Nutraceuticals and Functional Foods: Bridging Health and Food under a New Perspective

Nutraceuticals and Functional Foods: Bridging Health and Food under a New Perspective

Editor

Antonello Santini

MDPI • Basel • Beijing • Wuhan • Barcelona • Belgrade • Manchester • Tokyo • Cluj • Tianjin

Editor
Antonello Santini
University of Napoli Federico II,
Napoli, Italy

Editorial Office
MDPI
St. Alban-Anlage 66
4052 Basel, Switzerland

This is a reprint of articles from the Special Issue published online in the open access journal *Foods* (ISSN 2304-8158) (available at: https://www.mdpi.com/journal/foods/special_issues/SI_Nutraceuticals_Functional_Foods).

For citation purposes, cite each article independently as indicated on the article page online and as indicated below:

LastName, A.A.; LastName, B.B.; LastName, C.C. Article Title. *Journal Name* **Year**, *Volume Number*, Page Range.

ISBN 978-3-0365-7854-5 (Hbk)
ISBN 978-3-0365-7855-2 (PDF)

Contents

About the Editor

Antonello Santini

Antonello Santini is a Professor of Food Chemistry and Analysis of Food and Nutraceuticals in the Department of Pharmacy at the University of Napoli Federico II, Napoli, Italy. He has a Ph.D in Chemical Sciences and is also a Professor of Food Chemistry in the Agriculture Department of the same university. His research interests include, but are not limited to, addressing food chemistry and developing food supplements and nutraceuticals formulations. The recovery of by-products or waste from the agro food area to recover valuable bioactive compounds in view of a sustainable approach and the eco-compatible realization of added-value supplements and nutraceutical formulations, also including, where possible, unconventional novel foods and nano/micro formulations, is among his main research interests. He has published more than 200 scientific papers in peer-reviewed, reputed scientific journals. He has been a Member of the European Food Safety Authority EFSA, ERWG, Parma, Italy; a Member of the Italian Authority for Food Safety (CNSA), Italian Ministry of Health, Rome Italy; and a Member of Managing Board, Italian Chemistry Society (SCI) Division of Teaching (DD-SCI), Rome, Italy.

Preface to "Nutraceuticals and Functional Foods: Bridging Health and Food under a New Perspective"

This Special Issue, entitled "Nutraceuticals and Functional Foods: Bridging Health and Food Under a New Perspective", aims to approach the current state-of-the-art research on nutraceuticals and functional foods. The main issue in this field of research is the sustainability and recovery of bioactive substances from vegetal- or animal-origin byproducts to project and realize novel food supplements and nutraceuticals. Low environmental impact, safety, new food sources and analytical methodologies are of growing interest in the research area of food.

The perspective approach addresses the mechanism of action of nutraceuticals, safety and functional foods and nutraceuticals' mechanisms of action, revealing new possibilities for their use as tools in a complementary proactive approach to certain health issues to prevent the onset of health conditions or to be used in subjects who do not qualify for a conventional therapeutical approach. The areas involved in this perspective range from food chemistry and analysis to nutrition and from safety to sustainability; new therapeutical approaches and novel techniques of analysis and formulation are also involved, which require a wide inter- and multi-disciplinary approach. The overall assessment of these aspects creates new challenges for research and also impacts sustainability, health and safety. This Special Issue may also consider the bridging of health and food in their different declinations from a new perspective.

Antonello Santini
Editor

foods

MDPI

Editorial

Nutraceuticals and Functional Foods: Is It Possible and Sustainable for Bridging Health and Food?

Antonello Santini

Department of Pharmacy, University of Napoli Federico II, Via D. Montesano 49, 80131 Napoli, Italy; asantini@unina.it; Tel.: +39-81-2539317

This editorial is part of the Special Issue entitled "Nutraceuticals and Functional Foods: Bridging Health and Food Under a New Perspective". Current state-of-the-art research on nutraceuticals and functional foods is focusing on the recovery and re-use of bioactive substances from vegetal or animal byproducts. The following aspects of this research must be considered: (i) low environmental impact; (ii) safety of novel extraction strategies; and (iii) new sources (e.g., byproducts from the agro-food area) and new methodologies. Moreover, the byproducts deriving from transformation and manufacturing in the agro-food industry, which are often disposed of or used as feed, must adhere to good manufacturing practices and procedures [1]. Future research in this area includes: (i) the use of non-harmful solvents for the recovery; (ii) zero waste at the end of the processes; (iii) assessing the mechanism of action of nutraceuticals and functional foods. This will present new possibilities for their use as tools of proactive medicine. The expected endpoint is to prevent the onset of health conditions before a pharmacological therapy must be adopted, especially in subjects who do not qualify for a conventional therapeutical approach. This is becoming high desirable, considering the growing demand from the population of safe and all-natural remedies. This may be seen as a sort of "back-to-the-past" approach; however, the original pure holistic approach is presently developing into a more and more scientifically substantiated one. This Special Issue reports research and results of studies addressing some of what is described above. A multidisciplinary and multitarget involvement of different types of expertise is required. The Special Issue ranged from food chemistry to nutrition and from safety to sustainability and new therapeutical approaches, exploiting the role played by nutraceuticals and functional foods.

The health potential of nutraceuticals and functional food is triggering interest in research worldwide on the assessment of their mechanisms of action, which also involves possible interactions with physiological processes, with other molecules or pharmaceuticals, or with foodstuffs themselves. In vitro studies as well as studies on animals and humans must be conducted to exploit and assess these molecules' targets and mechanisms of action. The aim is to optimize safety, efficacy, and the appropriate formulation for their delivery. This aspect is of utmost importance: once the appropriate organ target is selected, the nutraceuticals and functional food must be allowed to reach it and provide the expected beneficial health effect. Based on this premise, new fields, new applications and emerging areas have been explored in this Special Issue, with an eye open to the predictable future development ahead. For example, there is growing interest in the development of nanonutraceuticals and nanoformulations with better bioavailability, supporting their specific beneficial health effects [2]. The paper by Cicero et al. [3] evaluated mineral and microbiological analyses of spices and aromatic herbs. Spices and aromatic herbs have been documented as a rich source of bioactive compounds that are used for their health benefits and for flavoring and coloring food. Nonetheless, they represent biological hazards and contain chemical substances of concern. The proposed paper gives a snapshot of the research aimed at monitoring the compliance of various spices and aromatic herbs from in the market of a non-European country according to the current European Union and WHO

Citation: Santini, A. Nutraceuticals and Functional Foods: Is It Possible and Sustainable for Bridging Health and Food?. *Foods* **2022**, *11*, 1608. https://doi.org/10.3390/foods11111608

Received: 25 May 2022
Accepted: 26 May 2022
Published: 30 May 2022

regulations. The results show the safety of the tested spices, which is an important step for their safe use. Lo Vecchio et al. [4] reported the full characterization of *Rhus coriaria* L., a fruit from Sicily which is a potential source of fiber (33.21 ± 1.02%) and unsaturated fatty acids and linoleic and α-linolenic acids (30.82 ± 1.21% and 1.85 ± 0.07%, respectively). Along with the high content in phenolic, anthocyanin, and minerals, its non-toxicity has been confirmed through tests of its extract on zebrafish embryos. This fruit is interesting, since it has also antimicrobial activity, as confirmed in the reported study against multidrug-resistant microorganisms, *Escherichia coli* and *Klebsiella pneumoniae* strains, isolated from poultry. D'Imperio et al. [5] studied the biofortification process of rocket and purslane with Zn. A bioaccessibility study of this nutrient using an in vitro gastrointestinal digestion process confirmed that it is possible to obtain a high content of Zn biofortified rocket and purslane by adopting an appropriate mineral plant nutrition solution enriched with this nutrient. The paper stressed, in fact, the growing interest in the research on these baby leaf vegetables. Biofortification could then be used to reduce the negative impact of mineral malnutrition with a relevant impact on health.

On the same area of interest, the paper by Buturi et al. [6] reported the relevance, novelty and importance of the biofortification of the vegetables with minerals as a tool to improve the human diet. It has been evidenced that the biofortification of vegetables can be a promising strategy to increase the content of specific compounds. The aim is to make them a functional food or source of nutraceuticals in order to formulate new food supplements. The possible interactions occurring at crop level, as well as the bioavailability of different minerals for the consumer, have been taken into consideration in the paper using a wide approach including the quantification of bioavailable fractions.

Santarcangelo et al. [7] reported that the long-ripened Parmigiano Reggiano Italian cheese has a beneficial content of minerals, indicating that it is an interesting food source, of selenium and chromium, among the others, which have beneficial health properties according to European Regulation 432/2012. The characteristics and health-promoting properties of green banana flours have been reported in the study of Khoza et al. [8]. The high content of phenolic and flavonoids and the antioxidant activity evidenced how the flours from Grand Naine and FHIA-01 GBF banana cultivars could potentially be used as raw materials for functional bakery products as well as for the fortification of snacks.

An alternative food source has been described by Mohamad Nasir et al. [9], who focused on edible bird nests, which are consumed as a Chinese traditional food for their health and medicinal purposes due to their high nutritional value. The study proposes a full characterization of edible bird nests, confirming their high nutritional value. This unconventional food matrix has been evaluated, evidencing how the characteristics and the nutrient-extraction methods may influence the availability of bioactive protein and peptides. The study also stressed the potential use of edible bird nests for their beneficial use, taking advantage of both the composition and content of biological active substances and the nutritional properties made available to human consumption.

Benchagra et al. [10] studied the beneficial antioxidant effect of Moroccan pomegranate against oxidative stress processes. The purpose of this study was to characterize the phenolic, flavonoids, and anthocyanin contents of different parts (namely peel and aril) of the *Sefri* variety of pomegranate. The results show that peel extract was richer in these compounds, especially in Punicalagin A and B. As a result, it was remarked that this Moroccan variety of pomegranate has protective effects against the development of metabolic disorder, cancer, atherosclerosis, and cardiovascular disease. Based on these results, the study of *Sefri* pomegranate extracts could open a new frontier in the fields of food preservation and health supplements.

Krobthong et al. proposed a novel approach aimed at developing safe strategies using natural hypolipidemic products and studied the possibility of the use of nano-liposomal Linzghi hydrolysate as novel functional ingredients in the treatment and prevention of obesity [11]. They studied Lingzhi, an edible fungus, as a potential lipid-suppression stimulant. Their results indicated the use of Lingzhi as a functional anti-hyperlipidemic ingredient. Excessive lipid accumulation is a serious health condition, and this result represents a good insight for the exploration of further lipid accumulation in adipocyte cells. The nano-liposomal Linzghi hydrolysate might serve as a novel functional ingredient in the treatment and prevention of obesity, and also indicates the current interest in new sources for nano-formulation of supplements and nutraceuticals.

Another novel and unconventional approach to food supplements was outlined by Gómez-Fernández et al. [12]. In their paper, the authors evaluated the effects on consumers' acceptability of a milk chocolate reformulation with alternative sugar sweeteners, probiotics, and ω-3 polyunsaturated fatty acids using milk chocolate as a carrier. The impact could be relevant as a potential functional food for the diabetic population. The authors concluded that the complete assessment of the health benefits of reformulated milk chocolates requires further in vitro, in vivo, and clinical studies. The topic is nonetheless triggering interest, since chocolate can be a candidate for the delivery of bioactive compounds. Due to its acceptable taste for the consumer, it may be a good carrier to formulate new nutraceuticals and functional foods.

Mlcek et al.'s paper describes the use of some edible flowers as foods for their beneficial health and nutritional properties [13]. This topic, although not much explored, is presently triggering interest in edible flowers as a possible food source and also for food supplements and nutraceutical development. The authors' contribution included the peculiarities of some ornamental edible flowers that represent a novel source of nutraceutical substances with valuable biological properties. The nutritional, chemical, and sensory characteristics and the antioxidant efficacy were explored in this paper, opening up a new area of interest towards flowers, which have so far been considered only as ornamental and beautiful to see in a garden or as a decoration on food preparations. Nonetheless, they may also have nutraceutical interest as a novel food.

The beneficial health-promoting properties of Amaranth, a pseudo-cereal crop, were described in the paper by Baraniak et al. [14], who evidenced the dual aspects of this plant that have been known centuries, i.e., its use as a functional food and as a medicine. Amaranth, in fact, has valuable biological properties, being rich in many phytochemicals and having wide pharmacological activity. Indeed, amaranth-based preparations are used in recipes for dietary supplements, functional food, and medicinal products. The authors concluded that the growth in the knowledge regarding this plant could trigger interest in research to promote its use in the development of innovative technologies in foods, nutraceuticals, and cosmetics industries.

In short, and as a final remark, the contributions presented in this Special Issue provide a snapshot of the current and growing interest in the research into the development of novel food supplements and nutraceuticals in view of building more and new knowledge to bridge food, supplements, nutraceuticals, and health in a coordinated and sustainable proactive approach.

Funding: This research received no external funding.

Conflicts of Interest: The authors declare no conflict of interest.

References

1. Souto, E.B.; Silva, G.F.; Dias-Ferreira, J.; Zielinska, A.; Ventura, F.; Durazzo, A.; Lucarini, M.; Novellino, E.; Santini, A. Nanopharmaceutics: Part I—Clinical Trials Legislation and Good Manufacturing Practices (GMP) of Nanotherapeutics in the EU. *Pharmaceutics* **2020**, *12*, 146. [CrossRef] [PubMed]
2. Durazzo, A.; Nazhand, A.; Lucarini, M.; Atanasov, A.G.; Souto, E.B.; Novellino, E.; Capasso, R.; Santini, A. An Updated Overview on Nanonutraceuticals: Focus on Nanoprebiotics and Nanoprobiotics. *Int. J. Mol. Sci.* **2020**, *21*, 2285. [CrossRef] [PubMed]
3. Cicero, N.; Gervasi, T.; Durazzo, A.; Lucarini, M.; Macrì, A.; Nava, V.; Giarratana, F.; Tardugno, R.; Vadalà, R.; Santini, A. Mineral and Microbiological Analysis of Spices and Aromatic Herbs. *Foods* **2022**, *11*, 548. [CrossRef] [PubMed]
4. Lo Vecchio, G.; Cicero, N.; Nava, V.; Macrì, A.; Gervasi, C.; Capparucci, F.; Sciortino, M.; Avellone, G.; Benameur, Q.; Santini, A.; et al. Chemical Characterization, Antibacterial Activity, and Embryo Acute Toxicity of Rhus coriaria L. Genotype from Sicily (Italy). *Foods* **2022**, *11*, 538. [CrossRef] [PubMed]

5. D'Imperio, M.; Montesano, F.F.; Serio, F.; Santovito, E.; Parente, A. Mineral Composition and Bioaccessibility in Rocket and Purslane after Zn Biofortification Process. *Foods* **2022**, *11*, 484. [CrossRef] [PubMed]
6. Buturi, C.V.; Mauro, R.P.; Fogliano, V.; Leonardi, C.; Giuffrida, F. Mineral Biofortification of Vegetables as a Tool to Improve Human Diet. *Foods* **2021**, *10*, 223. [CrossRef] [PubMed]
7. Santarcangelo, C.; Baldi, A.; Ciampaglia, R.; Dacrema, M.; Di Minno, A.; Pizzamiglio, V.; Tenore, G.C.; Daglia, M. Long-Aged Parmigiano Reggiano PDO: Trace Element Determination Targeted to Health. *Foods* **2022**, *11*, 172. [CrossRef] [PubMed]
8. Khoza, M.; Kayitesi, E.; Dlamini, B.C. Physicochemical Characteristics, Microstructure and Health Promoting Properties of Green Banana Flour. *Foods* **2021**, *10*, 2894. [CrossRef] [PubMed]
9. Mohamad Nasir, N.N.; Mohamad Ibrahim, R.; Abu Bakar, M.Z.; Mahmud, R.; Ab Razak, N.A. Characterization and Extraction Influence Protein Profiling of Edible Bird's Nest. *Foods* **2021**, *10*, 2248. [CrossRef] [PubMed]
10. Benchagra, L.; Berrougui, H.; Islam, M.O.; Ramchoun, M.; Boulbaroud, S.; Hajjaji, A.; Fulop, T.; Ferretti, G.; Khalil, A. Antioxidant Effect of Moroccan Pomegranate (*Punica granatum* L. Sefri Variety) Extracts Rich in Punicalagin against the Oxidative Stress Process. *Foods* **2021**, *10*, 2219. [CrossRef] [PubMed]
11. Krobthong, S.; Yingchutrakul, Y.; Visessanguan, W.; Mahatnirunkul, T.; Samutrtai, P.; Chaichana, C.; Papan, P.; Choowongkomon, K. Study of the Lipolysis Effect of Nanoliposome-Encapsulated Ganoderma lucidum Protein Hydrolysates on Adipocyte Cells Using Proteomics Approach. *Foods* **2021**, *10*, 2157. [CrossRef] [PubMed]
12. Gómez-Fernández, A.R.; Faccinetto-Beltrán, P.; Orozco-Sánchez, N.E.; Pérez-Carrillo, E.; Marín-Obispo, L.M.; Hernández-Brenes, C.; Santacruz, A.; Jacobo-Velázquez, D.A. Sugar-Free Milk Chocolate as a Carrier of Omega-3 Polyunsaturated Fatty Acids and Probiotics: A Potential Functional Food for the Diabetic Population. *Foods* **2021**, *10*, 1866. [CrossRef]
13. Mlcek, J.; Plaskova, A.; Jurikova, T.; Sochor, J.; Baron, M.; Ercisli, S. Chemical, Nutritional and Sensory Characteristics of Six Ornamental Edible Flowers Species. *Foods* **2021**, *10*, 2053. [CrossRef]
14. Baraniak, J.; Kania-Dobrowolska, M. The Dual Nature of Amaranth—Functional Food and Potential Medicine. *Foods* **2022**, *11*, 618. [CrossRef]

foods

Review

Mineral Biofortification of Vegetables as a Tool to Improve Human Diet

Camila Vanessa Buturi [1], Rosario Paolo Mauro [1,*], Vincenzo Fogliano [2], Cherubino Leonardi [1] and Francesco Giuffrida [1]

[1] Dipartimento di Agricoltura, Alimentazione e Ambiente (Di3A), University of Catania, Via Valdisavoia, 5-95123 Catania, Italy; camila.buturi@phd.unict.it (C.V.B.); cherubino.leonardi@unict.it (C.L.); francesco.giuffrida@unict.it (F.G.)

[2] Department of Agrotechnology and Food Sciences, Wageningen University & Research, P.O. Box 16, 6700 AA Wageningen, The Netherlands; vincenzo.fogliano@wur.nl

* Correspondence: rosario.mauro@unict.it

Abstract: Vegetables represent pillars of good nutrition since they provide important phytochemicals such as fiber, vitamins, antioxidants, as well as minerals. Biofortification proposes a promising strategy to increase the content of specific compounds. As minerals have important functionalities in the human metabolism, the possibility of enriching fresh consumed products, such as many vegetables, adopting specific agronomic approaches, has been considered. This review discusses the most recent findings on agronomic biofortification of vegetables, aimed at increasing in the edible portions the content of important minerals, such as calcium (Ca), magnesium (Mg), iodine (I), zinc (Zn), selenium (Se), iron (Fe), copper (Cu), and silicon (Si). The focus was on selenium and iodine biofortification thus far, while for the other mineral elements, aspects related to vegetable typology, genotypes, chemical form, and application protocols are far from being well defined. Even if agronomic fortification is considered an easy to apply technique, the approach is complex considering several interactions occurring at crop level, as well as the bioavailability of different minerals for the consumer. Considering the latter, only few studies examined in a broad approach both the definition of biofortification protocols and the quantification of bioavailable fraction of the element.

Keywords: fresh consumed vegetables; agronomic biofortification; mineral

Citation: Buturi, C.V.; Mauro, R.P.; Fogliano, V.; Leonardi, C.; Giuffrida, F. Mineral Biofortification of Vegetables as a Tool to Improve Human Diet. *Foods* **2021**, *10*, 223. https://doi.org/10.3390/foods10020223

Academic Editor: Antonello Santini

Received: 22 December 2020
Accepted: 19 January 2021
Published: 21 January 2021

1. Introduction

Many nutritional recommendations for human well-being and disease prevention have highlighted dietary styles based on the growing consumption of fresh fruits and vegetables and the reduction of simple carbohydrates, sodium, and saturated and trans-fats consumption [1].

In order to maintain a good health, people require several mineral nutrients that must be included in the diet. The essentiality of minerals can be demonstrated by the fact that vitamins cannot be absorbed solely or work in the absence of specific minerals, which are necessary in many physicochemical processes [2].

Deficiencies of specific mineral elements affect, in both underdeveloped areas and industrialized countries, up to two-thirds of the world's population [3–5] and the insufficient intake can cause severe damage to people's health [6]. For instance, in Europe and Central Asia, malnutrition problems related to diets with low micronutrient contents are increasing the number of women and children with anemia. In fact, iron and iodine deficiency disorders are the most common forms of malnutrition [7]. Besides, a recent study conducted in South Italy showed that the population has low intake of calcium and potassium [8].

Food, mainly plant-based, is the source of all important minerals, therefore it is important to keep on a regular basis a good and balanced diet that can provide the adequate proportion of minerals [9]. The enrichment of food with health related compounds and mineral elements could, however, be considered a strategy to fight undernourishment or to face with specific nutritional need [10].

In the case of not processed food, such as vegetables, the only option to enhance the nutrient content of products in preharvest using improved genotypes or adopting specific agronomical techniques [11].

The increasing interest in the enrichment of fresh consumed vegetables with mineral elements has encouraged intensive research activity focusing on the elaboration of suitable application protocols. This review describes developments in agronomic biofortification of vegetables with reference to some mineral elements often lacking or not adequately present in human diets, i.e., calcium (Ca), magnesium (Mg), iodine (I), zinc (Zn), selenium (Se), iron (Fe), copper (Cu), and silicon (Si). After synthetically considering the role in human nutrition and in plant physiology, this review aims to discuss the most successful agronomic strategies to increase the amount of the considered minerals in the edible portion of vegetables.

2. The Role of Vegetables for Human Health and How Biofortification Can Have an Impact

Plant foods make up a substantial part of the human diet and they provide most of calories, nutrients, and bioactive compounds necessary to keep a healthy status and prevent diseases. Vegetables are one of the pillars of a plant-based good diet, providing in particular dietary fiber, phytochemicals (such as vitamins, antioxidants), and minerals [12,13]. Minerals are considered essential nutrients: they are not synthesized by humans and must be obtained from the diet. Humankind evolved thanks to the dietary assumption of a significant number of vegetables and their insufficient intake is one of the reasons for many noncommunicable diseases, which are spread in Westernized societies. As an example, potassium, calcium, selenium, and iodine obtained through a vegetable-rich diet, can contribute to maintaining good blood pressure, bone strength, hormonal production, heart, and mental health [14,15]. In a recent study carried out in the UK, a data analysis from more than 40,000 people showed that changes in fruit and vegetable consumption may not only benefit physical health in the long-run, but also mental well-being in short term [16,17], besides the general population these benefits were also observed in cancer survivors [18]. On the other hand, vegetables play an important role in the economy, fighting poverty, hunger, and undernutrition, since they can be locally cultivated and consumed in a high diversity of shapes, sizes, colors, and tastes [12,19,20].

Nonoptimal intake of micronutrients and undernutrition, the so called hidden hunger, can be particularly severe for people following restricted diet for religious, ethical, or medical reasons [4,5,21]. Health authorities have established dietary reference intakes (DRI) based on recommended daily allowances (RDA) and tolerable upper levels (UL). As general principle, strategies that address vitamin or mineral deficiency must aim to achieve the DRI for each component without exceeding the UL [22].

However, the actual contribution of phytochemicals and minerals to human diet is not only related to their concentration in a certain plant tissue. The micronutrients must be released from the food matrix during the passage in the gastrointestinal track, absorbed into the blood and transported to their target tissues [23]. In fact, only the fraction released from the plant tissue become eventually available for absorption. This fraction is indicated as bioaccessible and to increase the bioaccessibility of plant phytochemicals and minerals is a promising target of agronomical strategies to improve the nutritional quality of vegetables [24].

Vegetable consumption should increase in the coming years for sustainability and health reasons. To deal with the rise of global population, more sustainable food sources will be needed [25]. According to Schreinemachers et al. [15], the most important vegetables in the current global economy are tomatoes, cucurbits (pumpkins, squashes, cucumbers, and gherkins), alliums (onions, shallots, and garlic), chilies, spinach, potatoes, carrots, and brassicas, therefore, it makes sense to focus the biofortification efforts on these species.

3. Biofortification of Vegetables

The approaches to address micronutrient malnutrition are different; medical supplementation and product fortification are the most commonly adopted. Fortification is the process of food enrichment with nutrients, adopting different methods during processing. However, in some contexts, fortification is challenged due to poor investments, infrastructure, and delivery system [26]. Under these conditions, an alternative strategy is to adopt new genotypes, characterized by improved compositional profiles, or to tailor specific agronomic techniques aimed to enhance the content of specific health effective compounds in widespread crops.[26]. While this can be considered an option for products which are transformed before they are used (e.g., staple foods), for fresh consumed products, such as vegetables, biofortification is the only choice to improve the content of health-related compounds in the edible portion.

Among the different strategies to obtain biofortified vegetables there are agronomic and genetic approaches, the latter can be done either through conventional breeding or transgenic methods [27,28]. The objectives are to increase in the edible portion the minerals content or other specific health related compounds. Transgenic programs involve biotechnology studies that allow to genetically modify a species, to obtain a plant with targeted characteristics (i.e., higher content of specific nutrients). Even though this approach could be cost effective in the long run, it is the least employed methodology today because the phase of research and development is still very slow and expensive. In any case, in developed countries the higher prices involved in the production of biofortified vegetables is countered by the achievement of a premium product with a superior nutritional quality, that can satisfy the new consumers' demand willing to pay for a healthier way of eating [29]. In addition, some countries have restrictive laws, that forbid genetically modified organisms (GMOs). Along the same lines, there is the option to cross different genotypes, with the aim to introduce in new cultivars desirable traits naturally occurring in plants. This genetic approach (traditional breeding) has been performed for decades and can allow to create new varieties with a higher content of certain nutrients. In this case, the limitation is to find the desired characteristics in the available genetic resources [30]. On the other hand, breeding programs, even when effective, may eliminate their effect due to the high renewal rate of cultivars coming from the large number of new introductions made by the vegetable seed industry [31].

Biofortification programs carried out through the agronomic approach are the best option, since they involve simple techniques to accumulate or to stimulate the production of specific compounds at plant level. A substantial part of the biofortification research that has been carried out in the last decades focused the attention on specific compounds such as vitamins and amino acids, rather than minerals [4,13,32–34]. A variety of biofortified products with vitamins or their precursor include banana, mango, sweet potato, wheat, and cauliflower [35]. In the same line, biofortification with amino acids proved to be effective in producing high lysine-rice, using the double strategy of maximizing its biosynthesis and minimizing its catabolism [36]. Besides, evidence shows that sulfur fertilization on wheat, barley, and potato can increase the sulfur-containing amino acids (SAAs) methionine and cysteine content in its edible part. In the same way, the application of nitrogen plus potassium has potential in increasing carotenoid content in carrots [37].

However, besides the increase of the content of some specific compounds (e.g., antioxidants) with controlled doses of stressors [38], agronomic biofortification consists in increasing or optimizing the application of mineral elements to the crop in order to increase the corresponding content in the edible organs. In this case the focus is on setting up the form of the mineral, the concentration, and the application form; indeed, certain mineral forms or quantities can cause indirect effects, damaging or compromising a crop [5,27].

4. Agronomic Mineral Biofortification

The production of vegetables is carried out in very diversified agronomic contexts as regards crop cycles, soil conditions, or growth environments. Agronomic approaches to increase the concentration of minerals in edible organs generally rely on the supply of mineral fertilizers and/or improvement of the mobilization and solubilization of mineral elements in the rhizosphere [27]. Vegetable crops are generally grown in agro-systems characterized by a high degree of intensification of the production processes and in which the supply of nutrients is increasingly based on the use of fertigation, soilless cultivation, and foliar fertilization. These alternatives offer different opportunities to implement targeted biofortification programs. In the case of the application of mineral elements by fertigation on soil cultivated crops, some interference may derive from element availability for the plant (phytoavailability), therefore selecting mineral forms and concentrations may have a relevant importance [27,32]. One alternative strategy to overcome the low mineral phytoavailability into the soil is the cultivation through hydroponic systems (soilless cultivation). The possibility of optimizing limited water resources and cultivating in the absence of suitable agricultural soils, has led to a considerable spread of soilless cultivation systems. It has been observed, for example, that hydroponic cultures can be among the best options to increase the nutrient content of plant tissues [39,40]. In the case of minerals not readily translocated to the edible tissues, such as for crops grown on soil and/or for minerals with scarce mobility, another alternative is the use of foliar fertilization [41].

4.1. Calcium

In human health, calcium (Ca) is required in several systems, like musculoskeletal, nervous and cardiac. It is essential to maintain good bones, teeth, and mineral homeostasis. It also acts as a cofactor in many enzyme reactions and contributes to the function of the parathyroid gland [42]. The RDA of Ca ranges between 1000 and 1300 mg day^{-1}. The UL for adults is 2500 mg day^{-1} [43]. Calcium is an important nutrient for plant metabolism, involved in structural functions of cell, acting as a counter-cation for organic and inorganic anions trafficking across the tonoplast and as an intracellular, cytosolic messenger [44]. It is one of most abundant nutrients in the earth's crust, with an average concentration of about 36.4 g kg^{-1}[45]. Ca^{2+} concentration in the soil solution is usually enough (0.1–20 mM) to meet the plants' demands or, in neutral and calcareous soils, to exceed their requirement, thus leading to Ca accumulation in the vicinity or inside the roots [44]. However, some Ca-deficient conditions can sometimes be encountered, especially in highly weathered tropical soils or in saline/sodic soils. Calcium is absorbed as divalent cation by the root apex and/or regions of lateral shoot initiation [46], where Casparian band between endodermal cells is absent or disrupted, and/or the endodermal cells surrounding the stele are not suberized [47]. Once inside the plant, Ca moves primarily through the xylem [46] with the water flow driven by transpiration [48,49], either as Ca^{2+} or complexed with organic acids [50]. However, Ca^{2+} movement inside the xylem vessels cannot be explained simply in terms of mass flow, as Ca^{2+} ions are also absorbed by adjacent cells and are complexed to nondiffusible anions in the xylem walls [48]. Due to its slow phloematic mobility, this element is present at lower concentrations in mostly phloem-fed organs (e.g., young leaves, fruits, and tubers) than in the older leaves (≈10-times less). Considering the mineral partitioning inside the plant, leafy vegetables can play a primary role in the dietary intake of Ca, so being possible targets for Ca biofortification [51]. This last point should be addressed at increasing the Ca content of the edible portions, without adversely impacting both plant growth and production costs [27]. Most plant species can accumulate high Ca contents in leaf laminae (up to 100 g kg^{-1} DW) without any symptoms of toxicity, because Ca exceeding plant's needs is detoxified by sequestering as insoluble Ca oxalate and deposited either in the cell wall or stored inside the vacuole [44,47]. Depending on the plant species, tissues, and growing conditions, Ca concentration in plants varies between 1 and >50 mg kg^{-1}. However, some species may have insufficient detoxification mechanisms, so their growth can be severely depressed at high Ca tissue content [44]. Excessive Ca can cause toxicity symptoms such as the presence of yellow flecks formed by crystals of calcium oxalate and growth inhibition, the latter can be observed even in calcicole species (plants occurring in calcareous soils) when submitted to a soil solution with a concentration higher than 10 mM Ca [46]. Strategies for Ca biofortification should include (i) increasing Ca supply to cells; (ii) increasing Ca uptake into cells; (iii) removing compounds making Ca unavailable and/or (iv) increasing Ca storage at the cellular and/or tissue level [27,52,53]. The application of Ca fertilizers can increase its concentration mostly in leafy vegetables (Table 1), whereas for grain, seeds, and fruits, sound indications are still to be reached. In 21-day old *Brassica rapa* plants grown on soil, the increased Ca supply to roots (compost mix supplemented with 0.4 vs. 3.5 g $CaCl_2$ L^{-1}) significantly enhanced its concentration in leaves (0.75 and 25 g kg^{-1} DW, respectively). The result was not influenced by the different supply of Mg fertilizer [54]. To reduce the effects of different soil characteristics (e.g., minerals concentration, pH) on Ca availability, soilless cultivation on inert substrates or water (e.g., floating system) allows a better control of the ion concentration in the root environment. In some leafy vegetables, D'Imperio et al. [24] increased the Ca concentration by adding calcium phosphate and calcium chloride in the nutrient solution (from 100 to 200 mg L^{-1}), determining an increase of Ca concentration in leaves of basil (≈15%) and mizuna (≈12%), but not in tatsoi or endive (Table 2). Moreover, the biofortification process did not influence their oxalate content nor Ca bioaccessibility. A higher Ca content (up to 5-fold higher than control) in lettuce (*Lactuca sativa* L.) grown in a floating system was obtained by Borghesi et al. [55] with a nutrient solution containing 800 mg Ca L^{-1} (as $CaCl_2$), compared to the control with no Ca addition. However, the high salt content increased both the Cl concentration and electrical conductivity of the nutrient solution, so reducing the marketable quality and yield (−32%). Foliar applications of soluble Ca fertilizers are commonly made for several horticultural crops, to prevent Ca-deficiency disorders. However, only few experiments refer to Ca biofortification through foliar applications. Moreover, these applications are expected to have limited effects on Ca content of roots, tubers, and seeds, because of the typical translocation patterns of the element. In one of few experiments, Yuan et al. [56] observed a significant increase of Ca concentration in lettuce sprayed three times every 20 days with 120 mg L^{-1} of $CaCl_2$ compared to 60 and 180 mg L^{-1} (21.4% and 5.2%, respectively), although this effect was genotype-dependent. Overall, Ca biofortification of vegetables using Ca chloride proved to be effective in the majority of the studied leafy crops even if negative effects on yield cannot be excluded; besides, one of the main challenges is related to the presence of oxalate, which can partially limit Ca bioavailability.

Table 1. Response of some vegetable crops to biofortification [(1)].

Element	Crop	Scopus® Papers (n.)	Average Concentration [(2)] (mg kg^{-1} FW)		Average Increase	Application Dose to Roots or Leaves (mg L^{-1})	
			Min	Max		Min	Max
Ca	Basil	1	950	1100	0.2-fold	100	200
Ca	Endive	1	1020	1080	0.1-fold	100	200
Ca	Indian colza	2	928	3000	2.2-fold	6	1603
Ca	Lettuce	2	695	2683	2.9-fold	0	800
Ca	Mizuna	1	1250	1400	0.1-fold	100	200
Ca	Potato	1	144	245	0.7-fold	350	5200
Ca	Tatsoi	1	1100	1150	1.1-fold	100	200
Mg	Indian colza	2	290	1059	2.7-fold	4	486
Mg	Onion	1	652	1627	1.5-fold	0	150
I	Basil	2	1	287	>100-fold	0.1	127
I	Cabbage	3	0.1	2.5	34.4-fold	0.1	0.6
I	Carrot	7	0.1	7.8	>50-fold	1	50
I	Chinese cabbage	3	0.1	48.7	>100-fold	0.1	50
I	Cowpea	2	4	1566	>100-fold	0.7	15
I	Lettuce	18	2	42.0	17.9-fold	0.1	50
I	Mizuna	2	0	1.0	>50-fold	0.7	1.1
I	Mustard	2	0	0.4	41-fold	0.7	1.1

Table 1. *Cont.*

Element	Crop	Scopus® Papers (n.)	Average Concentration [(2)] (mg kg^{-1} FW)		Average Increase	Application Dose to Roots or Leaves (mg L^{-1})	
			Min	Max		Min	Max
I	Onion	1	0	1.0	>50-fold	0	5
I	Potato	3	0.1	0.7	11.3-fold	0.6	5
I	Spinach	8	4.5	22.4	4.0-fold	1	5.1
I	Tomato	5	0.1	12.0	>100-fold	1	634
Zn	Arugula microgreens	1	3.0	70	22.3-fold	0	10
Zn	Broccoli	1	9.4	133	13.2-fold	121	408
Zn	Cabbage	4	4.1	39.1	8.6-fold	5	260
Zn	Carrot	1	42.1	802	18.1-fold	2.8	303
Zn	Indian colza	1	2.5	167	>50-fold	0	32.7
Zn	Kale	1	5.8	167	27.8-fold	2.8	303
Zn	Lettuce	3	2.2	30.4	12.8-fold	5.2	60
Zn	Okra	1	3.0	5.0	0.7-fold	2.8	303
Zn	Onion	1	2.5	7.8	2.1-fold	0	10
Zn	Potato	3	2.7	4.9	0.8-fold	9.7	250
Zn	Red cabbage microgreens	1	2.5	75	29-fold	0	10
Zn	Red mustard microgreens	1	2.1	92	42.8-fold	0	10
Se	Basil	4	0	8.3	>100-fold	2	12
Se	Broccoli	1	1.1	19.2	15.7-fold	10	100
Se	Carrot	3	0.1	1.8	35-fold	0.3	3.9
Se	Chard	1	0	0.5	45-fold	0	10
Se	Cucumber	1	0	0.2	7.6-fold	0	30
Se	Endive	1	0.1	1.2	23.5-fold	0.3	0.6
Se	Garlic	2	0.1	6.1	>50-fold	0.1	15
Se	Indian mustard	1	0	0.5	>50-fold	0	50
Se	Lettuce	12	0.1	6.9	>100-fold	0.5	20
Se	Onion	3	0.4	17.7	49.5-fold	2.0	20
Se	Potato	4	0.1	1.6	16.6-fold	0.5	0.8
Se	Radish	4	0.3	18.2	>50-fold	1	23.7
Se	Spinach	2	0.1	2.2	21.1-fold	0.2	0.3
Se	Strawberry	1	0.5	3.0	5.2-fold	0	4
Se	Tomato	3	0.3	3.4	9.1-fold	5	20
Se	Turnip	1	0.4	10.6	24.3-fold	0.2	2
Fe	Arugula microgreens	1	4.9	111	21.6-fold	0	40
Fe	Lettuce	1	2.3	4.3	0.9-fold	0.8	112
Fe	Red cabbage microgreens	1	7.7	448	>50-fold	0	40
Fe	Red mustard microgreens	1	4.9	323	>50-fold	0	40
Fe	Sweet potato	1	185	253	0.4-fold	0	100
Cu	Spinach	1	0.5	3.0	4.5-fold	0	3
Si	Basil	1	41.2	293	6.1-fold	0	100
Si	Chard	1	500	1450	1.9-fold	0	2.5
Si	Chicory	2	17.2	95	4.5-fold	0	101
Si	Green bean	1	57	252	3.4-fold	0	101
Si	Kale	1	700	2800	3-fold	0	2.5
Si	Mizuna	1	20	110	4.5-fold	0	100
Si	Purslane	1	14.8	98	5.6-fold	0	100
Si	Strawberry	1	475	8075	16-fold	0	100
Si	Swiss chard	1	18	145	7.1-fold	0	100
Si	Tatsoi	1	18	70	2.9-fold	0	100

[(1)] The list reports the most representative horticultural crops. In this and in the following tables, data refer to research on Scopus® using "biofortification" and "vegetables" as keywords performed in November 2020. Papers which tested more than one species were counted more than one time. [(2)] Calculated in the edible portion.

Table 2. Chemical forms of each mineral used in the biofortification of some vegetable crops.

	Basil	*Brassica* spp.	Carrot	Lettuce	Onion	Potato	Radish	Spinach	Tomato
Ca	Calcium phosphate, calcium chloride	Calcium chloride		Calcium chloride		Calcium chloride, calcium nitrate			
Cu								Cu-EDTA [1]	
Fe		Iron sulphate		Fe-EDDHA [2]					
I	Potassium iodide, potassium iodate	Potassium iodide, potassium iodate	Potassium iodide, potassium iodate	Potassium iodide, potassium iodate	Potassiumiodide	Potassium iodide, potassium iodate		Potassium iodide, potassium iodate	Potassium iodide, potassium iodate
Mg		Magnesium chloride, magnesium sulphate			Magnesium sulphate				
Se	Sodium selenate	Sodium selenate, sodium selenite	Sodium selenate, sodium selenite	Sodium selenate, sodium selenite	Sodium selenate	Sodium selenate, sodium selenite	Sodium selenate, sodium selenite	Sodium selenate, sodium selenite	Sodium selenate, sodium selenite
Si	Potassium meta silicate	Potassium silicate, sodium silicate							
Zn		Zinc nitrate, Zinc sulphate	Zinc oxide, zinc sulphate, Zinc EDTA [1]	Zinc sulphate	Zn-AML [3], Zn-EDDHSA [4], Zn-EDTA [1], Zn-PHP [5], Zn-HEDTA [6], Zn-EDTA [1]-HEDTA [6], Zn-DTPA [7]-HEDTA [6]-EDTA [1], Zn-EDDS [8]	Zinc nitrate, zinc oxide, zinc sulphate, Zn-EDTA [1]			

[1] Ethylenediaminetetraacetic acid, [2] Ethylenediamine-N,N′-bis(2-hydroxyphenylacetic acid), [3] Aminolignosulfonate, [4] ethylenediamine-di-(2-hydroxy-5-sulfophenylacetate), [5] Polyhydroxyphenylcarboxylate, [6] N-2-hydroxyethyl-ethylenediaminetriacetate, [7] Diethylenetriaminepentaacetate, [8] Ethylenediamine disuccinate. The list includes those papers that reported the initial and final concentration of mineral element in the edible part of vegetables.

4.2. Magnesium

In human health, magnesium (Mg) is essential in maintaining muscle tone and blood pressure. It participates in glycemic control, neuromuscular function, and myocardial contraction. It is also involved in the energy metabolism, besides being a cofactor of many enzymes [57]. The RDA for Mg ranges between 320 and 420 mg day^{-1}. The UL for adults is 420 mg day^{-1} [43]. Magnesium is a divalent cation and it is essential in plants because of its ability to interact with strongly nucleophilic ligands, it participates in the processes of enzyme regulation, pH cellular, and cation–anion balance, besides being a key metal in chlorophyll structure [58]. Magnesium is relatively mobile in soils, where its average concentration can vary between 0.5 and 40 g kg^{-1} [59], with a worldwide average of 5 g kg^{-1}. In addition to passive diffusion, as it happens with others divalent cations, Mg is actively absorbed by roots through permeable cation channels [60]. Regarding leaf uptake, younger leaves are more likely to absorb Mg than the aged ones [61]. Mg^{2+} transporters in higher plants are thought to be derived from the CorA transport system, acting as a gate locking it when the Mg concentration in the cytosol is increasing or opening otherwise [44,62]. Concentration of Mg in the metabolic pool of leaves is supposed to be between 2 and 10 mM, while free Mg concentration is expected to be lower (around 0.4 mM). For an optimal growth, plants demand between 1.5 and 3.5 g kg^{-1} of Mg in vegetative fractions [44]. Even though toxicity with Mg is rare, concentrations above 20 mM proved to be phytotoxic, causing symptoms like coppery colored leaves, decrease in starch contents, and growth reduction [63]. In contrast to the translocation difficulties observed for Ca, Mg shows a high phloem mobility and the application of Mg fertilizer can efficiently increase its concentration in leaves, tubers, fruits, seeds, and grains [27,44], making Mg agronomic biofortification of vegetables a feasible option to fight cases of malnutrition. As indicated in Table 1, plants of Indian colza (*Brassica rapa* ssp. *trilocularis*) submitted to different Mg biofortification protocols, showed on average a 3.6-fold increase in Mg content of leaves, when compared to untreated plants. In one experiment, after growing Indian colza plants on peat with a low (0.20 g L^{-1}) or high (3.04 g L^{-1}) Mg chloride ($MgCl_2$) concentration, leaf content increased up to 12 mg Mg kg^{-1} DW. However, the increase was 50% lower when plants received simultaneously a high dose of $CaCl_2$ (3.04 g), showing a possible negative interaction between Mg and Ca [54]. Similarly, Blasco et al. [64] submitted *Brassica rapa* plants to different nutrient solutions. When comparing the application of a low (4.86 mg L^{-1}) and a high (486.1 mg L^{-1}) dose of Mg (as $MgCl_2$) in the nutrient solution they noticed a 12-fold increase in the Mg content of shoots, passing from low to high dose. The same authors tested the interaction with other minerals and concluded that the Mg concentration in shoots increased with high Zn (500 μM) and low Ca (0.4 mM) supplies and decreased at high Ca (40 mM) supply. Another biofortification study of Mg was conducted applying doses of 0, 50, 100, 150, and 200 mg Mg dm^{-3} soil (as magnesium sulfate, $MgSO_4 \cdot 7H_2O$) on growth of onion plants (*Allium cepa* L.) [65]. The maximum Mg content in bulbs was obtained at the dose 150 mg dm^{-3}, i.e., almost 2-times higher than the untreated plants. However, this dose negatively affected the crop yield, and also caused a reduction in the uptake of Ca and potassium (K), showing that the antagonism between these minerals should be carefully evaluated. Therefore, the authors suggest using the Mg-100 dose, as it allowed to increase the Mg content of the bulbs (up to 1.4-fold, when compared to control), with a contextual increase in crop yield (up to 38%). There is evidence that fertilization of Mg via foliar spray can act to improve crop yield and quality [66,67]. The few studies on Mg biofortification show that both $MgSO_4 \cdot 7H_2O$ and $MgCl_2$ are effective in enhancing the element content in vegetables. However, Mg biofortification should be carefully managed considering its interaction with Ca, since high Ca content can inhibit Mg uptake by plants.

4.3. Iodine

Iodine (I) is essential for humans; it is required in the synthesis of the hormones thyroxine and triiodothyronine that are produced in the thyroid gland and are responsible for regulating growth and development, besides maintaining the basal metabolic rate [68]. The RDA of I is 150 μg day^{-1}, whereas the UL for adults is 1100 μg day^{-1} [69]. Typical I concentration in soils is between 0.5 and 20 mg kg^{-1}, and even though not essential to plant growth, it can be absorbed and translocated within the plant tissues. Plant leaves absorb I through stomata (60%) and leaf surface (40%), but I losses can occur too, due precipitation, wind, and tissue decay; the remaining can be partially transported via phloem to the other plant organs, including roots [70]. According to Smolèn et al. [71], leaves absorption occurs due the organophilic nature of I and its interaction with cuticular waxes or oxidation of I- (iodide) to I_2 (iodine), facilitating I penetration into the cuticle. It is known that root absorbs iodide better than elemental I or iodate, especially in plants grown in hydroponic systems. This I is majorly retained into the roots, but when in nutrient solution with concentrations higher than 0.01–10 μM it can also be translocated to the shoots [70,72]. In fact, I is efficiently transported into the xylem, transport in plants is analogous to chloride movement, I^- uptake being catalyzed by H+/anion symporters and released into the xylem by anion channels [27]. Concentration of I in plants can be zero or extremely low, about 30–100 μg kg^{-1} FW [73]. Depending on plant species, a nutrient solution with concentrations higher than 10–100 μM can already be phytotoxic and inhibit plant growth [70]. In general, the different I chemical forms present the following phytotoxicity order: (I_2) > (I^-) > iodate ($IO_3{}^-$) [74]. Horticultural crops are the best candidates for I biofortification, because of their ability to absorb and accumulate exogenous I into the edible fractions [75]. As reported in Table 1, once submitted to different biofortification protocols, leaf species such as basil and Chinese cabbage, showed an average I increase higher than 100-fold in their edible tissues, while cabbage, lettuce, mizuna, mustard, and spinach resulted in increases varying from 5 to 91 times. Average accumulation of I in vegetable fruit species was higher than 100-fold in both tomato and cowpea. Tuber species such as potato, showed a 13-fold average increased in I content, while root vegetables such as carrot presented a much higher average increase (greater than 100-fold). Biofortification of I through repeated foliar spray has been successfully performed in carrot and mustard plants [76,77]. Higher efficacy of lettuce iodine biofortification was noted Smolèn et al. [71] after foliar application, rather than adding the element to the nutrient solution. On the contrary, Caffagni et al. [75] demonstrated that, even though it is possible to enhance the I content of tomato fruits through KIO_3 foliar spray, better results were observed through fertigation with a 5 mM solution of KI; this allowed to achieve a 249-fold I increase in this vegetable. When grown in water culture, lettuce plants grown with 90 μg I L^{-1} as potassium iodide (KI) showed better biofortification results than plants submitted to the same amount of I as potassium iodate (KIO_3), with the result consisting of 30-times more I in leaves than untreated plants [78]. Low doses of I, such as 0.25 mg L^{-1} (KI) or 0.50 mg L^{-1} (KIO_3) in the nutrient solution are enough to achieve around 7 mg kg^{-1} DW of I in strawberry fruits, compared to 0 in the control, improving plant growth too [79]. Analogous results were observed in several leafy vegetables (e.g., broccoli raab, curly kale, mizuna or red mustard) when submitted to low doses of iodine (0.75 mg L^{-1}, 5.9 μM KIO_3) through fertigation, showing an increase ranging from 390 to 471 μg kg^{-1} FW [80]. However, high I levels (50 mg L^{-1}) in the nutrient solution, proved to increase the I content in carrot up to toxic amounts for humans (9 mg kg^{-1} FW) showing also phytotoxic effects on plants [76]. In addition, I biofortification should be carefully evaluated, since there is evidence that I can decrease Cu uptake by plants [73]. However, even though insufficient phloem loading and high volatilization rates could limit I accumulation, both K iodate and K iodide have successfully increased the I content in horticultural products.

4.4. Zinc

In human health, zinc (Zn) is essential for maintaining the structure and activity of many enzymes, besides playing a key role in the synthesis of nucleic acids and proteins. It acts in cell differentiation, glucose use, and insulin secretion [81]. The RDA of Zn ranges between 9 and 14 mg day^{-1}, whereas the UL for adults is 40 mg day^{-1} [69]. Zinc is essential in plant metabolism, as it plays a key role in chloroplast development and function through the Zn-dependent activity of SPP peptidase and repair of photosystem (PS I) I, besides participating in enzyme activation process such as RNA polymerases and superoxide dismutase, protein synthesis and metabolism of carbohydrate, lipid, and nucleic acid [82]. Although most of the world's cultivated soils contain enough Zn to sustain its accumulation in plants' edible portions (between 10 and 100 mg kg^{-1}), Zn phytoavailability is a factor often limiting its uptake by roots, so that it has been estimated that about one-fifth of the world's population actually suffers from Zn deficiency [83]. Under these conditions, agronomic strategies are aimed to improve the Zn phytoavailability into the soil, e.g., by correcting soil alkalinity, implementing more proper crop rotations, introducing beneficial soil microorganisms, or delivering phytoavailable Zn through the application of Zn-fertilizers to soil or foliage [83]. Zinc is absorbed by the plants from the soil solution primarily as Zn^{2+} (Strategy I plants) or complexed with organic ligands released by roots (phytometallophores), a mechanism which is restricted to cereals (Strategy II plants) [84]. Once inside the plant, xylem loading occurs either via symplast and apoplast, whereas in the xylem sap Zn is transported in its ionic form or in form of metal complexes with asparagine, histidine, organic acids, and nicotianamine [85]. Similarly, phloem Zn redistribution to various organs is thought to be affected either as divalent cation or in complexed forms with nicotianamine, malate, or histidine [27]. Due to its low phloematic mobility, Zn-supplied plants through the rhizosphere show a decreasing Zn concentration in the order shoot $\approx$ root > fruit, seed, tuber, thus showing a penalty on phloem-fed organs [86]. For this reason, root crops and leafy vegetables are thought to have the greater potential to increase dietary Zn uptake [83]. It must be pointed out that despite the low Zn phloematic mobility, Zn translocation through phloem for several plant species after application to foliage has been found to be nutritionally considerable for their growth and development, especially when cultivation occurs on substrates with low Zn phytoavailability [87]. Plants markedly differ in their ability to accumulate Zn in their tissues, but as a general rule, most crops require a leaf Zn concentration higher than 0.015–0.030 g kg^{-1} DW to reach their maximal yield. However, phytotoxicity symptoms are usually noticed at concentrations greater than 0.1–0.7 g kg^{-1} DM, depending on the species and exposure time [83]. When toxicity levels are attained, plants show an array of heavy metal stress responses such as growth and yield inhibition, leaf chlorosis and necrosis, restricted stomatal conductance and CO_2 fixation, changes in chlorophyll structure and concentration [88], so the higher threshold concentration actually represents a physiological limit to the biofortification achievements. Nonetheless Zn hyperaccumulation capacity has been observed in members of Brassicaceae, Caryophyllaceae, Polygonaceae, and Dichapetalaceae, whereas a greater Zn susceptibility has been noticed in the Linaceae, Poaceae, and Solanaceae [84]. Common inorganic Zn-fertilizers include $ZnSO_4$, ZnO, and synthetic chelates [27] such as Zn-EDTA, Zn-DTPA, or Zn-HEEDTA. When foliar applications are concerned, the Zn compounds used must be highly soluble and enter rapidly into the leaf apoplast, in order to promote Zn translocation to phloem-fed organs, so avoiding possible interferences with mesophyll metabolism [86]. Due to their ability to hyperaccumulate Zn, leafy *Brassicas* have been extensively studied in biofortification protocols (Table 1). In kale leaves (*Brassica oleracea* L. var. *acephala*), de Sousa Lima et al. [89] reported up to a 28-fold increase of Zn concentration by providing the crop with 300 mg Zn kg^{-1} soil. After applying 22.7 kg ha^{-1} of Zn (in the form of Zn sulphate, $ZnSO_4 \cdot 7H_2O$) to the soil, Mao et al. [90] observed a significant increase in the Zn concentration of the edible portions of canola (*Brassica napus* L.) and cabbage (*Brassica rapa* L. Chinensis Group) (by 25% and 200%, respectively). Zinc biofortification through foliar spray has been successfully performed in arugula (*Eruca sativa* L.) using 1.5 kg ha^{-1} of $ZnSO_4 \cdot 7H_2O$, with a resulting +94% increase of leaf Zn concentration [91]. Among non-Brassicas leafy vegetables, in a study conducted by Barrameda-Medina et al. [92] hydroponically cultured plants of lettuce (*Lactuca sativa* L.) supplemented with 100 μM $ZnSO_4 \cdot 7H_2O$ in the nutrient solution showed a 251% increase in leaf Zn concentration. Simultaneously biofortification programs must take into account that high Zn concentration on soil cultivated crops can negatively affect Fe absorption and improve the content of Mn and of amino acids [89]. In conclusion, Zn biofortification, especially in the form of sulphate is promising in increasing the mineral content in vegetable products.

4.5. Selenium

Selenium (Se) is an essential trace mineral, constituent of the selenoproteins responsible for important enzymatic functions. The function of selenoproteins in the human metabolism is most commonly connected to immunocompetence and cancer prevention, but it is known that Se functions go above that, as this mineral plays an important role in fertility and reproduction, brain functions, mood, thyroid health, and cardiovascular diseases [93]. The RDA of Se is 55 μg day^{-1}, and the UL for adults is set at 400 μg day^{-1} [69]. Selenium is not considered a micronutrient, but its appropriated use in plant nutrition can increase growth, stimulate seed germination, and contribute to protect several crops against pathogens and pests [94]. Soil concentration of Se is relatively low and it varies according to the type of rocks, being generally between 0.01 and 7 mg kg^{-1}, with a worldwide average of 0.4 mg kg^{-1} [95]. Plants take up Se inorganically both as selenite ($SeO_3{}^{2-}$) and selenate ($SeO_4{}^{2-}$) [96]. Plant absorption and transportation of Se are active processes [97]. Into the roots, due to its chemical similarity to sulfur (S), selenate moves through high-affinity sulphate transporters, while selenite movement is partially mediated by phosphate transporters [97,98]. Translocation of Se from root to the aerial parts of the plant is more likely to happen as selenate, since selenite is more prone to be accumulated in roots. Leaf surface can absorb volatile forms of Se from the atmosphere [99]. Foliar application of Se at late growth stages seems to optimize the uptake, translocation, and distribution of Se into the edible portions of plants, whereas selenate is more efficiently accumulated in plant tissues than selenite [100]. The tolerable Se content in most plant species is between 10 and 100 mg kg^{-1} DW [101] and phytotoxic effects due to Se excess can compromise plant growth through damages to photosynthetic apparatus, photosynthesis inhibition, and over-production of starch [102]. In addition, secondary accumulators, also called Se-indicator, as some vegetables of the Asteraceae, Brassicaceae, and Fabaceae family, when supplied with exogenous Se can accumulate up to 1 g kg^{-1} DW, being a good target for Se biofortification [101]. Skrypnik et al. [103] reported that Se biofortification of basil through foliar application of sodium selenite (Na_2SeO_3) at 10 μM (4 applications starting from the 7° day after transplanting) enhanced the Se concentration in leaves to up 10.74 mg kg^{-1} DW (more than 700-times higher than untreated plants). Moreover, five applications of Na_2SeO_4 (0.633 mM), as foliar spray, from the six-leaf phase, resulted in lettuce leaves enriched with up to 40 mg Se kg^{-1} DW, around 40 times greater than the control [71]. In another study, radish plants sprayed with 5 mg Se per plant 7 days before harvest, as sodium selenate, were able to produce roots with 346.5 mg kg^{-1} DW of Se, meaning that the consumption of 1–10 radishes is enough to fulfill the daily human requirement [104]. Meanwhile, da Silva et al. [105] found that fertigation of radish plants could be more efficient than foliar spray, after treating plants with a low dose of Na_2SeO_4 (3.6 mg of Se per pot). They obtained roots with approximately 50 mg Se kg^{-1} DW, while the leaf spray of the same chemical (0.36 mg of Se per pot, 93 mL per pot) resulted in plants with approximately 15 mg Se kg^{-1} DW. Lettuce appears to be a good candidate for Se biofortification, as demonstrated by do Nascimento da Silva et al. [106]. In this experiment, plants submitted to fertigation at 25 μM Se L^{-1} (as sodium selenate) resulted in lettuce leaves with as much as 39.4 mg Se kg^{-1} DW, around 40 times greater than the control. While, higher application rates of both sodium selenate (Na_2SeO_4) and selenite (Na_2SeO_3) reached numbers that exceeded the RDA of Se. Similarly, tomato plants fertigated with 5 mg L^{-1} of Se as sodium selenate, were enough to obtain a significant increase in Se concentration of fruits (35.8 mg kg^{-1} DW), twice the concentration in the untreated plants. At the same time this dose allowed to achieve good physiological responses on plants, such as increased enzyme activity of catalase, glutathione peroxidase, and superoxide dismutase in fruits [107]. Selenium biofortification was successfully implemented in many vegetable crops, using Na selenate or Na selenite. Besides, possible antioxidant and antisenescence effects of Se can improve shelf-life during postharvest storage [108]. However, because of the high toxicity of Se, especially in the form of selenate, attention must be made regarding agricultural workers and product safety.

4.6. Iron

In human health, the main function of iron (Fe) is related to the synthesis of hemoglobin and myoglobin besides being essential to many metabolic processes such as oxygen transport, deoxyribonucleic acid (DNA) synthesis, and electron transport, it is also required for energy production [109]. The RDA of Fe ranges between 8 and 18 mg day^{-1}, whereas the UL for adults is 45 mg day^{-1} [69]. Iron is a versatile, essential element in plant metabolism, whose biological functions are primarily based on the reversible redox reaction of Fe^{2+} (ferrous) and Fe^{3+} (ferric) ions, the ability to form octahedral complexes with various ligands and to change its redox potential in response to different environmental conditions. Due to this, Fe is involved in the transfer reactions at the base of life, since electron transfer chains of photosynthesis and respiration rely on iron–sulfur (S) clusters of the 2Fe–2S or 4Fe–4S type [110]. The concentration of this element in soil often exceeds plant requirements, being present at 20–40 mg kg^{-1} [111], but usually only a small amount of this is available for plant nutrition. Particularly in alkaline and calcareous soils, once applied through fertilization, Fe quickly becomes unavailable to roots absorption, because of precipitation, adsorption, and oxidation phenomena [112,113]. Plants have evolved two different strategies to acquire Fe from the growth substrate, based either on its reduction (Strategy I plants) or chelation with organic ligands (Strategy II plants) [114]. In nongraminaceous species (Strategy I plants), such as most of vegetable crops, organic acids and phenolic compounds released by roots chelate ferric Fe on the root surface (Fe^{3+}), which is subsequently reduced to its ferrous form (Fe^{2+}) to transport the element across the plasmalemma of root epidermal cells [27]. The Fe transportation within the plant occurs in chelated forms, mainly with citrate and malate in the xylem, and nicotianamine and its derivatives in the phloem [115]. This condition derives from the peculiarities of this metal, characterized by low solubility and high reactivity, so its transport inside the plant must be associated to proper chelating molecules controlling its redox states between ferrous and ferric forms [116]. The status of Fe into a plant is expressed by its quantity, redox state, speciation with chelating molecules, and its compartmentalization [117]. Chloroplasts represent the main pool of Fe within the cell, as they gather approximately 80–90% of cellular Fe [44]. This flows from the high Fe demand of the photosynthetic apparatus, and Fe-deficiency hampers the electron transfer between PSI and PSII, resulting in photooxidative damages [116]. Even though the range of Fe in leaves is between 50 and 150 mg kg^{-1} DW, Fe requirement is highly variable among species. For example, C4 species are more likely to require higher Fe amounts than C3 species; fast growing meristematic and expanding tissues need more Fe. On the other hand, Fe toxicity is reported in concentrations above 500 mg kg^{-1} DW, which can cause damages associated with formation of ROS, inducing the activity of antioxidative enzymes such as ascorbate peroxidase, besides damages to membrane and irreversible impairment of cellular structure, DNA, and proteins [44]. To improve Fe uptake agronomical solutions to make Fe available are acidification of soil [112] and/or use as Fe(III)-chelates synthetic fertilizers. Since the latter are expensive, their use is mainly restricted to soilless crops and to high added-value cash crops [117]. However, in the case of vegetable crops, the knowledge concerning Fe enrichment, and specifically biofortification, is still poor. One alternative to provide Fe to plants is the foliar spray even if, both adopting the chelated or the sulfate-salt form, a large fixation by cuticle can be observed [118]. Foliar spray of Fe sulphate heptahydrate ($FeSO_4 \cdot 7H_2O$) proved to be effective to increase Fe content both in leaves and sink organs of herbaceous crops [119,120]. In tomato, leaf spray with a 9 mM $FeSO_4$ solution increased by 3.8 times the Fe content in roots, mediated via phloem transport [121]. In a study conducted on potato, Kromann et al. [122] did not observe a positive relationship between Fe foliar spray with EDTA-chelated Fe and its concentration in tubers, thus the authors hypothesized that the limited effect was related to the Fe form used. As shown in Table 1, biofortification of vegetables with Fe through fertilization has been tested in few species. The use of EDDHA-chelated Fe up to 2.0 mM (112 mg L^{-1}) proved to be effective in soilless cultivation of lettuce in increasing the Fe content of the leaves from 2.31 mg kg^{-1} FW (control) to 4.30 mg kg^{-1} FW [123]. In addition, it has been reported that low doses of Fe can enhance the accumulation of secondary metabolites such as chlorogenic acid, β-carotene, violaxanthin, or neoxanthin, thus leading to improved functional profiles of vegetables [123]. However, the authors observed a yield reduction of about 25%, which increased proportionally with the amount of Fe added to the nutrient solution. Overall, Fe biofortification has not been investigated enough to draw a clear picture. Using sulphate or chelate forms only in some cases enhanced mineral content in the edible part of vegetables, however, the increase was coupled with a yield reduction. Concluding significant insolubilization in the soil, limited translocation into the plant and accumulation into edible organs and negative effects on yield are the main constraints in Fe biofortification.

4.7. Copper

In human health, copper (Cu) importance is related primarily to enzymes function, contributing also to maintain cardiovascular integrity, lung elasticity, normal development of connective tissue and nerve coverings, neovascularization; it has also neuroendocrine and immune functions and it is involved in the Fe metabolism too [124]. The RDA of Cu ranges between 1.0 and 1.6 mg day^{-1}, while the UL for adults is 10 mg day^{-1} [69]. Copper is a redox-active transition metal that under physiological conditions exists as Cu^{2+} and Cu^{+} [125]. In plants, it is essential to many physiological processes like photosynthesis, respiration, C and N metabolism, and protection against oxidative stress. It acts as co-factor of numerous proteins and in plants it is mainly present in complexed forms, the concentration of free Cu^{2+} and Cu^{+} in the cytoplasm being minimal [44]. The worldwide average Cu concentration in soils is 14 mg kg^{-1}, while in Europe the average concentration is 12 mg kg^{-1} [126]. Copper is mobile in soils and its absorption is directly related to its concentration in the soil solution [44]. Plants can absorb Cu in huge amounts by roots and in minor amounts by shoots and leaves [127]. Mechanisms involved in Cu uptake are supposedly similar to those of Fe. Copper chelate reductases are encoded by ferric reductase oxidases 4 and 5 and Cu reduction occurs at the roots (Strategy I plants) where Cu is absorbed and transported by proteins of the COPT family. Copper uptake from soil depends almost exclusively on the protein COPT1, while COPT2 could act in the processes of Cu and Fe homeostasis and phosphate metabolism [27,128]. Plants can also absorb Cu through leaves, as observed by Stepien and Wojtkowiak [129] that after treating wheat plants with a foliar fertilization of copper sulphate in the amount of 0.2 kg Cu ha^{-1} (1% $CuSO_4$ solution) obtained a 13% increase in the Cu content. On the other hand, the redox-active transition characteristic of Cu that makes it essential also contributes to its toxicity, since the reduction between Cu^{2+} and Cu^{+} catalyzes the production of toxic hydroxyl radicals that can damage DNA, cell membranes, and other biomolecules. Besides, damage to cell membranes can be reflected in low uptake of ions and water, so Cu toxicity can be indirectly expressed as growth inhibition and chlorosis caused by the generalized deficiency of nutrients and water [130]. Normally, crop species can tolerate a maximum of 20–30 mg kg^{-1} DW of Cu in leaves, but Cu-tolerant species can accumulate as much as 1000 mg kg^{-1} DW of Cu in leaves [44]. Moreover, foliar fertilization of Cu in maize should not exceed 100 g ha^{-1}, since at higher doses, between 200 and 600 g ha^{-1}, Barbosa et al. [131] noticed phytotoxic effects that caused growth and yield reduction up to 19% and 75%, respectively. In agriculture, Cu has been used for plant disease control for decades, a number of Cu formulations have been used as biocides to contain pathogens such as bacteria, fungi and in some cases, even invertebrates. In high concentrations, Cu interacts with nucleic acids, disrupting cell membranes of pathogens. In addition, direct application of Cu is used for seed treatment, to prevent seedling infections [132]. As shown in Table 1, among the few experiences in the biofortification of copper, Obrador et al. [133] conducted a study with spinach (*Spinacia oleracea* L.), var. 'Viroflay Esmeralda' applying eight different liquid fertilizers to the soil surface, with the irrigation water in a concentration ranging from 0 to 3 mg Cu kg^{-1} soil. Total Cu concentration in the dry matter of shoots increased by up to 450%, from 9.55 mg kg^{-1} (control treatment) to 52.51 mg kg^{-1} in the treatment where plants were submitted to 3 mg Cu kg^{-1} soil (as Cu-EDTA), a 4.54-fold increase (Table 1). However, at this dose they also noticed a 10% decrease in the dry matter yield. Instead, the dose 1 mg Cu kg^{-1} soil resulted in an increase in Cu content of 153% allowing also to obtain a yield increase of 71% when compared to the control. Regarding the chemical form, their results showed that the best fertilizers to increase Cu content in the edible part of spinach are Cu-DHE (Cu-diethylenetri-aminepentaacetate-N−2-hydroxyethyl-ethylenediamine-triacetate-ethylenediamine-tetraacetate) and especially Cu-EDTA. Curiously, in this study, even though the total concentration of Cu in spinach shoots was higher than the maximum concentration usually tolerated by plants, no visual phytotoxic symptoms and significant yield reductions were observed. In conclusion, Cu biofortification proved to be effective using different chelate forms and its potential as a biocide could benefit biofortification programs. In addition, when Cu biofortification is concerned attention must be made to the release of Cu in the soil substrate in relation to crop rotations and soil biological properties.

4.8. Silicon

Accumulating evidence from the last 30 years strongly suggests that silicon (Si) plays an essential role in bone formation and maintenance, improving the bone matrix quality and facilitating its mineralization. Increased intake of Si has been associated with increased bone mineral density and decreased osteoporosis [134]. Average daily dietary intake of Si is 20–50 mg for European population, the RDA has not been stablished; however, safe upper levels for humans have been recommended with a maximum range of 700–1750 mg day^{-1} [135]. Silicon is considered not essential for plant nutrition, but its inclusion in fertilization programs has proved to increase the crop tolerance to biotic and abiotic stressors [136], crop yield [137], or improve the absorption of macro- and microelements [138]. Silicon concentration in soil can vary depending on the type of soil. For example, alkaline soils containing sodium carbonate usually present a higher Si content. On average, the concentration of Si in soil is between 0.09 and 23.4 mg kg^{-1} [139]. If compared with other minerals, Si metabolism is still poorly understood. It seems that two main mechanisms of Si absorption coexist in plants, i.e., active and passive, whose relative contributions depend upon both plant species and external Si concentration [140]. This would explain the strong differences in Si concentration reported within tissues of different plants species [141]. In any case, Si is taken up by the roots as monosilicic acid with the involvement of channels belonging to the aquaporins' group, so the water flow resulting from leaf transpiration seems to play a determinant role in defining the rate of Si absorption and transport within the plant [142]. Once absorbed, monosilicic acid is subsequently translocated to the shoot through the xylem flow, where Si is concentrated thanks to transpiration and polymerized to silica (SiO_2), then deposited in the different tissues [143]. It has been reported in the Poaceae leaves that Si can be deposited both in mesophyll and epidermal cells, suggesting the coexistence of negative (transpiration-driven) and positive (though specific carriers) mechanisms controlling the Si accumulation process [144]. Plants markedly differ in their ability to accumulate Si in their various organs; concentrations ranging between 5 and 50 g kg^{-1} DW have been reported as critical for some species. The species with low mobilization capacity accumulate it in the roots and stems, while the species with high mobilization capacity accumulate Si in stems, leaves, fruits, and seeds [142]. Gao et al. [145] noticed that excessive Si supply (>2 mM) caused the formation of Si polymers on root surfaces, a feature that could affect nutrients uptake. In spite of the scarcity of available information, this aspect would deserve extensive study with reference to vegetable crops, due to their potential role as Si source in the human diet. Indeed, thanks to their usually low silicification capacity, vegetable crops are expected to contain high amounts of soluble

Si, which is theoretically more available to be assimilated after ingestion, so potentially being optimal candidates as Si source in the human diet [142]. As shown in Table 1, as regards the leafy vegetables, in a study concerning six crops grown in a greenhouse floating system, namely *Brassica rapa* L. (tatsoi and mizuna group), *Ocimum basilicum* L., *Portulaca oleracea* L., *Cichorium intybus* L. and *Beta vulgaris* L. ssp. *vulgaris*, D'Imperio et al. [146] found an increased Si content in plant tissues by providing them up to 100 mg L^{-1} Si (as potassium metasilicate) in the nutrient solution, with basil reaching the highest content of Si (293 mg kg^{-1} FW, expressed as SiO_2). Moreover, the authors found that Si became bioaccessible in all the considered species, in a range from 23% (basil) to 64% (chicory). In a different experiment concerning two leafy vegetables, namely chard (*Beta vulgaris* L. var. *cicla*) and kale (*Brassica oleracea* L. var. *acephala*) grown in a hydroponic system, De Souza et al. [147] compared the effects of two Si sources, namely potassium silicate and stabilized sodium potassium silicate with sorbitol, and four Si concentration in a foliar spray solution (from 0.00 to 2.52 g L^{-1}). They found that in both species, the Si concentration in leaves linearly increased in response to Si concentration in the foliar spray solution, with the best biofortification results obtained by spraying potassium silicate. In a study concerning the green bean (*Phaseolus vulgaris* L.) cultivated in a hydroponic system, Montesano et al. [148] found that biofortified pods (obtained by adding 3.6 mM of Si as potassium metasilicate to a standard nutrient solution) showed a 310% increase of Si (from 853.8 to 2496.3 mg kg^{-1} DW) when compared to unbiofortified ones. Moreover, they found that the bioaccessibility of Si in biofortified pods was higher than control pods (25.1% vs. 7.6%), even after cooking them by steaming or boiling. The Si biofortification protocol of strawberry fruits (*Fragaria* × *ananassa* Duchesne ex Rozier) was studied by Valentinuzzi et al. [149], who cultivated for 16 weeks in a hydroponic system provided with a standard nutrient solution, or with nutrient solutions enriched with 50 or 100 mg L^{-1} of Si (as Na_2SiO_3). The authors found that providing 100 mg L^{-1} of Si allowed to maximize the metalloid concentration in strawberry fruits (which increased from 6.44 up to 85 g kg^{-1} DW) without compromising crop yield. However, the they observed a decrease in total phenols and an increase in the content of flavonols in response to the highest Si supply. Overall, biofortification with Si using K silicate proved to effectively increase the mineral content in vegetables. In addition, its possible role as plant protector and its ability to improve the mineral status of the plant, both make Si a key element in biofortification programs.

5. Discussion and Future Trends

The evidence discussed above pointed out that biofortification should be contemplated as a promising strategy to face malnutrition in many circumstances. Biofortification can help to obtain products designed according to the needs of two categories of target consumers (Figure 1). The first concerns products enriched with minerals that can fulfil specific functional needs; this is the case of vegetables richer in one or more minerals to counter the deficiencies related to ordinary diet or new consumer habits. (e.g., vegans). Besides vitamins, in fact, vegan diets feature an inadequate content of calcium, potassium, iron, iodine, and magnesium [150]. A second target concerns products with premium quality or superfood aimed at improving health as a whole. This would satisfy the need of an increasing group of health-conscious consumers who look at plant-based foods, especially vegetables, as a sort of medicine to prevent the insurgence of chronic diseases.

Agronomic biofortification is comparatively simpler than other methods and potentially suitable for immediate results. However, the available studies on agronomic fortification of vegetables are of a considerable number only for few crops (e.g., lettuce, tomato, spinach, and *Brassica* spp.) and for few mineral elements (e.g., selenium, iodine). For these elements, aspects related to the form, application modality, concentration, and timing have been clarified for most important crops. For all the considered elements, and particularly for selenium and iodine, the biofortification adopting soilless crops or on soil fertigated crops have been mostly considered. In some cases the model describing the accumulation in relation to the application has also been described [151]. For some other mineral elements considered in this review, important as well in human nutrition (e.g., Fe), information is still lacking.

On the other hand, even when empirical evidence on biofortification showed a significant increase in the concentration of the mineral elements, the fortification is not economically worthwhile. In addition, an effective biofortification protocol is based on regular and frequent applications and a negative environmental impact cannot be excluded [32]. Besides, the step between biofortification and plant toxicity effects can be narrow and applications targeting the accumulation of essential micronutrients must be adjusted to avoid negative effects on plant growth [38].

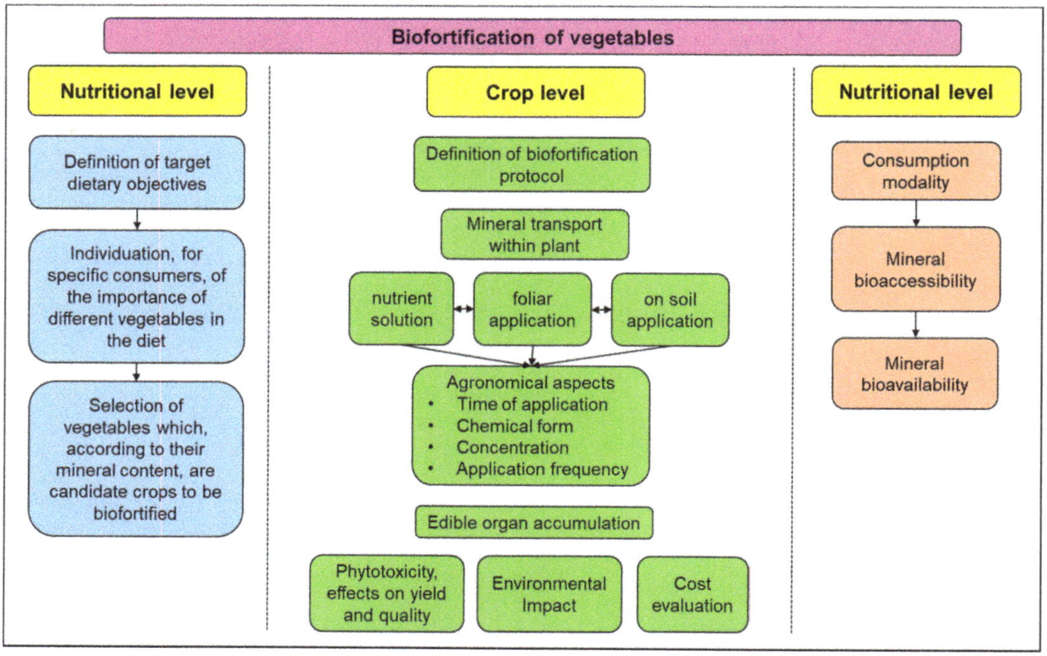

Figure 1. Key aspects to be considered in the agronomic mineral biofortification.

The application of biofortifying elements poses some problems related to the interaction with other factors at soil level (e.g., phytoavailability) and at plant level (e.g., competition with other elements) [152]. In many studies the traditional fertigation approach is adopted, rather than foliar spray, which can be more cost effective and environmentally friendly. Indeed, foliar fertilization represents the simplest and fastest method for the application of mineral elements used for the biofortification of vegetables; but, the effectiveness depends on the used plant organ and the mobility of the element inside the plant. To face some of these problems, technical innovations such as precision agriculture, soilless cultivation, etc., may help in defining more efficient biofortification protocols.

There are only few biofortified vegetable products already present on the market (e.g., selenium enriched potato, carrot and onion, 'Selenella' from Consorzio Patata Italiana di Qualità Soc. Cons., IT, iodine biofortified potato, 'Iodì' from the Pizzoli group, IT, selenium enriched brussels sprouts from Marks and Spencer, UK, etc.). It is clear that mostly iodine and selenium have been commercially considered as biofortification elements, probably because a more efficient accumulation system and for their lower toxicity at plant level. In the future, besides a broad choice of diversified vegetables, it is expected that the market will have biofortified products richer in more than one mineral. Therefore, research that comprises simultaneous biofortification is essential. In addition, further elements are being studied and are expected to be object of biofortification in the future (e.g., lithium, vanadium, etc.). In this regard, biofortification using Li-sulfate and Li-hydroxide was effective in increasing Li content in lettuce plants [153].

Based on the results in literature, biofortification is not expected to fully control mineral element deficiencies or eradicate them, but it complements other interventions to provide micronutrients to people. To be effective, a biofortification program should be based on very appropriate planning concerning: health and nutrition investigation, nutritional habits, design and validation of sustainable biofortification protocols, estimation of positive effect on health. Concerning biofortification protocols, the attention should be paid to those crops having an element content high enough to be conveniently targeted, and that prove to significantly benefit from mineral elements application.

In the reviewed literature most attention has been posed on the content of specific elements in plant edible portion but key concepts like bioaccessibility and bioavailability were seldomly considered. The first regards the nutrient fraction released from the food and available for absorption by the intestinal cells, while the latter expresses the amount of nutrients actually absorbed and therefore available for utilization in physiological functions [154,155]. While macronutrients (proteins, carbohydrates, and fats) are degraded and absorbed by specific and well-known biochemical mechanisms, phytochemicals and minerals are absorbable without biotransformation and often without a specific carrier [156,157]. The consequence of this poorly developed intestinal transport system is that the actual absorption of phytochemicals and minerals is deeply dependent on the food matrix. To modulate mineral bioavailability, attention should be devoted to those substances (e.g., vitamin C, β-carotene, oxalic acid, polyphenols, etc.) stimulating or inhibiting bioavailability [27,158]. Furthermore, some chemical bonds with other components in the food or the physical entrapping inside intact plant cell walls can dramatically decrease the bioaccessible and bioavailable fractions of phytochemicals and minerals [159].

6. Conclusions

In conclusion, the production of mineral-dense vegetables will deserve a prominent place in the coming years. Agronomic biofortification, even if it involves expensive experimental activities, represents the only strategy in the case of vegetables, for which genetic improvement programs would be rather complex and not very convenient due to the high rate of varietal turnover. The main challenges for agronomic biofortification in the immediate future will rely on the efficiency of fertilization process and bioavailability of minerals, the high cost of some specific chemical formulations, the possible yield losses due to biofortification-induced alterations of plant metabolism, and the potential environmental/health impact deriving from new agronomic protocols (as in the case, for example of copper and selenium). Deeper knowledge in these areas must be considered indispensable to achieve sound conclusions about the costs/benefits of biofortification.

The papers discussed in this review report promising results for several minerals and pillar vegetables in the human diet; however, the results are not entirely consistent and coherent. The future activities, beyond their specific scientific relevance, should be planned in a broader context, adopting an approach involving also farmers, traders, nutritionists, and educators. Evidence from research shows that farmers are willing to cultivate and commercialize biofortified crops and the few and selected products available in the market demonstrates that consumers are favorably eating them. Furthermore, nutrition specialists together with health educators can also have an impact on the population's eating habits and contribute to increase the consumption of the target vegetables. Such an approach, thanks also to the nutritional importance of vegetables, will certainly have a significant impact on improving human diet.

Author Contributions: Conceptualization, C.L. and F.G.; methodology, C.L., F.G., R.P.M. and C.V.B.; validation, R.P.M., F.G. and C.V.B.; data curation, C.V.B. and R.P.M.; writing—Original draft preparation, C.V.B. and R.P.M.; writing—Review and editing, V.F. and C.L.; supervision, C.L. All authors have read and agreed to the published version of the manuscript.

Funding: This research received no external funding.

Institutional Review Board Statement: Not applicable.

Informed Consent Statement: Not applicable.

Data Availability Statement: Data sharing not applicable.

Conflicts of Interest: The authors declare no conflict of interest.

References

1. Wang, L.; Sikand, G.; Wong, N.D. Nutrition, Diet Quality, and Cardiovascular Health. In *Molecular Basis of Nutrition and Aging: A Volume in the Molecular Nutrition Series*; Academic Press: Cambridge, MA, USA, 2016; ISBN 9780128018279.
2. Gupta, U.C.; Gupta, S.C. Sources and Deficiency Diseases of Mineral Nutrients in Human Health and Nutrition: A Review. *Pedosphere* **2014**. [CrossRef]
3. Bailey, R.L.; West, K.P.; Black, R.E. The epidemiology of global micronutrient deficiencies. *Ann. Nutr. Metab.* **2015**, *66*, 22–33. [CrossRef] [PubMed]
4. Hefferon, K.L. Nutritionally enhanced food crops; progress and perspectives. *Int. J. Mol. Sci.* **2015**, *16*, 3895–3914. [CrossRef] [PubMed]
5. O'Hare, T.J. Biofortification of vegetables for the developed world. *Acta Hortic.* **2015**, *1106*, 1–8. [CrossRef]
6. Tardy, A.-L.; Ballesta, A.A.; Yilmaz, G.C.; Dan, M.; Ramirez, D.M.; Lam, H.Y.; Azais-Braesco, V.; Pouteau, E. Adult's Dietary Intakes of Selected Vitamins & Minerals Essential for Energy Metabolism and Cognition: A Comparison Across Countries & Genders (FS10-04-19). *Curr. Dev. Nutr.* **2019**, *3*, 1501. [CrossRef]
7. FAO. *Europe and Central Asia: Regional Overview of Food Security and Nutrition*; Food and Agriculture Organization: Roma, Italy, 2018; ISBN 9789251311530.
8. Castiglione, D.; Platania, A.; Conti, A.; Falla, M.; D'Urso, M.; Marranzano, M. Dietary micronutrient and mineral intake in the mediterranean healthy eating, ageing, and lifestyle (Meal) study. *Antioxidants* **2018**, *7*, 79. [CrossRef]
9. Gharibzahedi, S.M.T.; Jafari, S.M. The importance of minerals in human nutrition: Bioavailability, food fortification, processing effects and nanoencapsulation. *Trends Food Sci. Technol.* **2017**, *62*, 119–132. [CrossRef]
10. Vlaic, R.A.; Mureşan, C.C.; Muste, S.; Mureşan, A.; Muresan, V.; Suharoschi, R.; Petruţ, G.; Mihai, M. Food Fortification through Innovative Technologies. In *Food Engineering*; IntechOpen: London, UK, 2019. [CrossRef]
11. Kyriacou, M.C.; Rouphael, Y. Towards a new definition of quality for fresh fruits and vegetables. *Sci. Hortic.* **2018**, *234*, 463–469. [CrossRef]
12. Fan, S. Ending hunger and undernutrition by 2025: The role of horticultural value chains. *Acta Hortic.* **2016**, *1126*, 9–20. [CrossRef]
13. Wang, G.; Xu, M.; Wang, W.; Galili, G. Fortifying horticultural crops with essential amino acids: A review. *Int. J. Mol. Sci.* **2017**, *18*, 1306. [CrossRef]
14. Fairweather-Tait, S.J.; Cashman, K. Minerals and trace elements. *World Rev. Nutr. Diet.* **2015**, *111*, 45–52. [CrossRef] [PubMed]
15. Schreinemachers, P.; Simmons, E.B.; Wopereis, M.C.S. Tapping the economic and nutritional power of vegetables. *Glob. Food Sec.* **2018**, *16*, 36–45. [CrossRef]
16. Ocean, N.; Howley, P.; Ensor, J. Lettuce be happy: A longitudinal UK study on the relationship between fruit and vegetable consumption and well-being. *Soc. Sci. Med.* **2019**, *222*, 335–345. [CrossRef]
17. Lalji, C.; Pakrashi, D.; Smyth, R. Can eating five fruit and veg a day really keep the doctor away? *Econ. Model.* **2018**, *70*, 320–330. [CrossRef]
18. Zhang, D.; Feng, Y.; Li, N.; Sun, X. Fruit and vegetable consumptions in relation to frequent mental distress in breast cancer survivors. *Support. Care Cancer* **2021**. [CrossRef]
19. Cicco, A. Eurostat Statistics Explained. Available online: https://ec.europa.eu/eurostat/statistics-explained/index.php/The_fruit_and_vegetable_sector_in_the_EU_-_a_statistical_overview (accessed on 20 January 2021).
20. Ogutu, S.O.; Ochieng, D.O.; Qaim, M. Supermarket contracts and smallholder farmers: Implications for income and multidimensional poverty. *Food Policy* **2020**, *95*, 101940. [CrossRef]
21. Sharma, A.; Verma, R.K. Biofortification: A Promising Approach toward Eradication of Hidden Hunger. In *Microbial Interventions in Agriculture and Environment: Volume 1: Research Trends, Priorities and Prospects*; Springer: Singapore, 2019; ISBN 9789811383915.
22. Sanahuja, G.; Farré, G.; Berman, J.; Zorrilla-López, U.; Twyman, R.M.; Capell, T.; Christou, P.; Zhu, C. A question of balance: Achieving appropriate nutrient levels in biofortified staple crops. *Nutr. Res. Rev.* **2013**, *26*, 235–245. [CrossRef]
23. Boland, M.J.; Golding, M.; Singh, H. *Food Structures, Digestion and Health*; Academic Press: Cambridge, MA, USA, 2014; ISBN 9780124046856.

24. D'Imperio, M.; Renna, M.; Cardinali, A.; Buttaro, D.; Serio, F.; Santamaria, P. Calcium biofortification and bioaccessibility in soilless "baby leaf" vegetable production. *Food Chem.* **2016**, *213*, 149–156. [CrossRef]
25. Ruini, L.F.; Ciati, R.; Pratesi, C.A.; Marino, M.; Principato, L.; Vannuzzi, E. Working toward Healthy and Sustainable Diets: The "Double Pyramid Model" Developed by the Barilla Center for Food and Nutrition to Raise Awareness about the Environmental and Nutritional Impact of Foods. *Front. Nutr.* **2015**. [CrossRef]
26. Govindaraj, M. Is Fortification or Bio Fortification of Staple Food Crops will Offer a Simple Solution to Complex Nutritional Disorder in Developing Countries? *J. Nutr. Food Sci.* **2015**, *5*, 2–5. [CrossRef]
27. White, P.J.; Broadley, M.R. Biofortification of crops with seven mineral elements often lacking in human diets–Iron, zinc, copper, calcium, magnesium, selenium and iodine. *New Phytol.* **2009**, *182*, 49–84. [CrossRef] [PubMed]
28. Siwela, M.; Pillay, K.; Govender, L.; Lottering, S.; Mudau, F.N.; Modi, A.T.; Mabhaudhi, T. Biofortified crops for combating hidden hunger in South Africa: Availability, acceptability, micronutrient retention and bioavailability. *Foods* **2020**, *9*, 815. [CrossRef] [PubMed]
29. Timpanaro, G.; Bellia, C.; Foti, V.T.; Scuderi, A. Consumer behaviour of purchasing biofortified food products. *Sustainability* **2020**, *12*, 6297. [CrossRef]
30. Gómez-Galera, S.; Rojas, E.; Sudhakar, D.; Zhu, C.; Pelacho, A.M.; Capell, T.; Christou, P. Critical evaluation of strategies for mineral fortification of staple food crops. *Transgenic Res.* **2010**, *19*, 165–180. [CrossRef] [PubMed]
31. Maynard, D.N. Enhancing the scope of vegetable cultivar evaluation in Florida. *HortTechnology* **2002**, *12*, 560–561. [CrossRef]
32. Carvalho, S.M.P.; Vasconcelos, M.W. Producing more with less: Strategies and novel technologies for plant-based food biofortification. *Food Res. Int.* **2013**, *54*, 961–971. [CrossRef]
33. Scarano, A.; Chieppa, M.; Santino, A. Looking at flavonoid biodiversity in horticultural crops: A colored mine with nutritional benefits. *Plants* **2018**, *7*, 98. [CrossRef]
34. Davies, K.; Espley, R. Opportunities and challenges for metabolic engineering of secondary metabolite pathways for improved human health characters in fruit and vegetable crops. *N. Z. J. Crop Hortic. Sci.* **2013**, *41*, 154–177. [CrossRef]
35. Shwetha, H.J.; Shilpa, S.; Arathi, B.P.; Raju, M.; Lakshminarayana, R. Biofortification of Carotenoids in Agricultural and Horticultural Crops. In *Vitamins and Minerals Biofortification of Edible Plants*; John Wiley & Sons: Hoboken, NJ, USA, 2020.
36. Yang, Q.Q.; Zhang, C.Q.; Chan, M.L.; Zhao, D.S.; Chen, J.Z.; Wang, Q.; Li, Q.F.; Yu, H.X.; Gu, M.H.; Sun, S.S.M.; et al. Biofortification of rice with the essential amino acid lysine: Molecular characterization, nutritional evaluation, and field performance. *J. Exp. Bot.* **2016**. [CrossRef]
37. Prasad, R.; Shivay, Y.S. Agronomic biofortification of plant foods with minerals, vitamins and metabolites with chemical fertilizers and liming. *J. Plant Nutr.* **2020**, *43*, 1534–1554. [CrossRef]
38. Rouphael, Y.; Kyriacou, M.C. Enhancing quality of fresh vegetables through salinity eustress and biofortification applications facilitated by soilless cultivation. *Front. Plant Sci.* **2018**, *9*, 1–6. [CrossRef] [PubMed]
39. Wiesner-Reinhold, M.; Schreiner, M.; Baldermann, S.; Schwarz, D.; Hanschen, F.S.; Kipp, A.P.; Rowan, D.D.; Bentley-Hewitt, K.L.; McKenzie, M.J. Mechanisms of selenium enrichment and measurement in brassicaceous vegetables, and their application to human health. *Front. Plant Sci.* **2017**, *8*, 1365. [CrossRef] [PubMed]
40. Li, R.; Li, D.W.; Liu, H.P.; Hong, C.L.; Song, M.Y.; Dai, Z.X.; Liu, J.W.; Zhou, J.; Weng, H.X. Enhancing iodine content and fruit quality of pepper (*Capsicum annuum* L.) through biofortification. *Sci. Hortic.* **2017**, *214*, 165–173. [CrossRef]
41. Niu, J.; Liu, C.; Huang, M.; Liu, K.; Yan, D. Effects of Foliar Fertilization: A Review of Current Status and Future Perspectives. *J. Soil Sci. Plant Nutr.* **2020**. [CrossRef]
42. Beto, J.A. The Role of Calcium in Human Aging. *Clin. Nutr. Res.* **2015**. [CrossRef]
43. Yates, A.; Suitor, C.; Schlicker, S.; Moats, S. *Dietary Reference Intakes for Calcium, Phosphorus, Magnesium, Vitamin D, and Fluoride*; The National Academies Press: Washington, DC, USA, 1997; ISBN 9780309063500.
44. Marschner, P. *Marschner's Mineral Nutrition of Higher Plants*, 3rd ed.; Academic Press: Cambridge, MA, USA, 2011; ISBN 9780123849052.
45. Lawlor, D.W.; Mengel, K.; Kirkby, E.A. Principles of plant nutrition. *Ann. Bot.* **2004**. [CrossRef]
46. White, P.J.; Broadley, M.R. Calcium in plants. *Ann. Bot.* **2003**, *92*, 487–511. [CrossRef]
47. Olle, M.; Bender, I. Causes and control of calcium deficiency disorders in vegetables: A review. *J. Hortic. Sci. Biotechnol.* **2009**, *84*, 577–584. [CrossRef]
48. Demidchik, V.; Shabala, S.; Isayenkov, S.; Cuin, T.A.; Pottosin, I. Calcium transport across plant membranes: Mechanisms and functions. *New Phytol.* **2018**, *220*, 49–69. [CrossRef]
49. Kerton, M.; Newbury, H.J.; Hand, D.; Pritchard, J. Accumulation of calcium in the centre of leaves of coriander (*Coriandrum sativum* L.) is due to an uncoupling of water and ion transport. *J. Exp. Bot.* **2009**, *60*, 227–235. [CrossRef]
50. Welch, R.M. Micronutrient Nutrition of Plants. *CRC Crit. Rev. Plant Sci.* **1995**, *14*, 49–82. [CrossRef]
51. Neeser, C.; Savidov, N.; Driedger, D. Production of hydroponically grown calcium fortified lettuce. *Acta Hortic.* **2007**, *744*, 317–322. [CrossRef]
52. Frossard, E.; Bucher, M.; Mächler, F.; Mozafar, A.; Hurrell, R. Potential for increasing the content and bioavailability of Fe, Zn and Ca in plants for human nutrition. *J. Sci. Food Agric.* **2000**, *80*, 861–879. [CrossRef]
53. Yang, J.; Punshon, T.; Guerinot, M.L.; Hirschi, K.D. Plant calcium content: Ready to remodel. *Nutrients* **2012**, *4*, 1120–1136. [CrossRef]

54. Rios, J.J.; Lochlainn, S.O.; Devonshire, J.; Graham, N.S.; Hammond, J.P.; King, G.J.; White, P.J.; Kurup, S.; Broadley, M.R. Distribution of calcium (Ca) and magnesium (Mg) in the leaves of Brassica rapa under varying exogenous Ca and Mg supply. *Ann. Bot.* **2012**, *109*, 1081–1089. [CrossRef]
55. Borghesi, E.; Carmassi, G.; Uguccioni, M.C.; Vernieri, P.; Malorgio, F. Effects of calcium and salinity stress on quality of lettuce in soilless culture. *J. Plant Nutr.* **2013**. [CrossRef]
56. Yuan, W.; Yuan, S.; Liu, Z.; Chen, L.; Qiu, Z. Effect of foliar application of $CaCl_2$ on lettuce growth and calcium concentrations with organic and conventional fertilization. *HortScience* **2018**, *53*, 891–894. [CrossRef]
57. Al Alawi, A.M.; Majoni, S.W.; Falhammar, H. Magnesium and Human Health: Perspectives and Research Directions. *Int. J. Endocrinol.* **2018**, *2018*, 1–17. [CrossRef]
58. Shaul, O. Magnesium transport and function in plants: The tip of the iceberg. *BioMetals* **2002**, *15*, 309–323. [CrossRef]
59. Gransee, A.; Führs, H. Magnesium mobility in soils as a challenge for soil and plant analysis, magnesium fertilization and root uptake under adverse growth conditions. *Plant Soil* **2013**, *368*, 5–21. [CrossRef]
60. Kuhn, A.J.; Schröder, W.H.; Bauch, J. The kinetics of calcium and magnesium entry into mycorrhizal spruce roots. *Planta* **2000**, *210*, 488–496. [CrossRef] [PubMed]
61. Oland, K.; Opland, T.R. Uptake of Magnesium by Apple Leaves. *Physiol. Plant.* **1956**, *9*, 401–411. [CrossRef]
62. Kobayashi, N.I.; Tanoi, K. Critical issues in the study of magnesium transport systems and magnesium deficiency symptoms in plants. *Int. J. Mol. Sci.* **2015**, *16*, 23076–23093. [CrossRef] [PubMed]
63. Guo, W.; Chen, S.; Hussain, N.; Cong, Y.; Liang, Z.; Chen, K. Magnesium stress signaling in plant: Just a beginning. *Plant Signal. Behav.* **2015**, *10*, e992287. [CrossRef]
64. Blasco, B.; Graham, N.S.; Broadley, M.R. Antioxidant response and carboxylate metabolism in Brassica rapa exposed to different external Zn, Ca, and Mg supply. *J. Plant Physiol.* **2015**, *176*, 16–24. [CrossRef]
65. Kleiber, T.; Golcz, A.; Krzesiński, W. Effect of magnesium nutrition of onion (*Allium cepa* L.). Part I. Yielding and nutrient status. *Ecol. Chem. Eng. S* **2012**, *19*, 97–105. [CrossRef]
66. Zlámalová, T.; Elbl, J.; Baroň, M.; Bělíková, H.; Lampíř, L.; Hlušek, J.; Lošák, T. Using foliar applications of magnesium and potassium to improve yields and some qualitative parameters of vine grapes (*Vitis vinifera* L.). *Plant Soil Environ.* **2015**, *61*, 451–457. [CrossRef]
67. Altarugio, L.M.; Loman, M.H.; Nirschl, M.G.; Silvano, R.G.; Zavaschi, E.; de Carneiro, L.M.S.; Vitti, G.C.; de Luz, P.H.C.; Otto, R. Yield performance of soybean and corn subjected to magnesium foliar spray. *Pesqui. Agropecu. Bras.* **2017**, *52*, 1185–1191. [CrossRef]
68. Zicker, S.; Schoenherr, B. Focus on nutrition: The role of iodine in nutrition and metabolism. *Compend. Contin. Educ. Vet.* **2012**, *2012*, 1–4.
69. Trumbo, P.; Yates, A.A.; Schlicker, S.; Poos, M. Dietary reference intakes: Vitamin A, vitamin K, arsenic, boron, chromium, copper, iodine, iron, manganese, molybdenum, nickel, silicon, vanadium, and zinc. *J. Am. Diet. Assoc.* **2001**, *101*, 3–294. [CrossRef]
70. Whitehead, D.C. The distribution and transformations of iodine in the environment. *Environ. Int.* **1984**, *10*, 321–339. [CrossRef]
71. Smoleń, S.; Kowalska, I.; Sady, W. Assessment of biofortification with iodine and selenium of lettuce cultivated in the NFT hydroponic system. *Sci. Hortic.* **2014**, *166*, 9–16. [CrossRef]
72. Dobosy, P.; Kröpfl, K.; Óvári, M.; Sandil, S.; Németh, K.; Engloner, A.; Takács, T.; Záray, G. Biofortification of green bean (*Phaseolus vulgaris* L.) and lettuce (*Lactuca sativa* L.) with iodine in a plant-calcareous sandy soil system irrigated with water containing KI. *J. Food Compos. Anal.* **2020**, *88*, 103434. [CrossRef]
73. Medrano-Macías, J.; Leija-Martínez, P.; González-Morales, S.; Juárez-Maldonado, A.; Benavides-Mendoza, A. Use of iodine to biofortify and promote growth and stress tolerance in crops. *Front. Plant Sci.* **2016**, *7*, 1146. [CrossRef]
74. Mackowiak, C.L.; Grossl, P.R.; Cook, K.L. Iodine toxicity in a plant-solution system with and without humic acid. *Plant Soil* **2005**, *269*, 141–150. [CrossRef]
75. Caffagni, A.; Pecchioni, N.; Meriggi, P.; Bucci, V.; Sabatini, E.; Acciarri, N.; Ciriaci, T.; Pulcini, L.; Felicioni, N.; Beretta, M.; et al. Iodine uptake and distribution in horticultural and fruit tree species. *Ital. J. Agron.* **2012**, *7*, 229–236. [CrossRef]
76. Signore, A.; Renna, M.; D'Imperio, M.; Serio, F.; Santamaria, P. Preliminary evidences of biofortification with iodine of "carota di polignano", an Italian carrot landrace. *Front. Plant Sci.* **2018**, *9*, 1–8. [CrossRef]
77. Golubkina, N.A.; Seredin, T.M.; Antoshkina, M.S.; Kosheleva, O.V.; Teliban, G.C.; Caruso, G. Yield, quality, antioxidants and elemental composition of new leek cultivars under organic or conventional systems in a Greenhouse. *Horticulturae* **2018**, *4*, 39. [CrossRef]
78. Voogt, W.; Holwerda, H.T.; Khodabaks, R. Biofortification of lettuce (*Lactuca sativa* L.) with iodine: The effect of iodine form and concentration in the nutrient solution on growth, development and iodine uptake of lettuce grown in water culture. *J. Sci. Food Agric.* **2010**, *90*, 906–913. [CrossRef]
79. Li, R.; Liu, H.P.; Hong, C.L.; Dai, Z.X.; Liu, J.W.; Zhou, J.; Hu, C.Q.; Weng, H.X. Iodide and iodate effects on the growth and fruit quality of strawberry. *J. Sci. Food Agric.* **2017**, *97*, 230–235. [CrossRef]
80. Gonnella, M.; Renna, M.; D'imperio, M.; Santamaria, P.; Serio, F. Iodine biofortification of four brassica genotypes is effective already at low rates of potassium iodate. *Nutrients* **2019**, *11*, 451. [CrossRef] [PubMed]
81. Roohani, N.; Hurrell, R.; Kelishadi, R.; Schulin, R. Zinc and its importance for human health: An integrative review. *J. Res. Med. Sci.* **2013**, *18*, 144. [PubMed]

82. Sharma, A.; Patni, B.; Shankhdhar, D.; Shankhdhar, S.C. Zinc—An Indispensable Micronutrient. *Physiol. Mol. Biol. Plants* **2013**, *19*, 11–20. [CrossRef]
83. White, P.J.; Pongrac, P.; Sneddon, C.C.; Thompson, J.A.; Wright, G. Limits to the biofortification of leafy brassicas with zinc. *Agricultures* **2018**, *8*, 32. [CrossRef]
84. Broadley, M.R.; White, P.J.; Hammond, J.P.; Zelko, I.; Lux, A. Zinc in plants. *New Phytol.* **2007**, *173*, 677–702. [CrossRef]
85. Gupta, N.; Ram, H.; Kumar, B. Mechanism of Zinc absorption in plants: Uptake, transport, translocation and accumulation. *Rev. Environ. Sci. Biotechnol.* **2016**, *15*, 89–109. [CrossRef]
86. White, P.J.; Broadley, M.R. Physiological limits to zinc biofortification of edible crops. *Front. Plant Sci.* **2011**, *2*, 1–11. [CrossRef]
87. Waters, B.M.; Sankaran, R.P. Moving micronutrients from the soil to the seeds: Genes and physiological processes from a biofortification perspective. *Plant Sci.* **2011**, *180*, 562–574. [CrossRef]
88. Tsonev, T.; Lidon, F.J.C. Zinc in plants—An overview. *Emirates J. Food Agric.* **2012**, *24*, 322–333.
89. De Sousa Lima, F.; Nascimento, C.W.A.; Da Silva Sousa, C. Zinc fertilization as an alternative to increase the concentration of micronutrients in edible parts of vegetables. *Rev. Bras. Ciencias Agrar.* **2015**, *10*, 403–408. [CrossRef]
90. Mao, H.; Wang, J.; Wang, Z.; Zan, Y.; Lyons, G.; Zou, C. Using agronomic biofortification to boost zinc, selenium, and iodine concentrations of food crops grown on the loess plateau in China. *J. Soil Sci. Plant Nutr.* **2014**, *14*, 459–470. [CrossRef]
91. Rugeles-Reyes, S.M.; Cecílio Filho, A.B.; López Aguilar, M.A.; Silva, P.H.S. Foliar application of zinc in the agronomic biofortification of arugula. *Food Sci. Technol.* **2019**, *39*, 1011–1017. [CrossRef]
92. Barrameda-Medina, Y.; Blasco, B.; Lentini, M.; Esposito, S.; Baenas, N.; Moreno, D.A.; Ruiz, J.M. Zinc biofortification improves phytochemicals and amino-acidic profile in Brassica oleracea cv. Bronco. *Plant Sci.* **2017**, *258*, 45–51. [CrossRef] [PubMed]
93. Rayman, M.P. Selenium and human health. *Lancet* **2012**, *379*, 1256–1268. [CrossRef]
94. Pandey, R. Mineral Nutrition of Plants. In *Plant Biology and Biotechnology: Plant Diversity, Organization, Function and Improvement*; Springer: Berlin/Heidelberg, Germany, 2015; ISBN 9788132222866.
95. Lopes, G.; Ávila, F.W.; Guilherme, L.R.G. Selenium behavior in the soil environment and its implication for human health. *Ciência Agrotecnol.* **2017**, *41*, 605–615. [CrossRef]
96. Sors, T.G.; Ellis, D.R.; Salt, D.E. Selenium uptake, translocation, assimilation and metabolic fate in plants. *Photosynth. Res.* **2005**, *86*, 373–389. [CrossRef]
97. Li, H.F.; McGrath, S.P.; Zhao, F.J. Selenium uptake, translocation and speciation in wheat supplied with selenate or selenite. *New Phytol.* **2008**, *178*, 92–102. [CrossRef]
98. White, P.J.; Bowen, H.C.; Parmaguru, P.; Fritz, M.; Spracklen, W.P.; Spiby, R.E.; Meacham, M.C.; Mead, A.; Harriman, M.; Trueman, L.J.; et al. Interactions between selenium and sulphur nutrition in Arabidopsis thaliana. *J. Exp. Bot.* **2004**, *55*, 1927–1937. [CrossRef]
99. Terry, N.; Zayed, A.M.; De Souza, M.P.; Tarun, A.S. Selenium in higher plants. *Ann. Rev. Plant Biol.* **2000**, *51*, 401–432. [CrossRef]
100. Deng, X.; Liu, K.; Li, M.; Zhang, W.; Zhao, X.; Zhao, Z.; Liu, X. Difference of selenium uptake and distribution in the plant and selenium form in the grains of rice with foliar spray of selenite or selenate at different stages. *Field Crop. Res.* **2017**, *211*, 165–171. [CrossRef]
101. White, P.J. Selenium accumulation by plants. *Ann. Bot.* **2016**, *117*, 217–235. [CrossRef] [PubMed]
102. Garousi, F.; Kovács, B.; Domokos-Szabolcsy, E.; Veres, S. Biological changes of green pea (*Pisum sativum* L.) by selenium enrichment. *Acta Biol. Hung.* **2017**, *68*, 60–72. [CrossRef] [PubMed]
103. Skrypnik, L.; Novikova, A.; Tokupova, E. Improvement of phenolic compounds, essential oil content and antioxidant properties of sweet basil (*Ocimum basilicum* L.) depending on type and concentration of selenium application. *Plants* **2019**, *8*, 458. [CrossRef] [PubMed]
104. Schiavon, M.; Berto, C.; Malagoli, M.; Trentin, A.; Sambo, P.; Dall'Acqua, S.; Pilon-Smits, E.A.H. Selenium biofortification in radish enhances nutritional quality via accumulation of methyl-selenocysteine and promotion of transcripts and metabolites related to glucosinolates, phenolics amino acids. *Front. Plant Sci.* **2016**, *7*, 1371. [CrossRef]
105. da Silva, D.F.; Cipriano, P.E.; de Souza, R.R.; Siueia Júnior, M.; da Silva, R.F.; Faquin, V.; de Souza Silva, M.L.; Guimarães Guilherme, L.R. Anatomical and physiological characteristics of *Raphanus sativus* L. submitted to different selenium sources and forms application. *Sci. Hortic.* **2020**. [CrossRef]
106. do Nascimento da Silva, E.; Aureli, F.; D'Amato, M.; Raggi, A.; Cadore, S.; Cubadda, F. Selenium Bioaccessibility and Speciation in Selenium-Enriched Lettuce: Investigation of the Selenocompounds Liberated after in Vitro Simulated Human Digestion Using Two-Dimensional HPLC-ICP-MS. *J. Agric. Food Chem.* **2017**, *65*, 3031–3038. [CrossRef]
107. Castillo-Godina, R.G.; Foroughbakhch-Pournavab, R.; Benavides-Mendoza, A. Effect of selenium on elemental concentration and antioxidant enzymatic activity of tomato plants. *J. Agric. Sci. Technol.* **2016**, *18*, 233–244.
108. Puccinelli, M.; Malorgio, F.; Pezzarossa, B. Selenium enrichment of horticultural crops. *Molecules* **2017**, *22*, 933. [CrossRef]
109. Abbaspour, N.; Hurrell, R.; Kelishadi, R. Review on iron and its importance for human health. *J. Res. Med. Sci.* **2014**, *19*, 164.
110. Hell, R.; Stephan, U.W. Iron uptake, trafficking and homeostasis in plants. *Planta* **2003**, *216*, 541–551. [CrossRef]
111. Cornell, R.M.; Schwertmann, U. The Iron Oxides. *Miner. Soil Environ.* **2003**, *1*, 379–435.
112. Shuman, L.M. Micronutrient Fertilizers. *J. Crop Prod.* **1998**. [CrossRef]
113. Rengel, Z.; Batten, G.D.; Crowley, D.E. Agronomic approaches for improving the micronutrient density in edible portions of field crops. *Field Crop. Res.* **1999**. [CrossRef]

114. Colombo, C.; Palumbo, G.; He, J.Z.; Pinton, R.; Cesco, S. Review on iron availability in soil: Interaction of Fe minerals, plants, and microbes. *J. Soils Sediments* **2014**, *14*, 538–548. [CrossRef]
115. Connorton, J.M.; Balk, J. Iron Biofortification of Staple Crops: Lessons and Challenges in Plant Genetics. *Plant Cell Physiol.* **2019**, *60*, 1447–1456. [CrossRef]
116. Kobayashi, T.; Nishizawa, N.K. Iron uptake, translocation, and regulation in higher plants. *Annu. Rev. Plant Biol.* **2012**, *63*, 131–152. [CrossRef]
117. Briat, J.F. Iron Nutrition and Implications for Biomass Production and the Nutritional Quality of Plant Products. In *The Molecular and Physiological Basis of Nutrient Use Efficiency in Crops*; John Wiley & Sons, Inc.: Hoboken, NJ, USA, 2011; ISBN 081381992X.
118. Ferrandon, M.; Chamel, A.R. Cuticular retention, foliar absorption and translocation of Fe, Mn and Zn supplied in organic and inorganic form. *J. Plant Nutr.* **1988**. [CrossRef]
119. Moosavi, A.A.; Ronaghi, A. Influence of foliar and soil applications of iron and manganese on soybean dry matter yield and iron-manganese relationship in a calcareous soil. *Aust. J. Crop Sci.* **2011**, *5*, 1550.
120. Niyigaba, E.; Twizerimana, A.; Mugenzi, I.; Ngnadong, W.A.; Ye, Y.P.; Wu, B.M.; Hai, J.B. Winter wheat grain quality, zinc and iron concentration affected by a combined foliar spray of zinc and iron fertilizers. *Agronomy* **2019**, *9*, 250. [CrossRef]
121. Carrasco-Gil, S.; Rios, J.J.; Álvarez-Fernández, A.; Abadía, A.; García-Mina, J.M.; Abadía, J. Effects of individual and combined metal foliar fertilisers on iron- and manganese-deficient Solanum lycopersicum plants. *Plant Soil* **2016**. [CrossRef]
122. Kromann, P.; Valverde, F.; Alvarado, S.; Vélez, R.; Pisuña, J.; Potosí, B.; Taipe, A.; Caballero, D.; Cabezas, A.; Devaux, A. Can Andean potatoes be agronomically biofortified with iron and zinc fertilizers? *Plant Soil* **2017**. [CrossRef]
123. Giordano, M.; El-Nakhel, C.; Pannico, A.; Kyriacou, M.C.; Stazi, S.R.; De Pascale, S.; Rouphael, Y. Iron biofortification of red and green pigmented lettuce in closed soilless cultivation impacts crop performance and modulates mineral and bioactive composition. *Agronomy* **2019**, *9*, 290. [CrossRef]
124. Bost, M.; Houdart, S.; Oberli, M.; Kalonji, E.; Huneau, J.F.; Margaritis, I. Dietary copper and human health: Current evidence and unresolved issues. *J. Trace Elem. Med. Biol.* **2016**, *35*, 107–115. [CrossRef] [PubMed]
125. Reed, S.T.; Martens, D.C. Copper and Zinc. In *Methods of Soil Analysis*; John Wiley & Sons, Ltd.: Hoboken, NJ, USA, 2018; pp. 703–722. ISBN 9780891188667.
126. Alloway, B.J. Sources of Heavy Metals and Metalloids in Soils. In *Heavy Metals in Soils: Trace Metals and Metalloids in Soils and their Bioavailability*; Alloway, B.J., Ed.; Springer: Dordrecht, The Netherlands, 2013; pp. 11–50. ISBN 978-94-007-4470-7.
127. Fu, L.; Chen, C.; Wang, B.; Zhou, X.; Li, S.; Guo, P.; Shen, Z.; Wang, G.; Chen, Y. Differences in copper absorption and accumulation between copper-exclusion and copper-enrichment plants: A comparison of structure and physiological responses. *PLoS ONE* **2015**, *10*, 1–18. [CrossRef] [PubMed]
128. Printz, B.; Lutts, S.; Hausman, J.-F.; Sergeant, K. Copper Trafficking in Plants and Its Implication on Cell Wall Dynamics. *Front. Plant Sci.* **2016**, *7*, 601. [CrossRef] [PubMed]
129. Stepien, A.; Wojtkowiak, K. Effect of foliar application of Cu, Zn, and Mn on yield and quality indicators of winter wheat grain. *Chil. J. Agric. Res.* **2016**, *76*, 220–227. [CrossRef]
130. Yruela, I. Copper in plants. *Brazilian J. Plant Physiol.* **2005**, *17*, 145–156. [CrossRef]
131. Barbosa, R.H.; Tabaldi, L.A.; Miyazaki, F.R.; Pilecco, M.; Kassab, S.O.; Bigaton, D. Absorção foliar de cobre por plantas de milho: Efeitos no crescimento e rendimento. *Cienc. Rural* **2013**, *43*, 1561–1568. [CrossRef]
132. Lamichhane, J.R.; Osdaghi, E.; Behlau, F.; Köhl, J.; Jones, J.B.; Aubertot, J.N. Thirteen decades of antimicrobial copper compounds applied in agriculture. A review. *Agron. Sustain. Dev.* **2018**, *38*, 28. [CrossRef]
133. Obrador, A.; Gonzalez, D.; Alvarez, J.M. Effect of inorganic and organic copper fertilizers on copper nutrition in Spinacia oleracea and on labile copper in soil. *J. Agric. Food Chem.* **2013**, *61*, 4692–4701. [CrossRef]
134. Price, C.T.; Koval, K.J.; Langford, J.R. Silicon: A review of its potential role in the prevention and treatment of postmenopausal osteoporosis. *Int. J. Endocrinol.* **2013**, *2013*, 1–7. [CrossRef] [PubMed]
135. Martin, K.R. The chemistry of silica and its potential health benefits. *J. Nutr. Health Aging* **2007**, *11*, 94. [PubMed]
136. Kaushik, P.; Saini, D.K. Silicon as a vegetable crops modulator—A review. *Plants* **2019**, *8*, 148. [CrossRef] [PubMed]
137. Epstein, E. Silicon. *Annu. Rev. Plant Biol.* **1999**, *50*, 641–664. [CrossRef]
138. Laane, H.M. The effects of foliar sprays with different silicon compounds. *Plants* **2018**, *7*, 45. [CrossRef]
139. Tubaña, B.S.; Heckman, J.R. Silicon in soils and plants. In *Silicon and Plant Diseases*; Springer: Cham, Switzerland, 2015; ISBN 9783319229300.
140. Ma, J.F.; Yamaji, N. Silicon uptake and accumulation in higher plants. *Trends Plant Sci.* **2006**, *11*, 392–397. [CrossRef]
141. Mvondo-She, M.A.; Marais, D. The investigation of silicon localization and accumulation in citrus. *Plants* **2019**, *8*, 200. [CrossRef]
142. López-Pérez, M.C.; Pérez-Labrada, F.; Ramírez-Pérez, L.J.; Juárez-Maldonado, A.; Morales-Díaz, A.B.; González-Morales, S.; García-Dávila, L.R.; García-Mata, J.; Benavides-Mendoza, A. Dynamic modeling of silicon bioavailability, uptake, transport, and accumulation: Applicability in improving the nutritional quality of tomato. *Front. Plant Sci.* **2018**. [CrossRef]
143. Raven, J.A. Cycling silicon—The role of accumulation in plants. *New Phytol.* **2003**, *158*, 419–421. [CrossRef]
144. Motomura, H.; Fujii, T.; Suzuki, M. Silica deposition in relation to ageing of leaf tissues in Sasa veitchii (Carrière) Rehder (Poaceae: Bambusoideae). *Ann. Bot.* **2004**, *93*, 235–248. [CrossRef]
145. Gao, X.; Zou, C.; Wang, L.; Zhang, F. Silicon decreases transpiration rate and conductance from stomata of maize plants. *J. Plant Nutr.* **2006**, *29*, 1637–1647. [CrossRef]

146. D'Imperio, M.; Renna, M.; Cardinali, A.; Buttaro, D.; Santamaria, P.; Serio, F. Silicon biofortification of leafy vegetables and its bioaccessibility in the edible parts. *J. Sci. Food Agric.* **2016**, *96*, 751–756. [CrossRef] [PubMed]
147. De Souza, J.Z.; De Mello Prado, R.; Silva, S.L.; Farias, T.P.; Neto, J.G.; Souza Junior, J.P. Silicon Leaf Fertilization Promotes Biofortification and Increases Dry Matter, Ascorbate Content, and Decreases Post-Harvest Leaf Water Loss of Chard and Kale. *Commun. Soil Sci. Plant Anal.* **2019**, *50*, 164–172. [CrossRef]
148. Montesano, F.F.; D'Imperio, M.; Parente, A.; Cardinali, A.; Renna, M.; Serio, F. Green bean biofortification for Si through soilless cultivation: Plant response and Si bioaccessibility in pods. *Sci. Rep.* **2016**, *6*, 1–9. [CrossRef] [PubMed]
149. Valentinuzzi, F.; Cologna, K.; Pii, Y.; Mimmo, T.; Cesco, S. Assessment of silicon biofortification and its effect on the content of bioactive compounds in strawberry (Fragaria × ananassa 'Elsanta') fruits. *Acta Hortic.* **2018**. [CrossRef]
150. Kowalska, K.; Brodowski, J.; Pokorska-Niewiada, K.; Szczuko, M. The change in the content of nutrients in diets eliminating products of animal origin in comparison to a regular diet from the area of middle-eastern Europe. *Nutrients* **2020**, *12*, 2986. [CrossRef]
151. White, P.J.; Broadley, M.R.; Gregory, P.J. Managing the nutrition of plants and people. *Appl. Environ. Soil Sci.* **2012**, *2012*, 1–13. [CrossRef]
152. Tran, B.T.T.; Cavagnaro, T.R.; Watts-Williams, S.J. Arbuscular mycorrhizal fungal inoculation and soil zinc fertilisation affect the productivity and the bioavailability of zinc and iron in durum wheat. *Mycorrhiza* **2019**, *29*, 445–457. [CrossRef]
153. da Silva, R.R.; de Faria, A.J.G.; Alexandrino, G.D.C.; Ribeiro, E.A.; dos Santos, A.C.M.; Deusdara, T.T.; do Nascimento, I.R.; Nascimento, V.L. Enrichment of lithium in lettuce plants through agronomic biofortification. *J. Plant Nutr.* **2019**, *42*, 2102–2113. [CrossRef]
154. Rousseau, S.; Kyomugasho, C.; Celus, M.; Hendrickx, M.E.G.; Grauwet, T. Barriers impairing mineral bioaccessibility and bioavailability in plant-based foods and the perspectives for food processing. *Crit. Rev. Food Sci. Nutr.* **2020**, *60*, 826–843. [CrossRef]
155. Fernández-García, E.; Carvajal-Lérida, I.; Pérez-Gálvez, A. In vitro bioaccessibility assessment as a prediction tool of nutritional efficiency. *Nutr. Res.* **2009**, *29*, 751–760. [CrossRef]
156. Basu, T.K.; Donaldson, D. Intestinal absorption in health and disease: Micronutrients. *Baillieres Best Pract. Res. Clin. Gastroenterol.* **2003**, *17*, 957–979. [CrossRef]
157. Jackson, A.D.; McLaughlin, J. Digestion and absorption. *Surgery* **2009**, *27*, 231–236.
158. Gupta, S.; Jyothi Lakshmi, A.; Prakash, J. In vitro bioavailability of calcium and iron from selected green leafy vegetables. *J. Sci. Food Agric.* **2006**, *86*, 2147–2152. [CrossRef]
159. Platel, K.; Srinivasan, K. Bioavailability of micronutrients from plant foods: An update. *Crit. Rev. Food Sci. Nutr.* **2016**. [CrossRef] [PubMed]

foods

MDPI

Review

The Dual Nature of Amaranth—Functional Food and Potential Medicine

Justyna Baraniak * and Małgorzata Kania-Dobrowolska

Department of Pharmacology and Phytochemistry, Institute of Natural Fibres and Medicinal Plants National Research Institute, Wojska Polskiego 71b Str., 60-630 Poznan, Poland; malgorzata.kania@iwnirz.pl
* Correspondence: justyna.baraniak@iwnirz.pl; Tel.: +48-61-665-95-50

Abstract: The beneficial health-promoting properties of plants have been known to mankind for generations. Preparations from them are used to create recipes for dietary supplements, functional food, and medicinal products. Recently, amaranth has become an area of increasing scientific and industrial interest. This is due to its valuable biological properties, rich phytochemical composition, and wide pharmacological activity. Amaranth is a pseudo-cereal crop with a dual character, combining the features of food and health-promoting product. This paper briefly and concisely reviews the current information on the chemical composition of amaranth, the value of its supplementation, the status of amaranth as a food ingredient as well as its key biological and pharmacological activities. The beneficial biological properties of amaranth preparations described in this paper may be an incentive to conduct further in-depth scientific research in this field and also to promote the development of innovative technologies in the food and cosmetics industry with the use of this plant.

Keywords: amaranth; pseudo-cereals; functional food; biological activity; pharmacological activity; health benefits

Citation: Baraniak, J.; Kania-Dobrowolska, M. The Dual Nature of Amaranth—Functional Food and Potential Medicine. *Foods* **2022**, *11*, 618. https://doi.org/10.3390/foods11040618

Academic Editor: Antonello Santini

Received: 24 January 2022
Accepted: 19 February 2022
Published: 21 February 2022

1. Introduction

In recent years, there has been noticed a growing interest in plant raw materials whose properties allow them to be used in both food and medicines. Various cereal grains are widely used in the food and beverage industry. There is a fairly broad group of plants that are classified as so-called pseudo-cereals. This means that the edible parts of these plants are the seeds and they usually are consumed in a similar way to cereals, being processed into flour. They also have similar nutritional values and taste to cereals. These are not typical cereals, but due to their similar composition and nutritional value mentioned above, they can be a good alternative. Pseudo-cereals have been the staple food of our ancestors for thousands of years, and all over the world. In different regions of the world, different pseudocereals predominate. Even today, pseudo-cereals still form the basis of nutrition in the poorest parts of the world. They have been increasingly appreciated in European countries for a long time. The best-known pseudo-cereals are amaranth, buckwheat, sorghum, millet, chia as well as khorasan. Actually, the most widely studied pseudocereals are quinoa, amaranth, chia, and buckwheat [1]. They present great potential as a natural source of a wide spectrum of biologically active compounds. Recent work suggests that first and foremost peptides and protein hydrolysates derived from these beneficial species for the human health are worth considering [1]. The first study on an amaranth protein deriving bioactive peptide with cholesterol esterase and pancreatic lipase inhibitory activities was published in 2021 by Ajayi and colleagues [2].

Due to climate change, the problem of world hunger, and changes in crop profiles in European and other countries around the world, it is becoming desirable to look for new plants with a high nutritional potential that can be combined with health benefits. Amaranth is a plant with valuable qualities as food and additionally with many valuable health properties. Additional important advantages of this plant are satisfactory yield performance, drought resistance, and enhanced photosynthesis. The excellent nutritional value of amaranth [3], the diverse chemical composition of amaranth seeds and leaves, the wide spectrum of biological activity, health-promoting properties, and the pharmacological activity of the plant have aroused the interest of researchers in recent years. This has resulted in a significant increase in the number of scientific studies on the properties and potential use of preparations from this plant. The PubMed database was used to locate publications with the most important data describing the nutritional and pharmacological activity of amaranth preparations.

2. Plant Characteristics

Amaranth has been well known since the time of the Aztecs, Mayans, and Incas [4]. In the 16–17th centuries, it spread widely in various other countries as a cereal, vegetable, weed, or crop. Amaranth seeds were used as food, but also as a sacred plant. It was used in many religious and ritual ceremonies [5]. It is a valuable plant whose potential is still not sufficiently exploited. This should be clearly emphasized because it has a huge economic value due to the various benefits it can bring to producers, food processors, and consumers. Amaranth is a member of the *Amaranthaceae* family comprising about 70 species of annual plants [4,6–9]. In many countries, *Amaranthus* species are cultivated for use as cereals, vegetables, or ornamentals, a few species are considered weeds. A review of the current literature suggests that mainly *Amaranthus cruentus, Amaranthus hypochondriacus*, and *Amaranthus caudatus* are grown for food purposes [4,9,10]. *Amaranthus blitum* Linn., *Amaranthus gangeticus* Linn., *Amaranthus mangostanus* Linn., *Amaranthus tricolor* Linn. Are cultivated all over India as a vegetable. Amaranth leaves are used in salads and to prepare other dishes, in African countries amaranth leaves are sometimes recommended for medicinal purposes [9]. Other species of amaranth, such as *A. viridis, A. tricolor, A. retroflexus*, and *A. hybridus* are known mainly as a vegetable. These species of *Amaranthus* grow very well in hot and humid regions of our globe. In Poland, amaranth is cultivated for seeds as a source of lipids and proteins for the production of flour, flakes, confectionery, expanded grains and bread, pasta, and noodles [10]. *Amaranthus cruentus* is the most widely grown species of this plant genus [5].

3. Chemical Composition of Amaranth

The main biological compounds found in amaranth are proteins, fats, carbohydrates, vitamins, and minerals [8]. The protein content (~18%) of amaranth seeds is higher than that of traditional cereals and varies according to the variety of the plant, the climate, and soil conditions and the method of fertilization [7,10]. Among proteins, albumins are the largest fraction. Protein contains all the essential amino acids required by the body [6], especially a lot of lysine and tryptophan. Starch is the main carbohydrate found in amaranth [7]. The amount of starch in amaranth seeds is approximately 45–65% [10]. An important group of compounds found in amaranth is the fiber fraction (high level)—its soluble (mainly pectins) and insoluble parts. The insoluble fraction consists of lignin, cellulose, and hemicelluloses, which have a beneficial effect on the digestive system. The amount of fiber in seeds, depending on the source of origin, averages 2–8% of dry weight [5]. The nutritional value of amaranth seed is mainly caused by lipids (~7%) [5] with a good ratio between saturated and unsaturated fatty acids and high protein content with the essential amino acids composition better than that in FAO/WHO standards [3,10]. Among unsaturated fatty acids, the most abundant are linoleic (~62%), oleic (~20%), linolenic (~1%), and arachidonic acid [5,11]. Amaranth contains saturated fatty acids (palmitic (~13%), stearic (~2.6%), arachidic (~0.7%), and myristic (~0.1%) in small amounts [5]. Among the lipid fraction of amaranth, tocopherols, tocotrienols, and sterols play an important biological role [12]. Squalene has been identified in the seeds and leaves of the plant, and they are also very rich in vitamins (especially the B group) and minerals [8]. The percentage content of squalene in oil derived from amaranth is 2–8% [6] or 6–8% [5,13], depending on the source and author. Amaranth seeds are a very good source of minerals, representing an average of 3.3% of their weight [10]. The levels of calcium, potassium, and magnesium are quite high, with iron, phosphorus present in the largest amount. Other minerals identified in amaranth include copper, zinc, sodium, chromium, manganese, nickel, lead, cadmium, and cobalt. The seeds and leaves of amaranth contain small amounts of polyphenols, saponins, hemagglutinins, phytin and nitrates (V), and oxalates. Astringent effect of amaranth also depends on the presence and activity of betacyans. Betacyans belong to the red or purple betalain pigments; the most known is betanidin. These compounds are identified in various species of amaranth [14]. Betalains have recently been recognized as highly bioactive natural compounds with potential human health benefits.

The structure of selected compounds from amaranth oil is presented in Figure 1.

(a) (b) (c) (d) (e) (f) (g)

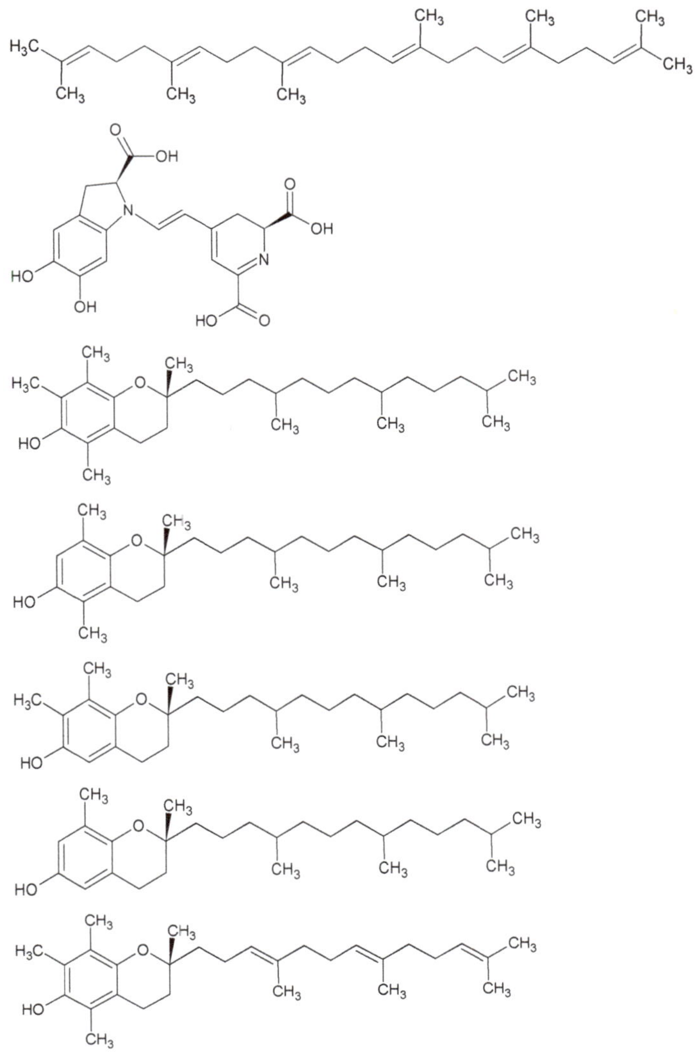

Figure 1. *Cont.*

(h)

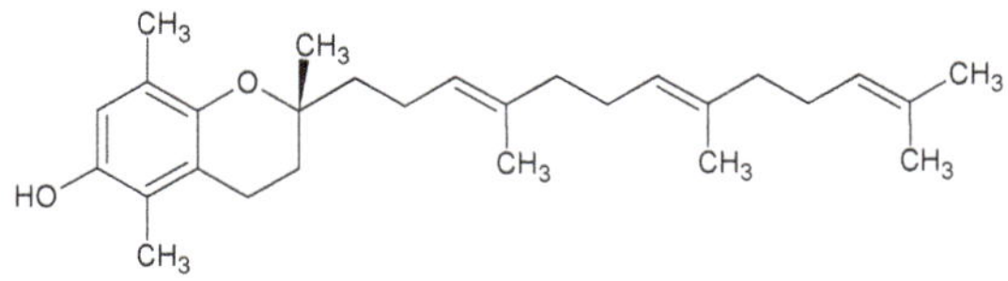

(i)

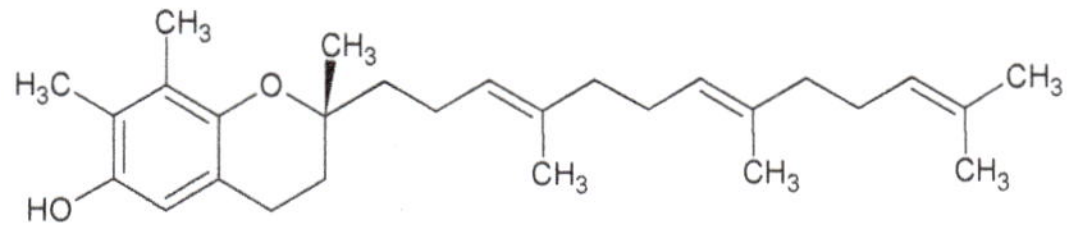

(j)

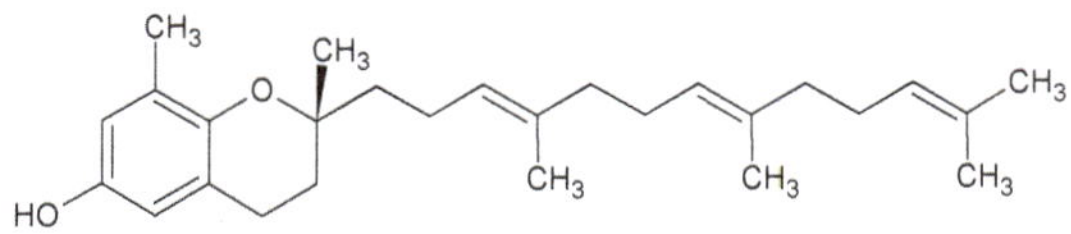

Figure 1. Structure of selected compounds from amaranth oil: (**a**) squalene; (**b**) betanidin; (**c**) α-; (**d**) β-; (**e**) γ-; (**f**) δ-tocopherol; (**g**) α-; (**h**) β-; (**i**) γ-; (**j**) δ-tocotrienol.

4. Supplementary Value of Amaranth

Amaranth seeds have a high nutritional value. The most important product obtained from amaranth is grain, which is a source of flour used in the baking industry [8]. Different plants such as millet, corn, sorghum, pseudocereals (amaranth), quinoa, and teff are the main components of a gluten diet [15]. The lack of gluten fraction makes the amaranth flour suitable for the production of dietetic food (gluten-free products) recommended for people who are allergic to gluten [8,10]. In a recent paper [16], technological and nutritional properties of an innovative gluten-free double-layered flat bread enriched with amaranth flour were examined. New formulations were developed in which rice flour (6%) and starch (6%) were partially replaced with amaranth [16]. Nowadays, such products of good quality are desirable because the number of people with celiac disease is increasing. Amaranth seeds are mainly used to produce flakes, flour, groats and muesli, and oil [7]. The high protein quality of amaranth means that it can be used alone or as a food fortifier in cereal grain mixtures. Recently, amaranth has been also used as a new alternative ingredient to compose functional cookies. The procedure basically relies on partial replacement of whole-wheat flour with formulations based on amaranth flour. The nutritional value of the fortified cookies (with amaranth flour) was found to be higher than that of traditional wheat flour cookies [17]. Oil pressed from amaranth seed is also very popular [12]. Amaranth oil is well known as a functional food [7]. A very important property (advantage) of amaranth oil is that it is highly resistant to oxidation.

During long-term storage, small changes in the fatty acid composition are observed [7]. It is worth noting that the iron content in Amaranth seed is much higher than in wheat, other seeds, spinach, and meat. For this reason, products with amaranth seeds can be an excellent dietary supplement for people with symptoms of anemia [6]. Amaranth and preparations from this plant are eaten in the form of soups, salads, puree, or tortillas [8]. Supplementation with amaranth oil contributes to lowering blood pressure, regulates lipid profile, manifests antioxidant and hepatoprotective effects. Preliminary results of the research indicate that amaranth oil may be used in the normalization of blood glucose levels [7]. Although amaranth is a valuable crop in terms of nutrient content, it is out of reach for many consumers due to its high price. The real problem is also the price of amaranth oil; it is very expensive and therefore long-term supplementation becomes impossible to implement for the average consumer.

Nowadays, with the promotion of healthy lifestyles and taking care of beauty, health-conscious consumers are increasingly choosing a healthy diet with proven health-promoting products and functional foods. An excellent example of such a product is amaranth oil, which, as a rich source of unsaturated fatty acids, tocopherols, and polyphenols, can be an excellent example of a functional food. Moreover, the extracts obtained from amaranth in the vegetation period and early flowering, due to their high content of hydroxycinnamic acid derivatives and rutin, can be a valuable source of antioxidants that can be exploited for the production of nutraceuticals or used as a functional food ingredient [3]. As it turns out, amaranth seeds can also be a source of iron, the amount of which may be important for preventing anemia. The study by Orsango and colleagues clearly presents the conclusion that the consumption of processed bread enriched with amaranth by children in underdeveloped countries decreased anemia prevalence and also increased mean hemoglobin concentration. An in-depth analysis showed that iron deficiency anemia risk was significantly decreased from 35% to 15% in a group of children treated with amaranth [18]. The most important values of amaranth for use as a food supplement are presented in Table 1.

Table 1. Key values of amaranth for use in food supplements.

Content	Source	Main Product Types
Lack of gluten	Seeds, flour	Food, food ingredient
High iron content	Any seed preparations	Food, food supplements
Rich source of unsaturated fatty acids and polyphenolic compound	Seeds, oil	Food, food supplements (nutricosmetics)
Valuable source of antioxidants	Oil	Food supplements (nutricosmetics)
Large amount of squalene	Oil	Food supplements (nutricosmetics)

5. Status of Amaranth as a Food or Food Ingredient

The seeds, oil, and leaves of this plant are used as food [3,8]. Amaranth seeds were consumed as early as the time of the Incan, Mayan, and Aztec Empires. According to the EU Novel Food Catalogue, in the case of *Amaranthus caudatus, Amaranthus cruentus, Amaranthus hypocondriacus* as food, only the use of grains from the plant is known in the EU. This product was present on the market as a food or food ingredient and was consumed to a significant degree before 15 May 1997, when the first regulation on novel food came into force. Thus, its access to the market is not subject to the Novel Food Regulation (EU) 2015/2283. However, other specific legislation may restrict the placing of this product as a food or food ingredient on the market in some Member States.

6. Biological and Pharmacological Activity

This plant has many valuable health benefits. Amaranth has been used as an astringent. This effect probably originates from the content of saponins, protoalkaloids, and betacyans [14]. According to PDR for Herbal Medicines, amaranth has been used for the treatment of diarrhea, ulcers, and in cases of pharyngitis. There are also reports on the use of the plant in excessive menstruation, skin problems such as acne and eczema, and as a mouthwash for sore mouths [19]. Saponins, protoalkaloids, and betacyans are responsible for the pharmacological activity of amaranth [14]. There are reports in the scientific literature regarding the beneficial activity of amaranth on the cardiovascular and nervous systems, hypoglycemic effect, antimicrobial activity, antioxidant activity. Amaranth is widely used in the pharmaceutical industry to produce medicinal products against atherosclerosis, stomach ulcers, tuberculosis, as well as antiseptic, antifungal, and anti-inflammatory preparations [6]. According to Khare 2004, the seeds of *Amaranthus hypochondriacus* L. in Unani medicine are considered as a spermatogenetic drug and tonic. A decoction is used in heavy menstrual bleeding, flowers are treated as remedium for diarrhea, dysentery, cough, and hemorrhages. *Amaranthus polygamus* Willd. is used as a spasmolytic, emmenagogue, galactagogue factor [20]. *Amaranthus spinosus* Linn. is taken to reduce heavy menstrual bleeding and in cases of excessive vaginal discharge, also as a diuretic medium. The whole plants of *Amaranthus blitum* Linn., *Amaranthus gangeticus* Linn., *Amaranthus mangostanus* Linn., and *Amaranthus tricolor* Linn. are considered as astringent, diuretic, demulcent, and cooling [20]. *Amaranthus* tricolor Linn. is placed and described in the Ayurvedic Pharmacopoeia of India. Amaranth seed oil exhibits hypolipemic, anti-atherosclerotic, hypotensive, and antioxidant activity [7]. Therefore, its consumption may lead to inhibition or delay in the development of diet-related diseases of civilization.

6.1. Hepatoprotective Activity

Various species of amaranth exhibit hepatoprotective activity. Information on such activity can be found in many scientific papers. Zeashan et al., (2009) demonstrated the hepatoprotective activity of whole plant extract, which was evaluated at 6, 7, 8, 9, and 10 mg/mL concentrations against CCl_4-induced toxicity in freshly isolated rat hepatocytes and HepG2 cells. Ethanolic extract of *Amaranthus spinosus* showed hepatoprotective activity in a dose-dependent manner [21]. In the study by Aneja et al., (2013), the hepatoprotective activity of aqueous extract of roots of *Amaranthus tricolor* Linn. was analyzed in paracetamol overdose-induced hepatotoxicity in a Wistar albino rat model [22]. The extract examined significantly prevented the physical, biochemical, histological, and functional changes induced by paracetamol in the liver of rats, thereby exhibiting hepatoprotective activity [22]. Other authors also mention the hepatoprotective activity of amaranth, which is attributed to the oil and extracts of the plant [7]. Enrichment of the diet with amaranth oil regulates the lipid profile and has a protective effect on the liver. Primarily, amaranth oil modulates physicochemical properties of lipids and cell membranes of hepatocytes. As a result, it stabilizes cell membranes and acts as a hepatoprotective agent [7]. Squalene is known to exhibit antioxidant and hepatoprotective properties, and also regulates cholesterol levels and helps remove toxic substances from the body [6]. Since a significant squalene content has been found in amaranth oil [13], this liver-protective activity is probably due to this.

6.2. Antioxidant Activity

Zeashan and colleagues documented the antioxidant activity of amaranth extract (obtained from the whole plant). In the study conducted by Zeashan et al., (2009), this extract showed significant antioxidant activity in the DPPH assay. In the next study by Lucero-Lopez et al., antioxidant properties of *Amaranthus hypochondriacus* seed extract were also examined. The study was conducted on the liver of rats sub-chronically exposed to ethanol. The results obtained in the experiments confirm the beneficial effect of the tested extract, which as a rich source of polyphenols, had a protective effect on the livers of rats [23]. Sarker and Oba's work characterized the phytochemical composition of *Amaranthus gangeticus* L. species. They particularly focused on the identification of phenolic compounds responsible for the antioxidant activity of these plants. Twenty-five different phenolic compounds were identified in the plant. Antioxidant components of *A. gangeticus* genotypes exhibited good radical scavenging activities [24]. In another study, the same authors presented chemical compounds found in amaranth *A. tricolor* (betaxanthins, betalains) that exhibit antioxidant activity [25]. In the study by Al-Mamun et al., the antioxidant activity of the methanol extract derived from the seed and stem of *A. hybridus* and *A. lividus* was tested. The DPPH radical scavenging assay showed that both extracts examined possessed significant dose-dependent antioxidant potential, exhibiting IC_{50} values of 28 ± 1.5 and 93 ± 3.23 μg/mL, respectively [26]. In a subsequent scientific paper, two polysaccharides from *A. hybridus* named AHP-H-1 and AHP-H-2 were characterized and examined as potential antioxidant factors. The results obtained in the study confirmed that the two polysaccharides purified from *A. hybridus* have strong antioxidant activity (DPPH radical scavenging activity and superoxide anion free radical scavenging activity) [27]. Kumari and colleagues confirmed the antioxidant properties of another amaranth species, *A. viridis*. Aqueous, chloroform, methanol, and hexane extracts were examined in several in vitro model systems. *A. viridis* exhibited dose-dependent effective antioxidant properties. Major components responsible for his antioxidant activity are gulonic and chlorogenic acids and also kaempferol [28]. In another paper describing the antioxidant activity of amaranth, the phenolic composition of the aerial part of *Amaranthus caudatus* was tested using ABTS+,

DPPH, and O_2 scavenging activity, ferric-reducing antioxidant power (FRAP), and Fe_2+ chelating ability methods. Different levels of antioxidant activity were observed depending on the stage of plant development and the content of biologically active substances (mainly a wide range of phenolic composition) responsible for generating such activity [3]. Studies focusing on the antioxidant capacity of amaranth over the period 2015–2020 were collected and summarized in Park et al.'s work. In this review, current knowledge on the antioxidant properties of different amaranth species was systematized and consolidated. These properties resulted not only from the presence of phenolic compounds but were also derived from hydrolysates and active peptides with superior antioxidant activity [8].

6.3. Anticancer Potential

Water extracts of two amaranth species (*A. lividus* and *A. hybridus*) were examined as anticancer factors. Female Swiss albino mice divided into a few groups were injected with EAC cells and received 25, 50, or 100 µg/mL/day/mouse of test extracts after 24 h of EAC cells injection. The measurement of cancer cells growth inhibition was conducted. Administration of *A. hybridus* and *A. lividus* extracts led to 45 and 43% growth inhibition of EAC cells [26]. The seed extract of *A. hybridus* possessed higher growth inhibitory activity than the stem extract of *A. lividus* and exhibited 14, 26, and 45% growth inhibition at 25, 50, and 100 µg/mL, respectively. In animals treated with amaranth extracts, morphological changes suggestive of apoptosis were also observed in EAC cells. Amaranth preparations can be considered as a potential target for cancer cure studies [26].

6.4. Antihyperglicemic and Hypolipidemic Activity

There are scientific papers in the databases on the sugar-lowering and cholesterol-lowering effects of amaranth-containing products. Methanolic extract of *Amarantus viridis* leaves (at the dose of 200 mg/kg and 400 mg/kg per day, 21 days) reduced blood sugar levels in streptozotocin-induced diabetic rats. The administration of the extract also reduced serum cholesterol and triglyceride levels [29]. Girija et al. investigated the anti-diabetic and anti-cholesterolemic activity of the methanol extract of leaves (200 and 400 mg/kg, for 21 days) from three species of amaranth: *A. caudatus, A. spinosus*, and *A. viridis* [30]. Experiments were conducted in streptozotocin-induced diabetic rats. Methanol extracts of all three species of amaranth showed significant glucose and cholesterol-lowering activity at a dose of 400 mg/kg [30]. Similar issues are presented in another paper published in 2011. Antihyperglycemic and hypolipidemic activity of the methanolic extract of leaves of *Amaranthus viridis* was investigated. Normal and streptozotocin-induced diabetic rats were fed with 200 mg/kg and 400 mg/kg of extract *per os* for 21 days. The authors of this study proved that the tested extract showed antiglycemic activity and improved the lipid profile in rats [29]. Studies on the activity of selected proteins from amaranth (*Amaranthus cruentus*) suggest hypocholesterolemic activity of this plant. Manolio Soares and colleagues showed that proteins from the plant affect the action of a key enzyme in cholesterol biosynthesis, 3-hydroxy-3-methyl-glutaryl-CoA reductase [31]. The hypolipemic effect of amaranth oil is associated with its significant squalene content. The mechanism of activity of squalene relies on the inhibition of HMG-CoA activity—a liver enzyme responsible for cholesterogenesis. Such activity has been demonstrated in both rat and clinical studies [7]. In another paper, the effects of consumption of the *Amaranthus mangostanus* on lipid metabolism and gut microbiota in high-fat diet-fed mice were examined. Amaranth powder supplementation significantly reduced the levels of triglycerides, total cholesterol, and phospholipids in the liver of rats and also downregulated the expression of a few lipogenesis-related genes [32]. Recent research findings suggest that the aqueous extract obtained from steamed red amaranth leaves might be used as a potent nutritional supplement to prevent diabetic retinopathy. Anti-glycative and anti-oxidative action of that extract against a high glucose-induced injury was examined in a human lens epithelial cell line HLE-B3 [33].

6.5. Neuroprotective and Antidepressant Action

An attempt was made to determine the neuroprotective effect of *A. lividus* L. and *A. tricolor* L. extracts against AGEs-induced cytotoxicity and oxidative stress. Advanced glycation end-products (AGEs) caused oxidative stress and cytotoxicity in neuronal cells. It was found that examined extracts protect human neuroblastoma SH-SY5Y cells against AGEs-induced cytotoxicity [34]. The authors suggest that amaranth may be useful for treating chronic inflammation associated with neurodegenerative disorders [34]. In another paper by the same authors, the neuroprotective effect of amaranth was again described. The methanol extracts of *A. lividus* and *A. tricolor* leaves were found to decrease cell toxicity and intracellular ROS production in human neuronal immortalized SH-SY5Y cells. Examined extracts decreased oxidative stress by suppressing gene expression of HMOX-1, RAGE, and RelA. Because of such activity and the high content of antioxidant substances, amaranth extracts may be a potential neuroprotective factor [35]. The methanol extract of *Amaranthus spinosus* (100 and 200 mg/kg, orally) was investigated for antidepressant activity. In the study, forced swimming test (FST) and tail suspension test (TST) models were used in experimental rats. The results of the tests prove the antidepressive potential of the methanol extract of this plant. The authors indicate that the mechanism of this activity has not yet been understood and its explanation requires further in-depth studies [36].

6.6. Anti-Inflammatory Activity

A. lividus and *A. tricolor* extracts possess anti-inflammatory activity and can reduce pro-inflammatory cytokine gene expression. An increased amount of proinflammatory cytokines, such as IL-1, IL-6, and TNF was observed [34]. In 2021, information on bioactive peptides with anti-inflammatory activity from germinated amaranth released by in vitro gastrointestinal digestion was described in the scientific literature for the first time [37].

6.7. Antimicrobial and Antiviral Effect

A new antimicrobial peptide with strong activity against *E. coli* was found in the medicinal plant *Amaranthus tricolor.* This peptide was selected after analysis of the protein fraction from *A. tricolor* and characterized as being highly antimicrobial [38]. The antimicrobial activity of ethanolic and aqueous extracts of *Amaranthus caudatus* was also examined in a study by Jimoh and colleagues [39]. *Streptococcus pyogenes, Staphylococcus aureus, Bacillus subtilis, Streptococcus pneumoniae, Escherichia coli,* and *Pseudomonas aeruginosa* were tested in this study. The used strains of fungi were: *Candida albicans, Penicillium chrysogenum, Candida glabrata,* and *Penicillium aurantiogriseum.* The ethanolic extract of amaranth showed stronger antimicrobial activity than the aqueous extract. The extracts also showed antifungal activity with an MIC in the range of 0.675 to 10 mg/mL [39]. A new application of amaranth seed oil (apolar fraction from *Amaranthus cruentus* L. seeds extract) as an agent against *Candida albicans* was examined by De Vita and colleagues. Amaranth oil in combination with an antifungal drug named terbinafine possesses synergic fungistatic and fungicidal activity and can be a potentially important ingredient of antifungal formulations [40]. In the next study, stem and seed methanol extracts of *A. lividus* and *A. hybridus* were examined as antimicrobial factors. In vitro susceptibility of five pathogenic bacteria (*E. coli, P. aeruginosa, B. subtilis, S. typhi, S. aureus*) was confirmed in the disk diffusion assay [26]. There have also been recent reports of the antiviral activity of amaranth. Chang and colleagues investigated the antiviral properties of betacyanin fractions from leaves of red spinach, *Amaranthus dubius* [41]. Betacyanin fractions from *A. dubius* inhibited DENV-2 in vitro. Betacyanin fractions exhibited antiviral activity against DENV-2 after virus adsorption to the host cells in a dose-dependent manner. For betacyanin fractions from red spinach, the IC50 value was 14.62 $\mu g\ mL^{-1}$, with an SI of 28.51. The authors point out that the mechanism of infectivity inhibition by the betacyanins must be confirmed by rigorous scientific studies [41]. In other experimental work, the antimicrobial activity of *A. tricolor* crude extract against *S. aureus* was assessed by disk diffusion, minimum inhibitory concentration (MIC) determinations, and growth curves. The authors of the experiment proved that the extract has antibacterial activity and the mechanism of this activity was connected with cell membrane depolarization, reduction in intracellular pH, decrease in bacterial protein content, DNA cleavage, and leakage of cytoplasm. The plant extract has the potential to be a good food preservative that improves meat quality [42]. The major biological effects of amaranth are summarized in Table 2.

Table 2. Key biological effects of amaranth.

Activity	Active Agent	References
Astringent	Saponins, protoalkaloids and betacyans	[14,20]
For skin problems	Naphthalene, squalene, sulfonates of *Amaranthus* spp.	[19]
Hypolipidemic, antihyperglycemic	Methanol extract of *A. viridis* leaves Methanol extract of *Amaranthus* spp. Proteins from *A. cruentus* squalene Leaves aqueous extracts	[29] [30] [31] [7] [33]
Action against microorganisms	*A. tricolor* isolated peptide Ethanolic, aqueous extract of *A. caudatus* Seed oil from *A. cruentus* Methanol extract of *A. lividus* and *A. hybridus* Betacyanins isolated from *A. dubius* *A. tricolor* crude extract	[38] [39] [40] [26] [41] [42]
Neuroprotective or antidepressant	*A. lividus, A. tricolor* extracts *A. lividus* and *A. tricolor* leaves methanol extract *A. spinosus* methanol extract	[34] [35] [36]
Anti-inflammatory	*A. lividus* and *A. tricolor* extracts Bioactive peptides	[34] [37]
Antioxidant	Whole plant extract Seed extract Phenolic compounds Betaxanthins, betalains Seed or stem methanol extract of *A. hybridus* AHP-H-1, AHP-H-2 polysaccharides from *A. hybridus* Different extracts of *A. viridis* Phenolic compounds *A. caudatus*	[21] [23] [24] [25] [26] [27] [28] [3]
Anticancer	Water extract of *A. lividus* Water extract of *A. hybridus*	[26]

7. Amaranth Seed Oil in the Cosmetics Industry

Due to its rich nutritional properties, some amaranth preparations are used in the cosmetics industry. Amaranth oil contains a large amount of unsaturated fatty acids, tocopherols, phytosterols, and squalene. These compounds are beneficial for hair and skin conditions. Amaranth seed oil may be used in the care of all skin types. It perfectly moisturizes, soothes irritations, accelerates wound healing, and has antimicrobial properties. It provides skin-nourishing and anti-aging effects. It contributes to the regeneration, nourishment, and strengthening of the epidermis and acts as an antioxidant. For example, innovative sunscreen formulations based on nanostructured lipid carriers (NLCs) which act as delivery systems for antioxidant and anti-UV bioactives were examined by Lacatusu and colleagues [43]. Amaranth oil and pumpkin seed oil were fitted in the lipid NLCs core, forming new delivery systems that were able to simultaneously entrap UVA and UVB filters and an antioxidant. It is an innovative and non-invasive design of herbal cosmetic formulations with superior photoprotection and enhanced antioxidant properties [43]. Amaranth seed oil is mostly found in skin creams and lotions, and is used as an ingredient in shampoos and shower gels. Amaranth oil as a natural, rich source of tocopherols, protects hair from the harmful effects of sunlight, is an effective way to solve problems associated with greasy hair, strengthens hair, and protects it from excessive hair loss. In addition, oil is also used in beauty clinics. It is usually used in body massages, baths, and relaxation treatments.

8. Future Remarks

Amaranth is characterized by many advantages (Figure 2). Therefore, it would be valuable to design a far-sighted cultivation strategy of this plant and to prepare a kind of global campaign promoting the advantages of amaranth, intended for food producers, cosmetics and pharmaceutical companies, and dietary supplement manufacturers. It is also extremely important to promote work leading to the development of new technologies and support research and development activities reflected in the production of food and cosmetics from this plant. This should lead to rapid and effective commercialization of these technologies and their introduction to the market in the form of specific products. In addition, new in-depth scientific research on all types of biological activities of amaranth preparations on human health is still needed.

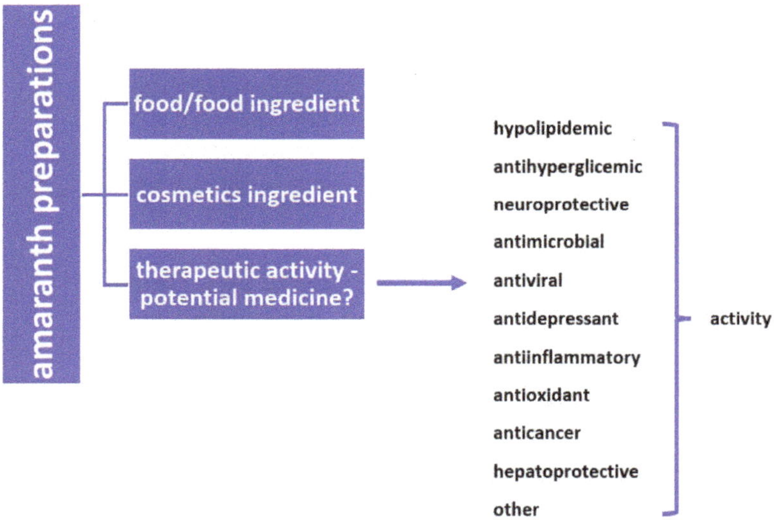

Figure 2. Multiple benefits of amaranth.

9. Conclusions

The properties of amaranth combine the characteristics of a health-promoting food and a raw material with potential therapeutic activity. Thus, all amaranth products might be utilized as natural agents in the pharmaceutical and food industries. The excellent nutritional value of amaranth and its health-promoting qualities should induce food manufacturers to develop new technologically innovative food products, especially functional foods. In addition, it is necessary to conduct detailed studies on the pharmacological activity of this plant. This will allow to determine the therapeutic doses of this raw material, which can be used in the formulation of medicinal products intended for application in the treatment of specific diseases. Ongoing statistics show that more than 60% of currently used anticancer drugs are related to plant products as their source.

Amaranth is a valuable plant with two faces—it has been a food for centuries, and at the same time in the future, it can be used to produce plant medicines. Amaranth may find wide application in the prevention and treatment of some civilization diseases, such as ischemic heart disease, allergy, type II diabetes, and celiac disease. However, further in-depth activity studies of this plant and preparations obtained from it are required.

Amaranth may also be a key factor in reducing hunger in underdeveloped countries. Amaranth should be recognized as one of the extremely promising nutritional and healthy crops with a great potential to feed the global population. This potential is still underexploited. Moreover, amaranth preparations are successfully used in the cosmetics industry. This is due to the presence of biological compounds with beneficial nutritional potential in this plant.

Author Contributions: All of the authors contributed significantly to the research; J.B. conceptualization and original draft preparation; M.K.-D. writing—review and editing. All authors have read and agreed to the published version of the manuscript.

Funding: This research received no external funding.

Institutional Review Board Statement: Not applicable.

Informed Consent Statement: Not applicable.

Data Availability Statement: Not applicable.

Conflicts of Interest: The authors declare no conflict of interest.

References

1. Morales, D.; Miguel, M.; Garcés-Rimón, M. Pseudocereals: A novel source of biologically active peptides. *Crit. Rev. Food Sci. Nutr.* **2020**, *61*, 1537–1544. [CrossRef] [PubMed]
2. Fisayo Ajayi, F.; Mudgil, P.; Gan, C.Y.; Maqsood, S. Identification and characterization of cholesterol esterase and lipase inhibitory peptides from amaranth protein hydrolysates. *Food Chem. X.* **2021**, *12*, 100165. [CrossRef] [PubMed]
3. Karamac, M.; Gai, F.; Longato, E.; Meineri, G.; Janiak, M.A.; Amarowicz, R.; Peiretti, P.G. Antioxidant activity and phenolic composition of amaranth (*Amaranthus caudatus*) during plant growth. *Antioxidants* **2019**, *8*, 173. [CrossRef]
4. Kulczyński, B.; Gramza-Michałowska, A.; Grdeń, M. Amarantus—Wartość odżywcza i właściwości prozdrowotne. *Bromat. Chem. Toksykol.* **2017**, *1*, 1–7.
5. Kaźmierczak, A.; Bolesławska, I.; Przysławski, J. Szarłat—Jego wykorzystanie w profilaktyce i leczeniu wybranych chorób cywilizacyjnych. *Nowiny Lekarskie* **2011**, *80*, 192–198.
6. Szwejkowska, B.; Bielski, S. Wartość prozdrowotna nasion szarłatu (*Amaranthus cruentus* L.). *Postepy Fitoter.* **2012**, *4*, 240–243.
7. Moszak, M.; Zawada, A.; Grzymisławski, M. Właściwości oraz zastosowanie oleju rzepakowego i oleju z amarantusa w leczeniu zaburzeń metabolicznych związanych z otyłością (The properties and the use of rapeseed oil and amaranth oil in the treatment of metabolic disorders related to obesity). *Forum Zaburzeń Metab.* **2018**, *9*, 53–64.
8. Park, S.J.; Sharma, A.; Lee, H.J. A review of recent studies on the antioxidant activities of a third-millennium food: *Amaranthus* spp. *Antioxidants* **2020**, *9*, 1236. [CrossRef]
9. Aderibigbe, O.R.; Ezekiel, O.O.; Owolade, S.O.; Korese, J.K.; Sturm, B.; Hensel, O. Exploring the potentials of underutilized grain amaranth (*Amaranthus* spp.) along the value chain for food and nutrition security: A review. *Crit. Rev. Food Sci. Nutr.* **2022**, *62*, 656–669. [CrossRef]
10. Januszewska-Jóźwiak, K.; Synowiecki, J. Charakterystyka i przydatność składników szarłatu w biotechnologii żywności. *Biotechnologia* **2008**, *3*, 89–102.
11. Gajewska, R.; Lebiedzińska, A.; Malinowska, E.; Szefer, P. Ocena jakości zdrowotnej szarłatu (amarantusa). *ROCZN PZH* **2002**, *53*, 141–147.
12. Jamka, M.; Morawska, A.; Krzyzanowska-Jankowska, P.; Bajerska, J.; Przysławski, J.; Walkowiak, J.; Lisowska, A. Comparison of the effect of amaranth oil vs. rapeseed oil on selected atherosclerosis markers in overweight and obese subjects: A randomized double-blind cross-over trial. *Int. J. Environ. Res. Public Health* **2021**, *18*, 8540. [CrossRef] [PubMed]
13. Obiedzińska, A.; Waszkiewicz-Robak, B. Oleje tłoczone na zimno jako żywność funkcjonalna. *Żywność. Nauka Technol. Jakość* **2012**, *1*, 27–44.
14. *PDR for Herbal Medicines*, 2nd ed.; Medical Economics Company: Montvale, NJ, USA, 2000; pp. 75–76.
15. Woomer, J.S.; Adedeji, A.A. Current applications of gluten-free grains—A review. *Crit. Rev. Food Sci. Nutr.* **2021**, *61*, 14–24. [CrossRef]
16. Piga, A.; Conte, P.; Fois, S.; Catzeddu, P.; Del Caro, A.; Sanguinetti, A.M.; Fadda, C. Technological, nutritional and sensory properties of an innovative gluten-free double-layered flat bread enriched with amaranth flour. *Foods* **2021**, *10*, 920. [CrossRef]
17. Uriarte-Frías, G.; Hernández-Ortega, M.M.; Gutiérrez-Salmeán, G.; Santiago-Ortiz, M.M.; Morris-Quevedo, H.J.; Meneses-Mayo, M. Pre-Hispanic Foods Oyster Mushroom (*Pleurotus ostreatus*), Nopal (*Opuntia ficus*-indica) and Amaranth (*Amaranthus* sp.) as New Alternative Ingredients for Developing Functional Cookies. *J. Fungi* **2021**, *7*, 911. [CrossRef]
18. Orsango, A.Z.; Loha, E.; Lindtjørn, B.; Engebretsen, I.M.S. Efficacy of processed amaranth-containing bread compared to maize bread on hemoglobin, anemia and iron deficiency anemia prevalence among two-to-five year-old anemic children in Southern Ethiopia: A cluster randomized controlled trial. *PLoS ONE* **2020**, *15*, e0239192. [CrossRef]

19. Esiyok, D.; Ötles, S.; Akcicek, E. Herbs as a food source in Turkey. *Asian Pac. J. Cancer Prev.* **2004**, *5*, 334–339.
20. Khare, C.P. Indian Herbal Remedies. In *Rational Western Therapy, Ayurvedic and Other Raditional Usage, Botany*; Springer: Berlin/Heidelberg, Germany, 2004.
21. Zeashan, H.; Amresh, G.; Singh, S.; Rao, C.V. Hepatoprotective and antioxidant activity of *Amaranthus spinosus* against CCl4 induced toxicity. *J. Ethnopharmacol.* **2009**, *125*, 364–366. [CrossRef]
22. Aneja, S.; Vats, M.; Aggarwal, S.; Sardana, S. Phytochemistry and hepatoprotective activity of aqueous extract of *Amaranthus tricolor* Linn. roots. *J. Ayurveda Integr. Med.* **2013**, *4*, 211–215. [CrossRef]
23. Lucero-Lopez, V.R.; Razzeto, G.S.; Gimenez, M.S.; Escudero, N.L. Antioxidant properties of *Amaranthus hypochondriacus* seeds and their effect on the liver of alcohol-treated rats. *Plant Foods Hum. Nutr.* **2011**, *66*, 157–162. [CrossRef] [PubMed]
24. Sarker, U.; Oba, S. Polyphenol and flavonoid profiles and radical scavenging activity in leafy vegetable *Amaranthus gangeticus*. *BMC Plant Biol.* **2020**, *20*, 499. [CrossRef] [PubMed]
25. Sarker, U.; Oba, S. Leaf pigmentation, its profiles and radical scavenging activity in selected *Amaranthus tricolor* leafy vegetables. *Sci. Rep.* **2020**, *10*, 18617. [CrossRef] [PubMed]
26. Al-Mamun, M.A.; Husna, J.; Khatun, M.; Hasan, R.; Kamruzzaman, M.; Hoque, K.M.F.; Reza, M.A.; Ferdousi, Z. Assessment of antioxidant, anticancer and antimicrobial activity of two vegetable species of *Amaranthus* in Bangladesh. *Complementary Altern. Med.* **2016**, *16*, 157. [CrossRef]
27. Tang, Y.; Xiao, Y.; Tang, Z.; Jin, W.; Wang, Y.; Chen, H.; Yao, H.; Shan, Z.; Bu, T.; Wang, X. Extraction of polysaccharides from *Amaranthus hybridus* L. by hot water and analysis of their antioxidant activity. *PeerJ* **2019**, *7*, e7149. [CrossRef]
28. Kumari, S.; Elancheran, R.; Devi, R. Phytochemical screening, antioxidant, antityrosinase, and antigenotoxic potential of *Amaranthus viridis* extract. *Indian J. Pharmacol.* **2018**, *50*, 130–138. [CrossRef]
29. Krishnamurthy, G.; Lakshman, K.; Pruthvi, N.; Chandrika, P.U. Antihyperglicemic and hypolipidemic activity of methanolic extract of *Amaranthus viridis* leaves in experimental diabetes. *Indian J. Pharmacol.* **2011**, *43*, 450–454. [CrossRef]
30. Girija, K.; Lakshman, K.; Udaya, C.; Sachi, G.S.; Divya, T. Anti-diabetic and anti-cholesterolemic activity of methanol extracts of three species of Amaranthus. *Asian Pac. J. Trop. Biomed.* **2011**, *1*, 133–138. [CrossRef]
31. Manólio Soares, R.A.; Mendonça, S.; Andrade de Castro, L.I.; Cardoso Corrêa Carlos Menezes, A.C.; Gomes Arêas, J.A. Major peptides from amaranth (*Amaranthus cruentus*) protein inhibit HMG-CoA reductase activity. *Int. J. Mol. Sci.* **2015**, *16*, 4150. [CrossRef]
32. Yang, Y.; Fukui, R.; Jia, H.; Kato, H. Amaranth supplementation improves hepatic lipid dysmetabolism and modulates gut microbiota in mice fed a high-fat diet. *Foods* **2021**, *10*, 1259. [CrossRef]
33. Hsiao, L.W.; Tsay, G.J.; Mong, M.C.; Liu, W.H.; Yin, M.C. Aqueous extract prepared from steamed red amaranth (*Amaranthus gangeticus* L.) leaves protected human lens cells against high glucose induced glycative and oxidative stress. *J. Food Sci.* **2021**, *86*, 3686–3697. [CrossRef] [PubMed]
34. Amornrit, W.; Santiyanont, R. Effect of *Amaranthus* on advanced glycation end-products induced cytotoxicity and proinflammatory cytokine gene expression in SH-SY5Y cells. *Molecules* **2015**, *20*, 17288. [CrossRef] [PubMed]
35. Amornrit, W.; Santiyanont, R. Neuroprotective effect of *Amaranthus lividus* and *Amaranthus tricolor* and their effects on gene expression of RAGE during oxidative stress in SH-SY5Y cells. *Genet. Mol. Res.* **2016**, *15*, gmr15027562. [CrossRef]
36. Ashok Kumar, B.S.; Lakshman, K.; Velmurugan, C.; Sridhar, S.M.; Gopisetty, S. Antidepressant activity of methanolic extract of *Amaranthus spinosus*. *Basic Clin. Neurosci. Winter* **2014**, *5*, 11–17.
37. Sandoval-Sicairos, E.S.; Milán-Noris, A.K.; Luna-Vital, D.A.; Milán-Carrillo, J.; Montoya-Rodríguez, A. Anti-inflammatory and antioxidant effects of peptides released from germinated amaranth during in vitro simulated gastrointestinal digestion. *Food Chem.* **2021**, *1*, 128394. [CrossRef] [PubMed]
38. Moyer, T.B.; Heil, L.R.; Kirkpatrick, C.L.; Goldfarb, D.; Lefever, W.A.; Parsley, N.C.; Wommack, A.J.; Hicks, L.M. PepSAVI-MS reveals a proline-rich antimicrobial peptide in *Amaranthus tricolor*. *J. Nat. Prod.* **2019**, *82*, 2744–2753. [CrossRef] [PubMed]
39. Jimoh, M.O.; Afolayan, A.J.; Lewu, F.B. Toxicity and antimicrobial activities of *Amaranthus caudatus* L. (Amaranthaceae) harvested from formulated soils at different growth stages. *J Evid Based Complementary Altern. Med.* **2020**, *25*, 1–11. [CrossRef]
40. De Vita, D.; Messore, A.; Toniolo, C.; Frezza, C.; Scipione, L.; Bertea, C.M.; Micera, M.; Di Sarno, V.; Madia, V.N.; Pindinello, I.; et al. Towards a new application of amaranth seed oil as an agent against *Candida albicans*. *Nat. Prod. Res.* **2021**, *35*, 4621–4626. [CrossRef]
41. Chang, Y.J.; Pong, L.Y.; Hassan, S.S.; Choo, W.S. Antiviral activity of betacyanins from red pitahaya (*Hylocereus polyrhizus*) and red spinach (*Amaranthus dubius*) against dengue virus type 2 (GenBank accession no. MH488959). *Access Microbiol.* **2020**, *2*, 1–6. [CrossRef]
42. Guo, L.; Wang, Y.; Bi, X.; Duo, K.; Sun, Q.; Yun, X.; Zhang, Y.; Fei, P.; Han, J. Antimicrobial activity and mechanism of action of the *Amaranthus tricolor* crude extract against *Staphylococcus aureus* and potential application in cooked meat. *Foods* **2020**, *9*, 359. [CrossRef]
43. Lacatusu, I.; Arsenie, K.L.V.; Badea, G.; Popa, O.; Oprea, O.; Badea, N. New cosmetic formulations with broad photoprotective and antioxidative activities designed by amaranth and pumpkin seed oils nanocarriers. *Ind. Crops Prod.* **2018**, *123*, 424–433. [CrossRef]

foods MDPI

Review

Nutraceuticals and Food-Grade Lipid Nanoparticles: From Natural Sources to a Circular Bioeconomy Approach

Cristina Blanco-Llamero [1,2], Joel Fonseca [1], Alessandra Durazzo [3], Massimo Lucarini [3], Antonello Santini [4,*], Francisco J. Señoráns [2] and Eliana B. Souto [1,5,*]

1 Department of Pharmaceutical Technology, Faculty of Pharmacy, University of Porto, Rua de Jorge Viterbo Ferreira, 228, 4050-313 Porto, Portugal; cristina.blanco@uam.es (C.B.-L.); jfonseca@ff.up.pt (J.F.)
2 Healthy Lipids Group, Departmental Section of Food Sciences, Faculty of Sciences, Autonomous University of Madrid, 28049 Madrid, Spain; javier.senorans@uam.es
3 CREA-Research Centre for Food and Nutrition, Via Ardeatina 546, 00178 Rome, Italy; alessandra.durazzo@crea.gov.it (A.D.); massimo.lucarini@crea.gov.it (M.L.)
4 Department of Pharmacy, University of Napoli Federico II, Via D. Montesano 49, 80131 Napoli, Italy
5 REQUIMTE/UCIBIO, Faculty of Pharmacy, University of Porto, Rua de Jorge Viterbo Ferreira, 228, 4050-313 Porto, Portugal
* Correspondence: asantini@unina.it (A.S.); ebsouto@ff.up.pt (E.B.S.)

Abstract: Nutraceuticals have gained increasing attention over the last years due to their potential value as therapeutic compounds formulated from natural sources. For instance, there is a wide range of literature about the cardioprotective properties of omega-3 lipids and the antioxidant value of some phenolic compounds, which are related to antitumoral activity. However, the value of nutraceuticals can be limited by their instability under gastric pH and intestinal fluids, their low solubility and absorption. That is why encapsulation is a crucial step in nutraceutical design. In fact, pharmaceutical nanotechnology improves nutraceutical stability and bioavailability through the design and production of efficient nanoparticles (NPs). Lipid nanoparticles protect the bioactive compounds from light and external damage, including the gastric and intestinal conditions, providing a retarded delivery in the target area and guaranteeing the expected therapeutic effect of the nutraceutical. This review will focus on the key aspects of the encapsulation of bioactive compounds into lipid nanoparticles, exploring the pharmaceutical production methods available for the synthesis of NPs containing nutraceuticals. Moreover, the most common nutraceuticals will be discussed, considering the bioactive compounds, their natural source and the described biological properties.

Keywords: nutraceuticals; lipids; solid lipid nanoparticles; nanostructured lipid carriers; food-grade ingredients

Citation: Blanco-Llamero, C.; Fonseca, J.; Durazzo, A.; Lucarini, M.; Santini, A.; Señoráns, F.J.; Souto, E.B. Nutraceuticals and Food-Grade Lipid Nanoparticles: From Natural Sources to a Circular Bioeconomy Approach. *Foods* **2022**, *11*, 2318. https://doi.org/10.3390/foods11152318

Academic Editor: Aimin Shi

Received: 12 June 2022
Accepted: 1 August 2022
Published: 3 August 2022

1. Introduction

In recent decades, nanoparticles have increasingly become a subject of interest for researchers, mainly due to the use of materials and production methods that are easy to scale and safe, especially the ones that avoid solvents. These properties make these systems especially attractive for the synthesis of nutraceuticals. A nutraceutical is formed by nutrient compounds, each of which has a pharmaceutical and standardized value, with physiological benefits for human health, performance and well-being, and which is obtained from natural sources. In fact, the description of bioactive compounds includes lowering the risk of developing specific medical diseases, and these compounds include, for instance, bioactive peptides, polyphenols, omega-3 PUFA, probiotics, carotenoids, etc. However, nutraceuticals derived from natural sources are quickly oxidized and unstable, which limits their utilization. On the other hand, the bioactive ingredients present in nutraceuticals are often unable to achieve their potential outcomes due to limited aqueous solubility, leading to a poor bioavailability profile and interaction with gastro-intestinal fluids. Therefore, their encapsulation in order to increase their bioavailability and adsorption is instrumental to improve their therapeutic potential. Recent progress in the field of nutraceutical delivery has incorporated nanotechnology to overcome the drawbacks accompanying nutraceuticals. Nano-based carrier systems provide several benefits, including an undesirable taste, odor and color masking, providing a pH-triggered controlled release, improved stability, improved shelf life, preservation of volatile ingredients, protection against gastric conditions and pH before reaching the target and protection for the ingredients from different environmental parameters, including oxygen, heat, water and light.

Nanotechnology can improve the bioavailability of nutraceuticals by their encapsulation using nanocarriers for proper delivery to their target, or by the transformation of these compounds into a nanoparticle form. This comprehensive review focuses on issues associated with nutraceuticals and nano-scale formulation approaches, describing recent nanodelivery systems used to encapsulate different nutraceuticals or bioactive compounds with biological value and therapeutic properties. In addition, the advantages and disadvantages of the most common encapsulation techniques and nanodelivery systems used to encapsulate nutraceuticals will be explored, taking into account the main challenges related to their stability.

2. Sources of Nutraceuticals with Biological Value

Bioactive compounds obtained from natural sources are a widespread and heterogeneous group with highly different biological properties described to be potential alternative therapeutic tools useful in both the prevention or treatment of some diseases. Their therapeutic value has already been proven effective by numerous clinical trials. Most common bioactive compounds can be classified as bioactive lipids (omega-3 fatty acids, oleic acid), carotenoids, bioactive peptides and phenolic compounds, among others [1,2]. Table 1 shows the different more common bioactive compounds, including their natural sources and their proven biological value.

Table 1. Source of common nutraceuticals and biological properties, including bioactive lipids, carotenoids, phenolic compounds and bioactive peptides.

Bioactive Compound	Sources	Biological Value	References
Lipids			
Omega-3 EPA and DHA	Fish, krill, microalgae including *Nannochloropsis gaditana*, *Isochrysis galbana*, *Tetraselmis chuii* and *Phaedodactylum tricornutum*	• Anti-inflammatory properties • Antitumoral effect on in vitro studies on breast, prostate and colon cancer • Action during nursing and pregnancy associated with tissue growth, visual and neuronal function development at a dosage of 450 mg of DHA and EPA per day • Cardiovascular effect due to the inhibition of the atheroma plaque formation, prevention of arrythmias and antithrombotic effect at a dosage of 250 mg per day • High daily dosage of EPA and DHA (6 g) are associated with a low risk of AMD and neurological disease such as TDAH or depression • Evidence for the treatment of autoimmune diseases such as rheumatoid arthritis and psoriasis, as well as inflammatory intestinal diseases such as Crohn's disease or ulcerative colitis	[1,3–14]
Oleic acid (18:1 Ω -9)	Olive oil	• Prevention and treatment of cardiovascular disease, lowering the blood pressure and the synthesis of TXB_2	[15–19]
Conjugated linoleic acid (CLA)	Beef, milk, lamb	• Increases β oxidation of fatty acids • Modulates the release of adipokines and cytokines • Modulates the metabolism of adipocytes • Increases spending of energy	[17,18]
Carotenoids			
Lutein and zeaxanthin	Yolk egg, marigold flower, green vegetables, especially spinach, fruits and microalgae including *Chlorella vulgaris*, *Scenedesmus almeriensis* and *Nannochloropsis gaditana*	• Antioxidant properties • Lutein and zeaxanthin are mainly associated with retinal and neurological health at a dosage from 14 to 40 mg per day • Reduction in vision loss in AMD patients after supplementation • Retinal protection from light exposure and UV radiation, reduction in the associated oxidative stress	[1,2,8,9,20–29]
Astaxanthin	Crustaceous, fish including salmon and microalgae including *Haematococcus pluvialis*	• 10 times more antioxidant than b-carotene	[20,21,23,30]

Table 1. *Cont.*

Bioactive Compound	Sources	Biological Value	References
b-carotene (pro-vitamin A)	Carrots, microalgae including *Dunaliella salina*	• Vision function • Antioxidant	[1,2,8,20,21,23,31]
Lycopene	Tomato and derivates	• Is the dietary carotenoid with the strongest antioxidant effect • Association with lower risk of prostate cancer due to its antioxidant activity, the induction of the apoptosis, the inhibition of the cellular growth, decrease in IGF-1 and IGF-BP-3, induction of phase II enzymes, modulation of androgenic metabolism	[20]
Fucoxanthin	Brown algae and microalgae including *Phaedodactylum tricornutum*	• Antidiabetic and antiobesity properties due to the stimulation of lipolysis and inhibition of lipogenesis, increase in b-oxidation of FA, inhibition of adipocytes differentiation in murine models	[1,20,21,23,32–34]
Phenolic compounds			
• Flavonoids: anthocyanins, flavanols, catechins, gallocatechins • Phenolic acids: caffeic acid, vanillin acid • Lignans • Stilbenes: Resveratrol	Fruits (grapes, red fruits, citric fruits), vegetables (soy, rosemary, salvia) coffee, tea, cocoa, olive oil	• All phenolic compounds present antioxidant, anti-inflammatory and antitumoral effects • The antioxidant properties of the phenolic compounds, including catechin and quercetin, are associated with a reduced risk of cardiovascular disease due to the inhibition of LDL oxidation, the antihypertensive effect, anti-inflammatory effect and regulation of the immune response, platelet antiaggregant • Resveratrol is associated with an anticancerogenic effect in the prevention and treatment due to the induction of the apoptosis of damaged cells, inhibition of angiogenesis in the tumoral tissues • Isoflavones found in soy are associated with an antiestrogenic effect due to their interaction with 17-b-estradiol receptors • Curcumin is reported to be a potent anti-inflammatory agent	[35–39]

Table 1. *Cont.*

Bioactive Compound	Sources	Biological Value	References
Protein compounds			
Bioactive peptides	Milk, soy, meat, eggs, algae, fish, wine, cereals	• Antihypertensive, antithrombotic, antioxidant, antiproliferative, anti-inflammatory, apiaceous, hypocholesterolemic, antithrombotic, mineral fixative effects • Inhibition of ECA: NWGPLV (soy), LKP, IKP, LRP (fish), IKW, LKW (meat), lactochinins and casoquinine (milk), ookinin, KVREGTTY (egg) • Immunomodulation: IAP, immunopeptides (wheat), YPK (broccoli), GYPMYPLR (rice) and TTMPLW (milk) • Opioids: exorphins A4, A5, B4, B5, C (wheat), casomorphins, lactoferroxin, casoxins (milk) • Antimicrobials: f 109–200 (egg), lactoferricin (milk) • Antithrombotic: K-CN, casoplatelins (milk) • Chelator of metals: casein phosphopeptides (milk) • Hypocholesterolemic: LPYPR (soy), IIAEK (milk) • Antioxidants: MY (fish), MHIRL, YVEEL, WYSLAMAASDI (milk)	[40–44]

Omega-3 fatty acids are long-chain polyunsaturated fatty acids, and their main bioactive forms are: docosahexaenoic acid, DHA (22:6 Ω-3), eicosapentaenoic acid, EPA (20:5 Ω-3) and α-linolenic acid, ALA (18:3 Ω-3). In fact, ALA is the precursor of the bioactive fatty acids EPA and DHA; nevertheless, its conversion rate in the human body has been described as low; thus, the bioactive forms of EPA and DHA must be obtained from natural sources. The traditional sources of omega-3 fatty acids, including EPA and DHA, have been fish and krill oil. This approach, however, supports the overexploiting of the ocean and abusive fishing. Therefore, new alternative biomasses are needed to obtain EPA and DHA. Microalgae are a promising source of bioactive compounds, especially omega-3 fatty acids, as they are the only nonanimal source of the bioactive form of omega-3. One of the main advantages of microalgae as omega-3 producers is their ability to grow in wastewater without competing with terrestrial plants for arable lands. Microalgae have a grow rate 2–3 times higher than that of terrestrial plants, and they are able to accumulate a wide range of bioactive compounds in their cells depending on the microalgae species. Among them, *Nannochloropsis gaditana*, *Isochrysis galbana*, *Tetraselmis chuii* and *suecia* and *Phaedodactylum tricornutum* are the main species producing EPA and DHA [1,3,4,45–48]. Omega-3 lipids have been proven to be related to the prevention and treatment of some world-recognized diseases, such as cardiometabolic disease and age-related macular degeneration, with promising results [5–11,49].

Carotenoids are a widespread group constituted by more than 400 different types of substances, which have been described in natural sources such as vegetables, egg, fish, algae and microalgae. The most common carotenoids included in nutraceuticals are lutein, zeaxanthin, fucoxanthin, astaxanthin and beta-carotene, which have been proven to be related to different biological values in clinical trials [20,21,32–34]. For instance, lutein and zeaxanthin have been shown to be alternative and effective treatments for early AMD after supplementation, and they have been proven to influence neurological and visual development during pregnancy. Lycopene has been shown to be able to reduce the risk of prostate cancer due to its antioxidant activity and ability to realize the induction of the apoptosis, the inhibition of cellular growth, the decrease in IGF-1 and IGF-BP-3, the induction of phase II enzymes and the modulation of androgenic metabolism [2,20–24,33,50–52].

It is interesting to highlight the wide biological properties that this group of bioactive peptides shows depending on the peptide studied, including antihypertensive, antioxidant, antiproliferative, anti-inflammatory, apiaceous, hypocholesteremia, antithrombotic and mineral chelator functions [40–44,53]. On the other hand, phenolic compounds, including flavonoids (anthocyanins, flavanols, catechins, gallo catechins, etc.), phenolic acids (caffeic acid, vanillin acid, etc.), lignans and stilbenes, show mainly antioxidant and antitumoral activities related to the reduced risk of cardiovascular disease due to the inhibition of LDL oxidation, an antihypertensive effect, anti-inflammatory effect and regulation of the immune response, platelet antiaggregant. Isoflavones found in soy, on the other hand, are associated with an antiestrogenic effect due to their interaction with 17-b-estradiol receptors [35–39,54–56].

3. Challenges Encountered in Nutraceutical Stability

Due to the complex nature of the different available bioactive compounds and the heterogenicity of them all (including molecular weight, charge, thermosensitivity or polarity), there are important challenges related to the properties of each one that must be considered. First of all, the low solubility of most of the bioactive compounds described should be highlighted, including, for instance, bioactive lipids or carotenoids. Due to this fact, optimal formulations must be designed to encapsulate these compounds. Indeed, if the final purpose of using a nutraceutical is its inclusion in the food matrix, the dosage of the bioactives is crucial, whereas micro/nanoemulsions must be produced to make them dispersible in the often-aqueous food matrix and resistant enough to pass through the rest of the food process. Additionally, the nutraceutical must be delivered in the target area at a controlled velocity, which will depend on the nanoparticle's constituents. Regarding bioactive lipids, their low solubility, the crystallinity of some at room temperature and their thermosensitivity, which forbids the use of high temperatures in the process, should be highlighted. Additionally, the formulation should protect the bioactive from the stomach pH and digestion, maintaining the biological properties after these processes. This is also a crucial point for bioactive peptides, which should be protected from high temperatures, physiology pH and organic solvents, which may damage their structure and make them lose their biological value. Additionally, as some of them are bitter, they benefit from an adequate vehiculation system masking their flavor. Regarding phenolic compounds and phytochemicals, their solubility is affected by their polarity, functional groups, molecular weight, if it is oxidized or reduced, or if it is complexed to another molecule. Therefore, each compound should be studied deeply before choosing one production technique or another to preserve their biological value and to load it efficiently [57–62].

Taking all these aspects into account, the main key points that must be considered in the formulation of NPs including nutraceuticals are shown below (Figure 1).

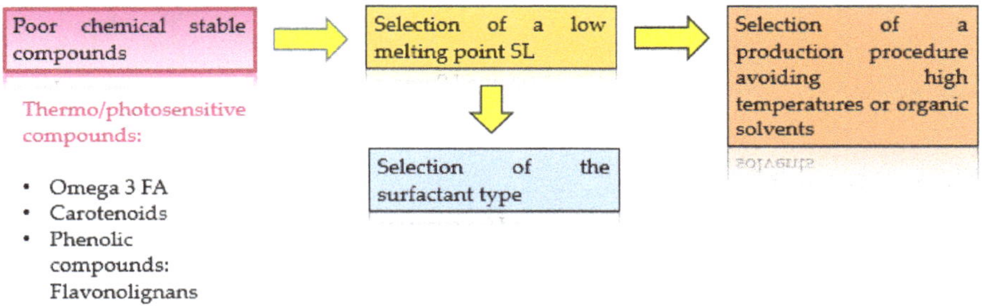

Figure 1. Diagram representing the key points to consider in nanoparticle formulation for loading unstable bioactive compounds.

The main principles of different nutraceuticals to be considered in the selection of wall materials and nanodelivery system production procedures are described below:

- Bioactive solubility: the selection of wall lipid materials relies on physiological tolerance, physiochemical structure, active principle' solubility and solid lipid–liquid lipid miscibility. Solubility of bioactive compounds in the lipidic matrix is one of the most important factors determining its loading capacity. Preliminary studies must be performed to study this factor with each compound and each solubility, taking into account their polarity, functional groups, molecular weight, if it is oxidized or reduced or if it is complexed to another molecule. Indeed, most of the bioactive compounds have low water solubility, including carotenoids, bioactive lipids and some phenolic compounds, whereas bioactive peptides are usually more hydrophilic [42–44].
- Wall materials' compatibility: in NLCs, liquid lipid and solid lipid molecules should have good miscibility and compatibility with each other. This prevents the formation of the solid lipid crystalline matrix, promoting an amorphous structure typically presented in NLCs. The ideal ratio between liquid lipids and solid lipids is reported to range from 70:30 up to a ratio of 99.9:0.1 [63].
- Bioactive stability: most of the bioactive compounds described are highly unstable during oxidation, including carotenoids, lipids and phenolic compounds. The lipid matrix plays a crucial role in the protection of the bioactive ingredients. Wall materials should be stable against chemical degradation, including oxidation and lipolysis. Medium-chain triglycerides (MCTs) are the most common oils used in NP production. They have a small molecular weight and are water-soluble. Moreover, MCTs' (e.g., Miglyol 812) digestion is faster than that of long-chain triglycerides (e.g., corn oil), and they have higher stability against oxidation. Additionally, they are generally recognized as safe (GRAS) by the US Food and Drug Administration (FDA) for direct addition into many foods, including beverages, as a carrier, solvent and emulsifier.

Oleic acid or PUFAs are commonly used in the food and pharmaceutical industries. However, their susceptibility to oxidation may cause damage to encapsulated compounds through production of free radicals; thus, their use should be limited to highly stable compounds, avoiding carotenoids or omega-3 lipids. Emulsifiers have been used for stabilization of the lipid dispersions by reduction in interfacial tension between the lipid phase and the aqueous phase during the production of the particles, leading to fine nanocarriers. It has been found that utilization of a mixture of emulsifiers can be more efficient in preventing particle aggregation. A combination of tween 80 and lecithin caused smaller particles with a lower PI and higher stability due to an increased zeta potential in comparison to their use separately. Additionally, compared to ionic surfactants, tween 80 has low toxicity and is approved for use in specific food products and is GRAS [64,65].

- Bioactive thermosensitivity: the method used for NP production should be selected based on the bioactive compound properties, thermosensitivity and water solubility, which are the main factors affecting its selection. Homogenization at high pressure is the most used method in food industry; however, the high temperatures employed in this technique have led to the development of cold homogenization, which could be more adequate for carotenoids, bioactive lipids and peptides [65–67].

4. Nanoparticles Used for the Loading of Nutraceuticals: Food-Grade Nanosystems

Overcoming the main drawbacks related to some bioactive compounds, such as low water solubility, poor chemical and oxidation stability associated with thermosensitivity and photosensitive compounds, gastric degradation and thus poor bioavailability, must be the focus. Many of these challenges, such as the low water solubility, gastric degradation and the poor bioavailability, can be overcome by encapsulating these compounds into nanoparticles as efficient delivery vehicles contributing to the industry's economy. A circular bioeconomy has become a model for commercial production that enhances reuse, recycle and recovery with a smaller environmental footprint in nutraceutical industries. Nutraceutical industry costs are mainly related to the challenges mentioned above, requiring large numbers of bioactive ingredients to achieve an adequate in vivo effect, so many natural biomasses are still unexploited [68,69]. Nutraceuticals' production using lipid nanoparticles as carriers allows for their obtention in one stage of production with higher encapsulation efficiencies, as was reported in previous works, providing economic advantages, decreasing the number of process steps and the dosage of nutrients by increasing their bioavailability [66,70–73]. NPs overcoming the main challenges related to nutraceuticals have been widely reported in the scientific literature. Recent works on NP nutraceutical release showed an increased water solubility and in vitro bioavailability of curcumin [72]. Improved oxidative stability has also been described in NP omega-3 works [74]. Thus, the encapsulation of bioactive compounds in nanocarriers allows for the incorporation of natural biomasses into the circular bioeconomy.

Interestingly, lipid NPs' structure can overcome the abovementioned poor bioavailability by different causing factors. The main critical steps in the oral absorption of nutrients could be the rate of dissolution and the rate of nutrient penetration across the bio membrane. An essential prerequisite for the absorption of a bioactive compound is its ability to exist in a stable aqueous solution. This fact depends on its aqueous solubility and dissolution rate. Lipid-based nanoparticles improve the bioavailability due to their increased surface area. Additionally, by decreasing the particle size, the thickness of the diffusion layer is decreased, leading to faster transport and faster dissolution [57,63,64].

Going deeper, lipid NPs promote enhanced gastrointestinal (GI) absorption due to induced permeability changes caused by the surfactant, and due to the increased residence time in the stomach and upper small intestine owing to their lipidic nature and their adhesion to the intestinal underlying epithelium. The bioactive compound is protected from the harsh gastric conditions as it is encapsulated in the NPs, promoting their stability. On the other hand, NPs have low stability in acidic environments, which make them degradable by gastric lipases [75]. Thus, lipid NPs are transformed by lipase and colipase into micelles (consisting of bioactive and lipid monoglycerides), stimulating bile flow to form mixed micelles. Mixed micelles are absorbed by chylomicron formation into lymphatic vessels, avoiding the first-pass effect and enhancing drug bioavailability alongside the fat absorption process. Through systemic and lymphatic transport, they increase the concentration of nutrients in the systemic circulation. On the other hand, nanoparticulate systems were reported to improve oral drug bioavailability by intracellular uptake by M cells of Peyer's patches. Indeed, transient opening of tight junctions (gaps between two adjacent intestinal epithelial cells) has been reported due to the effect of highly lipophilic surfactants, improving paracellular absorption [63].

Moreover, bioactive compound release from lipid particles occurs by diffusion and simultaneous degradation of lipid particles in the body. Controlled release from NPs can lead to a prolonged half-life and slows down the enzymatic attack in systematic circulation. The degradation rate, and therefore the kinetics of compound release, depend on the type of lipid used. The shorter the fatty acids of triacylglycerols, the faster the degradation rate. Surfactants including lecithin or sodium cholate can play an important role in accelerating the degradation rate by inducing attachment of the lipase/colipase to the nanoparticle [65,76].

Other main drawbacks must be studied to achieve an adequate formulation. Therefore, in order to produce lipid NPs with thermosensitive compounds, including omega-3 FA, carotenoids and flavonolignans, the selection of a low-melting-point lipid, such as glyceryl monooleate, glycerin, Monosteol™ or Softisan™, and a production procedure avoiding high temperatures, is a crucial step. It is also important to avoid the use of organic solvents that could damage the stability of the bioactive compounds, and to work with photosensitive compounds, opaque lipids must be selected as formulation components. Lipid components, solids and liquids (in the case of NLCs), will be selected based on these points, and based on their miscibility with the bioactive compounds.

Once the lipid components are selected, the surfactant type should be chosen as a function of the former. The lipid and surfactant content in the formulation will be the main factors affecting the chemical properties of the nanoparticles developed. Higher numbers of solid lipids (SL) will lead to a larger particle size, although it will also depend on the surfactant content [70,77]. However, to increase the NPs' load, it is necessary to increase the SL rate, and thus, compromise conditions must be selected for each case. Frequently, lipid NPs are produced through the combined use of Ultra-Turrax with sonication to reduce particle size. Nevertheless, the addition of the sonication step can also increase the polydispersity index and particle size distribution. On the other hand, when the target administration is intravenous or ocular, the rheology of the NPs produced must be studied to explore the NPs' behavior under different conditions. Ideally, the viscosity must decrease with the stress applied, whereas the loss and modulus storage must increase, with the storage always being higher than the loss ($G' > G''$), to achieve an appropriate rheological behavior upon administration [25,58–60,62,70,77–85]. The lipid-based nanosystems solid lipid nanoparticles (SLNs) and nanostructured lipid carriers (NLCs) will be described, as well as the component types and the production procedures available. Table 2 lists the latest works on the production of NPs encapsulating nutraceuticals, highlighting the ones focused on lipid nanoparticles. Recent studies have also investigated the use of resistant starch or soy-protein-based polymeric nanoparticles for the encapsulation of ferulic acid and curcumin, respectively, showing higher stability of the nutraceutical loaded [60,84].

Table 2. Overview of the recent works on nanoparticles containing nutraceuticals: Type of nanoparticles (NPs), NP components, bioactive compound encapsulated and production procedure employed are detailed.

NPs	NPs Component	Nutraceutical	Production Procedure	Major Findings	References
NLC, SLN, LNE	Imwitor 900 K, medium-chain triglyceride (MCT) Lipoid® SPC-3, pure soybean phosphatidylcholine, Tween® 80 (polysorbate 80) and Span 20	Quercetin	HPH	Maximum bioaccessibility was observed with LNE compared to SLN and NLC Bioaccessibility is affected by lipid's physical state and composition	[70]
NLC	Palmitic acid (PA), polyoxyethylene sorbitan monolaurate (Tween 20), ethanol and acetone	Β-carotene	Solvent diffusion	Lipid phase and surfactant concentrations have an important effect on particle size Liquid lipid content in NLCs and temperature significantly affect β-carotene degradation	[31]
NLC	Fish oil (FO) with omega-3 fatty acid composition, carnauba wax (CW), glycerol stearate (GS), Poloxamer 407 Tween 80	Lutein	Melting emulsification coupled with the high-shear homogenization technique	Fish oil concentration was found to enhance the lutein entrapment efficiency In vitro lutein release from lipid nanocarriers is slower than that from nanoemulsion.	[25]
NLC	2.7% Q, 9.4% soy lecithin, 23.6% glyceryl tridecanoate, 6.7% glyceryl tripalmitate, 13.4% vitamin E acetate, 44.2% Kolliphor HS15 and an aqueous mixture containing 1% of NaCl in deionized water	Quercetin	Phase-inversion-based process	Q-NLCs decrease the viability of breast cancer cells and induce their apoptosis Q-NLCs increase cellular uptake of Q by breast cancer cells Q-NLCs enhance solubility and stability of Q in aqueous solution	[54]
SLN	Compritol 888 ATO, Pluronic F68 and 1,7-octadiene	Curcumin	High-shear homogenization and ultrasonication techniques	Solid lipid nanoparticles were stable for over 1 month at 4 ± 1 °C In vitro studies showed a good release of curcumin from lipid nanoparticles Formulation increases the amount of curcumin permeated by 2 orders of magnitude	[72]
SLN, NLC	glycerol monostearate (GMS), soy lecithin, Tween 80, glycerol distearate (GDS) (Precirol Ato 5), medium-chain triglyceride (MCT) (Labrafac Lipopile WL 1349)	Zeaxanthin	Homogenization at high speed and ultrasonication	Encapsulation efficiency and loading are higher for NLCs than SLNs	[85]
SLN	Witepsol H15 and Polysorbate 80 (Tween 80)	Rosmarinic acid	Hot-melt ultrasonication method	High association of rosmarinic acid was detected, and stable particles were obtained	[58]

Lipid nanoparticles are similar to polymeric ones in terms of the solid matrix structure that they share. However, lipid nanoparticles appear to avoid the problems of toxicity of some polymers and solvents used in their production. Lipid nanoparticles are based on biodegradable lipids and emulsifiers, either lipophilic or hydrophilic. Among lipids, those mostly used were tristearin and tripalmitin, whereas amongst the surfactant agents, soy lecithin and polysorbate 80 should be emphasized as lipophilic agents and poloxamer 188 and tween 80 as hydrophilic ones [31,54,70,80–83].

The other advantage of these kinds of systems is their ability to encapsulate either lipophilic or hydrophilic molecules, and the possibility of covering them with polymers to modify their characteristics, such as the use of polyethylene glycol or chitosan to modulate the permanence of the molecules in the mucous membranes of the organism. Indeed, lipid nanoparticles are versatile systems that can be administrated by different pathways, such as intravenous, oral, cutaneous, pulmonary, ocular and transdermal pathways. They also have limitations though, since during storage, the purest lipids tend to crystalize in a perfect crystal structure that may lead to the drug or bioactive being loaded. That is why a new generation of lipid nanoparticles appeared to solve this problem, the nanolipid carriers (NLCs), whose structure is just imperfect enough to avoid the expulsion of the loaded drug, which is achieved by formulations including not only solid lipids but liquid and solid lipids in combination, which increases the encapsulation efficiency and minimizes the expulsion of the bioactive compounds inside during the storage [25,58–60,62,70,77–86].

5. Production Procedures

The major objective of the encapsulation is creating coating-sensitive compounds or reducing side effects of some useful compounds applied in high concentrations; these compounds are located in the core and coated by suitable wall materials. Encapsulation techniques protect nutraceuticals or bioactive compounds from unbalanced and unfavorable conditions, including pH, light, moisture, heat, chemical and biological degradation, and oxygen during storage, processing and utilization. Wall materials, including lipids and surfactants, have a critical role in the encapsulation technique because of their important effects on target delivery, bioavailability, biocompatibility and protection of bioactive compounds. Additionally, these materials should be safe and do not have an impact on flavor, color, texture or other properties of foods. The most important properties of suitable wall materials include a low cost, low viscosity, film-forming capacity, high solubility, low hydroscope, high stability in the media of the target, high protection, abundance, nontoxicity and compatibility in food or drug formulations. Several techniques are used for encapsulating bioactive agents; the preferred encapsulation technique depends on the bioactive compound structure and its end use. The most common encapsulation techniques for lipid NPs include emulsification, homogenization at high pressure, microemulsion and emulsion–evaporation of the solvent, sometimes combined with sonication. The main production procedures are discussed below for their use in the food industry.

5.1. Homogenization at High Pressure (HPH)

The melted lipid is emulsified in an aqueous solution containing the surfactant at the same temperature by agitation at high speed or ultrasounds. The pre-emulsion is then subjected to high-pressure homogenization. As typical production conditions, 500 bar pressure and between 3 and 5 homogenization cycles are repeated. Finally, the nanoemulsion is cooled, the lipid phase solidifies, and the suspension of lipid nanoparticles is formed. It must be highlighted that increasing homogenization cycles may lead to particle coalescence, resulting in a bigger particle size. This technique is especially aimed at the encapsulation of lipophilic molecules since the hydrophilic ones diffuse in a large proportion to the aqueous phase during the homogenization phase, giving rise to a low encapsulation efficiency. One of the drawbacks of this technology is the exposure of the active ingredients to high temperatures, although for a very short time, this allows sensitive compounds to resist the process. Additionally, the high temperatures used in hot HPH may reduce the emulsifying capacity of most surfactants, therefore causing nanocarriers' instability [64,66,67,74].

For the encapsulation of thermosensitive compounds, a cold homogenization method was designed in which the molten lipid is rapidly cooled in dry ice, the solid form of carbon dioxide, or in liquid nitrogen. In this way, the fragility of the lipid is increased to facilitate the grinding process for obtaining microparticles. These are dispersed in the cold solution of the surfactant, and finally, the suspension is subjected to high-pressure homogenization at or below room temperature [87]. HPH is the most used production technique for nanocarriers encapsulating food ingredients due to the advantages that it has compared to other methods, including large-scale production, disuse of organic solvents and shorter production time.

5.2. Preparation Technique via Microemulsion

This method requires low energy and is based on the basic mechanism of microemulsions, which can be transformed into an ultrafine nanoemulsion after their rupture by adding a certain volume of water.

In the microemulsion formation, the lipid melts and the active substance or a drug is dissolved in it. Next, the surfactant, cosurfactant and water are added at a high temperature to form the microemulsion, which is poured over cold water, breaking into nanoparticles of emulsion, which crystallize to form lipid nanoparticles. As drawbacks of this process, we can point out the high concentration of the surfactant and cosurfactant which is required, the use of solvents to form the emulsion and the high dilution to which the particles are subjected, which leads to the final content in particles being below 1%. The temperature difference between the chilled water and the microemulsion extremely influences the particle size in this method. The faster the solidification, the smaller the particle sizes. Although this method is operated under mild conditions, it requires abundant surfactant and cosurfactant, which could be a disadvantage for its use in the food industry [63].

5.3. Solvent Emulsification–Evaporation Technique

In this method, very low or no energy is required, and it is widely used for the preparation of polymeric micro- and nanoparticles. The lipid material in this case is dissolved in a water-immiscible organic solvent, in which the active ingredient is also dissolved. This organic phase is emulsified with the aqueous phase containing the surfactant agent by means of mechanical agitation or an ultrasound probe. After evaporation of the solvent at reduced pressure, the dispersion of nanoparticles occurs after the precipitation of the lipid. The preparation of double emulsion in this technique allows the encapsulation of numerous compounds. As there is no heat involved, this method is suitable for heat-sensitive active compounds. The low energy required is another great advantage of this method. The main disadvantages of this technique are solvent-residue-associated toxicity and diluted particles. These disadvantages can be reduced by the selection of a food-grade solvent such as ethanol or ethyl acetate, making this method a good option for the encapsulation of food ingredients [63,77].

5.4. Solvent Emulsification–Diffusion Technique

This technique is similar to the previous one, differing only in the method of precipitation of the lipid from the emulsion. In this case, it is achieved by adding extra water to the aqueous phase, which causes immediate diffusion of the organic solvent, with the consequent precipitation of the lipid.

In the solvent emulsification–evaporation process, the lipid is dissolved in the water-immiscible solvent, and then it is emulsified in an aqueous phase containing the surfactant, followed by evaporation of the solvent under reduced pressure. Lipid precipitation occurs upon solvent evaporation, leading to nanocarriers' formation. Merits of this method include its lab-scale acceptability, higher stability and ability to obtain the smallest particle size, but its demerits are the use of toxic solvents, the increase in lipid content, which leads to an increase in the polydispersity index, and particle size distribution [31,78].

6. Conclusions and Future Prospects

Lipid nanoparticles are especially interesting for oral administration, for different reasons, including, in the first place, the mucoadhesive properties that they present due to their colloidal nature, and to which their ability to facilitate the release in the area of the intestine to which they adhere is attributed. On the other hand, there is the possibility that they are internalized by the intestinal cells, and the promoting effect of the absorption of the constituent lipid components must also be considered. Nanotechnology has wide applications in nutrition, food supplements, nutraceuticals and medical science [88–90]. The recent literature suggests that nanotechnology will overcome the current main challenges that bioactive compounds and nutraceuticals must face, such as their stability, low solubility, targeted delivery and prolonged release. Additionally, with regard to the food industry, new products must avoid problems related to their color, flavor or nutrient content. Accommodation of each health need could be achieved with the aid of pharmaceutical nanotechnology. In fact, it seems like a promising technology approach to reduce the dose levels and to achieve better and longer stability of the nutraceuticals. The formulations of the bioactives as nanostructured products will help in their superior characterization, improved patient acceptability and, above all, high reproducibility of their therapeutic effectiveness. Thus, a lot of nutraceuticals in nanosized forms have been developed in many works regarding the optimum production procedure or the most adequate wall materials for each nanoparticle, considering lipid types and surfactants. Additionally, many nutraceutical products containing NPs are commercially available on the market. Therefore, it can be concluded that nano-based carrier systems provide better means for enhancing the efficacy and availability of nutraceuticals having issues with solubility, stability and bioavailability.

Nevertheless, components of lipid NPs should be carefully selected since they will directly influence product stability and effectiveness. For future prospects, it should be remarked that studies on orally administered NPs are still very limited, and the molecular mechanisms by which they are absorbed through the intestinal lumen into the circulation should be better clarified by studying each lipid component. Although NPs possess great potential as delivery carriers, more preclinical and clinical studies are needed to better understand their behavior. Additionally, NPs have some related challenges, such as the need to improve their colloidal stability under harsh conditions, including food processing (heating, high pressure, drying, etc.) and the gastrointestinal environment (low pH, bile salt and digestive enzymes); studying interactions between bioactive compounds and nanoparticles for optimal encapsulation; and accepting the biological fate of these nanoparticles upon oral administration. Thus, further investigation on food nanotechnology is needed with regard to the in vivo and food processing stages.

Author Contributions: Methodology, formal analysis, investigation, resources and data curation: C.B.-L., J.F., A.D., M.L. and A.S. Writing of the original manuscript: C.B.-L., F.J.S. and E.B.S. Conceptualization, review and editing of the manuscript, project administration, supervision and funding acquisition: C.B.-L., A.S., F.J.S., A.S. and E.B.S. All authors have read and agreed to the published version of the manuscript.

Funding: This work received the funding support from Santander/Autonoma University of Madrid, granted to the first author.

Institutional Review Board Statement: Not applicable.

Informed Consent Statement: Not applicable.

Data Availability Statement: Not applicable.

Conflicts of Interest: The authors declare no conflict of interest.

References

1. Jacob-Lopes, E.; Maroneze, M.M.; Deprá, M.C.; Sartori, R.B.; Dias, R.R.; Zepka, L.Q. Bioactive food compounds from microalgae: An innovative framework on industrial biorefineries. *Curr. Opin. Food Sci.* **2018**, *25*, 1–7. [CrossRef]
2. Mitra, S.; Rauf, A.; Tareq, A.M.; Jahan, S.; Bin Emran, T.; Shahriar, T.G.; Dhama, K.; Alhumaydhi, F.A.; Aljohani, A.S.; Rebezov, M.; et al. Potential health benefits of carotenoid lutein: An updated review. *Food Chem. Toxicol.* **2021**, *154*, 112328. [CrossRef] [PubMed]
3. Chua, E.T.; Schenk, P.M. A biorefinery for Nannochloropsis: Induction, harvesting, and extraction of EPA-rich oil and high-value protein. *Bioresour. Technol.* **2017**, *244*, 1416–1424. [CrossRef]
4. Chew, K.W.; Yap, J.Y.; Show, P.L.; Suan, N.H.; Juan, J.C.; Ling, T.C.; Lee, D.-J.; Chang, J.-S. Microalgae biorefinery: High value products perspectives. *Bioresour. Technol.* **2017**, *229*, 53–62. [CrossRef]
5. Georgiou, T.; Neokleous, A.; Nicolaou, D.; Sears, B. Pilot study for treating dry age-related macular degeneration (AMD) with high-dose omega-3 fatty acids. *PharmaNutrition* **2014**, *2*, 8–11. [CrossRef]
6. Jiang, H.; Shi, X.; Fan, Y.; Wang, D.; Li, B.; Zhou, J.; Pei, C.; Ma, L. Dietary omega-3 polyunsaturated fatty acids and fish intake and risk of age-related macular degeneration. *Clin. Nutr.* **2021**, *40*, 5662–5673. [CrossRef]
7. Kaarniranta, K.; Salminen, A. NF-κB signaling as a putative target for ω-3 metabolites in the prevention of age-related macular degeneration (AMD). *Exp. Gerontol.* **2009**, *44*, 685–688. [CrossRef]
8. McCusker, M.M.; Durrani, K.; Payette, M.J.; Suchecki, J. An eye on nutrition: The role of vitamins, essential fatty acids, and antioxidants in age-related macular degeneration, dry eye syndrome, and cataract. *Clin. Dermatol.* **2016**, *34*, 276–285. [CrossRef] [PubMed]
9. Querques, G.; Souied, E.H. The role of omega-3 and micronutrients in age-related macular degeneration. *Surv. Ophthalmol.* **2014**, *59*, 532–539. [CrossRef]
10. Rezende, F.A.; Lapalme, E.; Qian, C.X.; Smith, L.E.; San Giovanni, J.P.; Sapieha, P. Omega-3 Supplementation Combined with Anti–Vascular Endothelial Growth Factor Lowers Vitreal Levels of Vascular Endothelial Growth Factor in Wet Age-Related Macular Degeneration. *Am. J. Ophthalmol.* **2014**, *158*, 1071–1078.e1. [CrossRef]
11. Tuo, J.; Ross, R.J.; Herzlich, A.A.; Shen, D.; Ding, X.; Zhou, M.; Coon, S.L.; Hussein, N.; Salem, N., Jr.; Chan, C.-C. A High Omega-3 Fatty Acid Diet Reduces Retinal Lesions in a Murine Model of Macular Degeneration. *Am. J. Pathol.* **2009**, *175*, 799–807. [CrossRef] [PubMed]
12. Fan, H.; Song, J.-T. Potential mechanisms of macular degeneration protection by fatty fish consumption. *Curr. Opin. Pharmacol.* **2022**, *63*, 102186. [CrossRef] [PubMed]
13. Vanthoor-Koopmans, M.; Wijffels, R.H.; Barbosa, M.J.; Eppink, M.H. Biorefinery of microalgae for food and fuel. *Bioresour. Technol.* **2013**, *135*, 142–149. [CrossRef] [PubMed]
14. Yen, H.-W.; Hu, I.-C.; Chen, C.-Y.; Ho, S.-H.; Lee, D.-J.; Chang, J.-S. Microalgae-based biorefinery—From biofuels to natural products. *Bioresour. Technol.* **2013**, *135*, 166–174. [CrossRef] [PubMed]
15. Atef, B.; Ishak, R.A.H.; Badawy, S.S.; Osman, R. Exploring the potential of oleic acid in nanotechnology-mediated dermal drug delivery: An up-to-date review. *J. Drug Deliv. Sci. Technol.* **2022**, *67*, 103032. [CrossRef]
16. Bowen, K.J.; Kris-Etherton, P.M.; Shearer, G.C.; West, S.G.; Reddivari, L.; Jones, P.J.H. Oleic acid-derived oleoylethanolamide: A nutritional science perspective. *Prog. Lipid Res.* **2017**, *67*, 1–15. [CrossRef] [PubMed]
17. Gong, M.; Hu, Y.; Wei, W.; Jin, Q.; Wang, X. Production of conjugated fatty acids: A review of recent advances. *Biotechnol. Adv.* **2019**, *37*, 107454. [CrossRef] [PubMed]
18. Gong, M.; Wei, W.; Hu, Y.; Jin, Q.; Wang, X. Structure determination of conjugated linoleic and linolenic acids. *J. Chromatogr. B* **2020**, *1153*, 122292. [CrossRef] [PubMed]
19. Pastor, R.; Bouzas, C.; Tur, J.A. Beneficial effects of dietary supplementation with olive oil, oleic acid, or hydroxytyrosol in metabolic syndrome: Systematic review and meta-analysis. *Free Radic. Biol. Med.* **2021**, *172*, 372–385. [CrossRef]
20. Gammone, M.A.; Riccioni, G.; D'Orazio, N. Carotenoids: Potential allies of cardiovascular health? *Food Nutr. Res.* **2015**, *59*, 26762. [CrossRef]
21. Pereira, A.G.; Otero, P.; Echave, J.; Carreira-Casais, A.; Chamorro, F.; Collazo, N.; Jaboui, A.; Lourenço-Lopes, C.; Simal-Gandara, J.; Prieto, M.A. Xanthophylls from the Sea: Algae as Source of Bioactive Carotenoids. *Mar. Drugs* **2021**, *19*, 188. [CrossRef] [PubMed]
22. Leung, I.Y.-F. Macular pigment: New clinical methods of detection and the role of carotenoids in age-related macular degeneration. *Optometry* **2008**, *79*, 266–272. [CrossRef] [PubMed]
23. Liu, C.; Hu, B.; Cheng, Y.; Guo, Y.; Yao, W.; Qian, H. Carotenoids from fungi and microalgae: A review on their recent production, extraction, and developments. *Bioresour. Technol.* **2021**, *337*, 125398. [CrossRef]
24. Loane, E.; Nolan, J.M.; O'Donovan, O.; Bhosale, P.; Bernstein, P.S.; Beatty, S. Transport and Retinal Capture of Lutein and Zeaxanthin with Reference to Age-related Macular Degeneration. *Surv. Ophthalmol.* **2008**, *53*, 68–81. [CrossRef]
25. Lacatusu, I.; Mitrea, E.; Badea, N.; Stan, R.; Oprea, O.; Meghea, A. Lipid nanoparticles based on omega-3 fatty acids as effective carriers for lutein delivery. Preparation and in vitro characterization studies. *J. Funct. Foods* **2013**, *5*, 1260–1269. [CrossRef]
26. Peng, M.-L.; Chiu, H.-F.; Chou, H.; Liao, H.-J.; Chen, S.-T.; Wong, Y.-C.; Shen, Y.-C.; Venkatakrishnan, K.; Wang, C.-K. Influence/impact of lutein complex (marigold flower and wolfberry) on visual function with early age-related macular degeneration subjects: A randomized clinical trial. *J. Funct. Foods* **2016**, *24*, 122–130. [CrossRef]

27. Richer, S.; Stiles, W.; Statkute, L.; Pulido, J.; Frankowski, J.; Rudy, D.; Pei, K.; Tsipursky, M.; Nyland, J. Double-masked, placebo-controlled, randomized trial of lutein and antioxidant supplementation in the intervention of atrophic age-related macular degeneration: The Veterans LAST study (Lutein Antioxidant Supplementation Trial). *Optom.—J. Am. Optom. Assoc.* **2004**, *75*, 216–229. [CrossRef]
28. Robman, L.; Vu, H.; Hodge, A.; Tikellis, G.; Dimitrov, P.; McCarty, C.; Guymer, R. Dietary lutein, zeaxanthin, and fats and the progression of age-related macular degeneration. *Can. J. Ophthalmol.* **2007**, *42*, 720–726. [CrossRef]
29. Tian, Y.; Kijlstra, A.; Webers, C.A.B.; Berendschot, T.T. Lutein and Factor D: Two intriguing players in the field of age-related macular degeneration. *Arch. Biochem. Biophys.* **2015**, *572*, 49–53. [CrossRef]
30. Kim, B.; Youn Lee, S.; Lakshmi Narasimhan, A.; Kim, S.; Oh, Y.-K. Cell disruption and astaxanthin extraction from Haematococcus pluvialis: Recent advances. *Bioresour. Technol.* **2022**, *343*, 126124. [CrossRef]
31. Hejri, A.; Khosravi, A.; Gharanjig, K.; Hejazi, M. Optimisation of the formulation of β-carotene loaded nanostructured lipid carriers prepared by solvent diffusion method. *Food Chem.* **2013**, *141*, 117–123. [CrossRef] [PubMed]
32. Maria, A.G.; Graziano, R.; Nicolantonio, D.O. Anti-Obesity Activity of the Marine Carotenoid Fucoxanthin. *Mar. Drugs* **2015**, *13*, 2196–2214. [CrossRef]
33. Liu, M.; Li, W.; Chen, Y.; Wan, X.; Wang, J. Fucoxanthin: A promising compound for human inflammation-related diseases. *Life Sci.* **2020**, *255*, 117850. [CrossRef]
34. Maeda, H. Nutraceutical Effects of Fucoxanthin for Obesity and Diabetes Therapy: A Review. *J. Oleo Sci.* **2015**, *64*, 125–132. [CrossRef]
35. Alara, O.R.; Abdurahman, N.H.; Ukaegbu, C.I. Extraction of phenolic compounds: A review. *Curr. Res. Food Sci.* **2021**, *4*, 200–214. [CrossRef] [PubMed]
36. Bondam, A.F.; Da Silveira, D.D.; Dos Santos, J.P.; Hoffmann, J.F. Phenolic compounds from coffee by-products: Extraction and application in the food and pharmaceutical industries. *Trends Food Sci. Technol.* **2022**, *123*, 172–186. [CrossRef]
37. Dorosh, O.; Rodrigues, F.; Delerue-Matos, C.; Moreira, M.M. Increasing the added value of vine-canes as a sustainable source of phenolic compounds: A review. *Sci. Total Environ.* **2022**, *830*, 154600. [CrossRef] [PubMed]
38. Maya-Cano, D.A.; Arango-Varela, S.; Santa-Gonzalez, G.A. Phenolic compounds of blueberries (*Vaccinium* spp.) as a protective strategy against skin cell damage induced by ROS: A review of antioxidant potential and antiproliferative capacity. *Heliyon* **2021**, *7*, e06297. [CrossRef]
39. Mikołajczak, N.; Tańska, M.; Ogrodowska, D. Phenolic compounds in plant oils: A review of composition, analytical methods, and effect on oxidative stability. *Trends Food Sci. Technol.* **2021**, *113*, 110–138. [CrossRef]
40. Islam, M.S.; Wang, H.; Admassu, H.; Sulieman, A.A.; Wei, F.A. Health benefits of bioactive peptides produced from muscle proteins: Antioxidant, anti-cancer, and anti-diabetic activities. *Process Biochem.* **2022**, *116*, 116–125. [CrossRef]
41. Manzoor, M.; Singh, J.; Gani, A. Exploration of bioactive peptides from various origin as promising nutraceutical treasures: In vitro, in silico and in vivo studies. *Food Chem.* **2022**, *373*, 131395. [CrossRef]
42. Mirzaei, M.; Shavandi, A.; Mirdamadi, S.; Soleymanzadeh, N.; Motahari, P.; Mirdamadi, N.; Moser, M.; Subra, G.; Alimoradi, H.; Goriely, S. Bioactive peptides from yeast: A comparative review on production methods, bioactivity, structure-function relationship, and stability. *Trends Food Sci. Technol.* **2021**, *118*, 297–315. [CrossRef]
43. Okoye, C.O.; Ezeorba, T.P.C.; Okeke, E.S.; Okagu, I.U. Recent Findings on the Isolation, Identification and Quantification of Bioactive Peptides. *Appl. Food Res.* **2022**, *2*, 100065. [CrossRef]
44. Sosalagere, C.; Kehinde, B.A.; Sharma, P. Isolation and functionalities of bioactive peptides from fruits and vegetables: A reviews. *Food Chem.* **2022**, *366*, 130494. [CrossRef] [PubMed]
45. Blanco-Llamero, C.; García-García, P.; Señoráns, F.J. Cross-Linked Enzyme Aggregates and Their Application in Enzymatic Pretreatment of Microalgae: Comparison Between CLEAs and Combi-CLEAs. *Front. Bioeng. Biotechnol.* **2021**, *9*, 794672. [CrossRef] [PubMed]
46. Blanco-Llamero, C.; García-García, P.; Señoráns, F.J. Combination of Synergic Enzymes and Ultrasounds as an Effective Pretreatment Process to Break Microalgal Cell Wall and Enhance Algal Oil Extraction. *Foods* **2021**, *10*, 1928. [CrossRef] [PubMed]
47. Blanco-Llamero, C.; Señoráns, F. Biobased Solvents for Pressurized Liquid Extraction of *Nannochloropsis gaditana* Omega-3 Lipids. *Mar. Drugs* **2021**, *19*, 107. [CrossRef]
48. Halim, R.; Papachristou, I.; Kubisch, C.; Nazarova, N.; Wüstner, R.; Steinbach, D.; Chen, G.Q.; Deng, H.; Frey, W.; Posten, C.; et al. Hypotonic osmotic shock treatment to enhance lipid and protein recoveries from concentrated saltwater Nannochloropsis slurries. *Fuel* **2021**, *287*, 119442. [CrossRef]
49. Van Leeuwen, E.M.; Emri, E.; Merle, B.M.J.; Colijn, J.M.; Kersten, E.; Cougnard-Gregoire, A.; Dammeier, S.; Meester-Smoor, M.; Pool, F.M.; de Jong, E.K.; et al. A new perspective on lipid research in age-related macular degeneration. *Prog. Retin. Eye Res.* **2018**, *67*, 56–86. [CrossRef]
50. Eisenhauer, B.; Natoli, S.; Liew, G.; Flood, V.M. Lutein and Zeaxanthin—Food Sources, Bioavailability and Dietary Variety in Age-Related Macular Degeneration Protection. *Nutrients* **2017**, *9*, 120. [CrossRef]
51. Johnson, E.J. Role of lutein and zeaxanthin in visual and cognitive function throughout the lifespan. *Nutr. Rev.* **2014**, *72*, 605–612. [CrossRef] [PubMed]
52. Li, L.H.; Lee, J.C.-Y.; Leung, H.H.; Lam, W.C.; Fu, Z.; Lo, A.C.Y. Lutein Supplementation for Eye Diseases. *Nutrients* **2020**, *12*, 1721. [CrossRef] [PubMed]

53. Tahmasian, A.; Broadbent, J.A.; Juhász, A.; Nye-Wood, M.; Le, T.T.; Bose, U.; Colgrave, M.L. Evaluation of protein extraction methods for in-depth proteome analysis of narrow-leafed lupin (*Lupinus angustifolius*) seeds. *Food Chem.* **2022**, *367*, 130722. [CrossRef]
54. Sun, M.; Nie, S.; Pan, X.; Zhang, R.; Fan, Z.; Wang, S. Quercetin-nanostructured lipid carriers: Characteristics and anti-breast cancer activities in vitro. *Colloids Surf. B Biointerfaces* **2014**, *113*, 15–24. [CrossRef] [PubMed]
55. Jabbehdari, S.; Handa, J.T. Oxidative stress as a therapeutic target for the prevention and treatment of early age-related macular degeneration. *Surv. Ophthalmol.* **2021**, *66*, 423–440. [CrossRef] [PubMed]
56. Li, Y.; Li, X.; Li, X.; Zeng, Z.; Strang, N.; Shu, X.; Tan, Z. Non-neglectable therapeutic options for age-related macular degeneration: A promising perspective from traditional Chinese medicine. *J. Ethnopharmacol.* **2022**, *282*, 114531. [CrossRef]
57. Ali, A.; Ahmad, U.; Akhtar, J.; Badruddeen; Khan, M.M. Engineered nano scale formulation strategies to augment efficiency of nutraceuticals. *J. Funct. Foods* **2019**, *62*, 103554. [CrossRef]
58. Campos, D.A.; Madureira, A.R.; Gomes, A.M.; Sarmento, B.; Pintado, M.M. Optimization of the production of solid Witepsol nanoparticles loaded with rosmarinic acid. *Colloids Surf. B Biointerfaces* **2014**, *115*, 109–117. [CrossRef]
59. Lydia, D.E.; Khusro, A.; Immanuel, P.; Esmail, G.A.; Al-Dhabi, N.A.; Arasu, M.V. Photo-activated synthesis and characterization of gold nanoparticles from Punica granatum L. seed oil: An assessment on antioxidant and anticancer properties for functional yoghurt nutraceuticals. *J. Photochem. Photobiol. B* **2020**, *206*, 111868. [CrossRef]
60. Liu, L.-L.; Li, X.-T.; Zhang, N.; Tang, C.-H. Novel soy β-conglycinin nanoparticles by ethanol-assisted disassembly and reassembly: Outstanding nanocarriers for hydrophobic nutraceuticals. *Food Hydrocoll.* **2019**, *91*, 246–255. [CrossRef]
61. Rashidi, L. Different nano-delivery systems for delivery of nutraceuticals. *Food Biosci.* **2021**, *43*, 101258. [CrossRef]
62. Vozza, G.; Khalid, M.; Byrne, H.J.; Ryan, S.M.; Frias, J.M. Nutraceutical formulation, characterisation, and in-vitro evaluation of methylselenocysteine and selenocystine using food derived chitosan: Zein nanoparticles. *Food Res. Int.* **2019**, *120*, 295–304. [CrossRef] [PubMed]
63. Elmowafy, M.; Al-Sanea, M.M. Nanostructured lipid carriers (NLCs) as drug delivery platform: Advances in formulation and delivery strategies. *Saudi Pharm. J.* **2021**, *29*, 999–1012. [CrossRef]
64. Azar, F.A.N.; Pezeshki, A.; Ghanbarzadeh, B.; Hamishehkar, H.; Mohammadi, M. Nanostructured lipid carriers: Promising delivery systems for encapsulation of food ingredients. *J. Agric. Food Res.* **2020**, *2*, 100084. [CrossRef]
65. Zardini, A.A.; Mohebbi, M.; Farhoosh, R.; Bolurian, S. Production and characterization of nanostructured lipid carriers and solid lipid nanoparticles containing lycopene for food fortification. *J. Food Sci. Technol.* **2018**, *55*, 287–298. [CrossRef] [PubMed]
66. Pereira, I.; Zielińska, A.; Ferreira, N.R.; Silva, A.M.; Souto, E.B. Optimization of linalool-loaded solid lipid nanoparticles using experimental factorial design and long-term stability studies with a new centrifugal sedimentation method. *Int. J. Pharm.* **2018**, *549*, 261–270. [CrossRef] [PubMed]
67. Vieira, R.; Severino, P.; Nalone, L.A.; Souto, S.B.; Silva, A.M.; Lucarini, M.; Durazzo, A.; Santini, A.; Souto, E.B. Sucupira Oil-Loaded Nanostructured Lipid Carriers (NLC): Lipid Screening, Factorial Design, Release Profile, and Cytotoxicity. *Molecules* **2020**, *25*, 685. [CrossRef] [PubMed]
68. Huy, M.; Vatland, A.K.; Kumar, G. Nutraceutical productions from microalgal derived compounds via circular bioeconomy perspective. *Bioresour. Technol.* **2022**, *347*, 126575. [CrossRef]
69. Sheldon, R.A. CLEAs, Combi-CLEAs and 'Smart' Magnetic CLEAs: Biocatalysis in a Bio-Based Economy. *Catalysts* **2019**, *9*, 261. [CrossRef]
70. Aditya, N.P.; Macedo, A.S.; Doktorovova, S.; Souto, E.B.; Kim, S.; Chang, P.-S.; Ko, S. Development and evaluation of lipid nanocarriers for quercetin delivery: A comparative study of solid lipid nanoparticles (SLN), nanostructured lipid carriers (NLC), and lipid nanoemulsions (LNE). *LWT-Food Sci. Technol.* **2014**, *59*, 115–121. [CrossRef]
71. Dieng, S.M.; Anton, N.; Bouriat, P.; Thioune, O.; Sy, P.M.; Massaddeq, N.; Enharrar, S.; Diarra, M.; Vandamme, T. Pickering nano-emulsions stabilized by solid lipid nanoparticles as a temperature sensitive drug delivery system. *Soft Matter* **2019**, *15*, 8164–8174. [CrossRef] [PubMed]
72. Righeschi, C.; Bergonzi, M.C.; Isacchi, B.; Bazzicalupi, C.; Gratteri, P.; Bilia, A.R. Enhanced curcumin permeability by SLN formulation: The PAMPA approach. *LWT* **2016**, *66*, 475–483. [CrossRef]
73. Wiedenmann, V.; Oehlke, K.; Van der Schaaf, U.; Koivula, H.M.; Mikkonen, K.S.; Karbstein, H.P. Emulsifier Composition of Solid Lipid Nanoparticles (SLN) Affects Mechanical and Barrier Properties of SLN-Protein Composite Films. *J. Food Sci.* **2019**, *84*, 3642–3652. [CrossRef] [PubMed]
74. Shahparast, Y.; Eskandani, M.; Rajaei, A.; Khosroushahi, A.Y. Preparation, Physicochemical Characterization and Oxidative Stability of Omega-3 Fish Oil/α-Tocopherol-co-Loaded Nanostructured Lipidic Carriers. *Adv. Pharm. Bull.* **2019**, *9*, 393–400. [CrossRef] [PubMed]
75. Fathi, M.; Mozafari, M.R.; Mohebbi, M. Nanoencapsulation of food ingredients using lipid based delivery systems. *Trends Food Sci. Technol.* **2012**, *23*, 13–27. [CrossRef]
76. Bandi, S.P.; Bhatnagar, S.; Venuganti, V.V.K. Advanced materials for drug delivery across mucosal barriers. *Acta Biomater.* **2021**, *119*, 13–29. [CrossRef] [PubMed]
77. Fangueiro, J.F.; Andreani, T.; Egea, M.A.; Garcia, M.L.; Souto, S.B.; Souto, E.B. Experimental factorial design applied to mucoadhesive lipid nanoparticles via multiple emulsion process. *Colloids Surf. B Biointerfaces* **2012**, *100*, 84–89. [CrossRef]

78. Anton, N.; Benoit, J.-P.; Saulnier, P. Design and production of nanoparticles formulated from nano-emulsion templates—A review. *J. Control. Release* **2008**, *128*, 185–199. [CrossRef] [PubMed]
79. Fangueiro, J.F.; Gonzalez-Mira, E.; Martins-Lopes, P.; Egea, M.A.; Garcia, M.L.; Souto, S.B.; Souto, E.B. A novel lipid nanocarrier for insulin delivery: Production, characterization and toxicity testing. *Pharm. Dev. Technol.* **2013**, *18*, 545–549. [CrossRef] [PubMed]
80. González-Mira, E.; Egea, M.A.; García, M.L.; Souto, E.B. Design and ocular tolerance of flurbiprofen loaded ultrasound-engineered NLC. *Colloids Surf. B Biointerfaces* **2010**, *81*, 412–421. [CrossRef] [PubMed]
81. Gonzalez-Mira, E.; Egea, M.A.; Souto, E.B.; Calpena, A.C.; García, M.L. Optimizing flurbiprofen-loaded NLC by central composite factorial design for ocular delivery. *Nanotechnology* **2011**, *22*, 045101. [CrossRef] [PubMed]
82. Haider, M.; Abdin, S.M.; Kamal, L.; Orive, G. Nanostructured Lipid Carriers for Delivery of Chemotherapeutics: A Review. *Pharmaceutics* **2020**, *12*, 288. [CrossRef] [PubMed]
83. Müller, R.H.; Petersen, R.D.; Hommoss, A.; Pardeike, J. Nanostructured lipid carriers (NLC) in cosmetic dermal products. *Adv. Drug Deliv. Rev.* **2007**, *59*, 522–530. [CrossRef]
84. Noor, N.; Gani, A.; Jhan, F.; Shah, M.A.; Ashraf, Z.U. Ferulic acid loaded pickering emulsions stabilized by resistant starch nanoparticles using ultrasonication: Characterization, in vitro release and nutraceutical potential. *Ultrason. Sonochem.* **2022**, *84*, 105967. [CrossRef] [PubMed]
85. Osanlou, R.; Emtyazjoo, M.; Banaei, A.; Hesarinejad, M.A.; Ashrafi, F. Preparation of solid lipid nanoparticles and nanostructured lipid carriers containing zeaxanthin and evaluation of physicochemical properties. *Colloids Surf. A Physicochem. Eng. Asp.* **2022**, *641*, 128588. [CrossRef]
86. Dubey, S.K.; Pradhan, R.; Hejmady, S.; Singhvi, G.; Choudhury, H.; Gorain, B.; Kesharwani, P. Emerging innovations in nano-enabled therapy against age-related macular degeneration: A paradigm shift. *Int. J. Pharm.* **2021**, *600*, 120499. [CrossRef]
87. Hanifiyah, I.A.; Rosita, N.; Purwanti, T. Production Method of Nanostructured Lipid Carrier (NLC): Hot and Cold Homogenization Against NLC-Coenzyme Q10 Characteristics. *J. Comput. Theor. Nanosci.* **2021**, *18*, 26–31. [CrossRef]
88. Durazzo, A.; Nazhand, A.; Lucarini, M.; Atanasov, A.G.; Souto, E.B.; Novellino, E.; Capasso, R.; Santini, A. An Updated Overview on Nanonutraceuticals: Focus on Nanoprebiotics and Nanoprobiotics. *Int. J. Mol. Sci.* **2020**, *21*, 2285. [CrossRef] [PubMed]
89. Yeung, A.W.K.; Souto, E.B.; Durazzo, A.; Lucarini, M.; Novellino, E.; Tewari, D.; Wang, D.; Atanasov, A.G.; Santini, A. Big impact of nanoparticles: Analysis of the most cited nanopharmaceuticals and nanonutraceuticals research. *Curr. Res. Biotechnol.* **2020**, *2*, 53–63. [CrossRef]
90. Souto, E.B.; Silva, G.F.; Dias-Ferreira, J.; Zielinska, A.; Ventura, F.; Durazzo, A.; Lucarini, M.; Novellino, E.; Santini, A. Nanopharmaceutics: Part I—Clinical Trials Legislation and Good Manufacturing Practices (GMP) of Nanotherapeutics in the EU. *Pharmaceutics* **2020**, *12*, 146. [CrossRef] [PubMed]

foods

MDPI

Article

Mineral and Microbiological Analysis of Spices and Aromatic Herbs

Nicola Cicero [1,2], Teresa Gervasi [1,*], Alessandra Durazzo [3], Massimo Lucarini [3], Antonio Macrì [1], Vincenzo Nava [1], Filippo Giarratana [4], Roberta Tardugno [2], Rossella Vadalà [1] and Antonello Santini [5,*]

1 Department of Biomedical and Dental Sciences and Morphofunctional Imaging, University of Messina, 98168 Messina, Italy; ncicero@unime.it (N.C.); macri.anto@outlook.it (A.M.); vnava@unime.it (V.N.); rosvadala@tiscali.it (R.V.)
2 Science4life, Spinoff Company, University of Messina, 98168 Messina, Italy; roberta.tardugno@gmail.com
3 CREA—Research Centre for Food and Nutrition, Via Ardeatina 546, 00178 Rome, Italy; alessandra.durazzo@crea.gov.it (A.D.); massimo.lucarini@crea.gov.it (M.L.)
4 Department of Veterinary Sciences, University of Messina, 98168 Messina, Italy; fgiarratana@unime.it
5 Department of Pharmacy, University of Napoli Federico II, Via D. Montesano 49, 80131 Napoli, Italy
* Correspondence: tgervasi@unime.it (T.G.); asantini@unina.it (A.S.)

Abstract: Spices and aromatic herbs have always had great historical importance in human nutrition. Their use has been documented for centuries as a rich source of bioactive compounds; they have been used for their health benefits and also for flavoring or coloring food. However, despite the many health properties linked to the use of spices and aromatic herbs, these can represent biological hazards and can contain chemical substances of concern. Certainly, monitoring potential health hazards in spices and aromatic herbs includes microbiological safety and also the content of inorganic substances: both represent a key step. This research aims at monitoring the compliance of various spices and aromatic herbs from a non-European country market (namely: black cumin seeds, Iranian Tokhme Sharbati, clove buds, Shahjeera, Abbaszadeh saffron, organic fenugreek, whole black pepper, cinnamon, Abthul Ahmar (Asario), Ajwan seeds, whole coriander seeds, black sesame seeds, Sabja seeds) with the current European Union (EU) and WHO regulations, when available, regarding mineral and microbiological parameters. In particular, microbiological assays using rapid and conventional methods, and trace mineral determination by inductively coupled plasma mass spectrometry (ICP-MS) were performed. Results show the safety of the tested spices, given that the microbiological parameters were within the legal microbiological criteria set by the European Commission Regulation (EC) No. 2073/2005 and its amendment Regulation (EC) No. 1441/2007. With reference to potentially toxic Cd, Pb, As, Hg, these were within the limits set by the European Commission Regulation (EC) No. 1881/2006 and its amendments, Regulation (EU) No. 1317/2021 and Regulation (EU) No. 1323/2021, and WHO. According to EU regulations, for Pb content, 2 samples out of 16 showed values different from the set limits.

Keywords: spices; aromatic herbs; minerals; chemical analysis; microbiological analysis

Citation: Cicero, N.; Gervasi, T.; Durazzo, A.; Lucarini, M.; Macrì, A.; Nava, V.; Giarratana, F.; Tardugno, R.; Vadalà, R.; Santini, A. Mineral and Microbiological Analysis of Spices and Aromatic Herbs. *Foods* **2022**, *11*, 548. https://doi.org/10.3390/foods11040548

Academic Editor: Amin Mousavi Khaneghah

Received: 10 January 2022
Accepted: 11 February 2022
Published: 15 February 2022

1. Introduction

Spices and aromatic herbs have always had great historical significance in human nutrition and in holistic approaches to health issues. In fact, they were used in ancient times not only in the food sector, i.e., to flavor and aromatize dishes, but also in the medical field, in many religious rituals, and for the preservation of food. They are documented as rich source of bioactive compounds linked to health benefits [1]. However, vegetals can be contaminated by microorganisms and can accumulate heavy metals, pesticide residues, and other potentially toxic substances from the environment depending on environmental factors, e.g., soil characteristics and absorbability, water, air, plant genotypes, and anthropogenic activities [2–9].

Acute or chronic poisonings may occur following heavy metal intake through food. Their bioaccumulation may lead to diverse toxic effects on a variety of body tissues and organs. Heavy metals disrupt cellular events including growth, proliferation, differentiation, damage-repairing processes, and apoptosis [10].

Thus, spices and aromatic herbs can be biological and chemical threats when used as food ingredients or for medicinal uses [11–13]. The conservation of spices and aromatic herbs is carried out in most cases via the dehydration process, through physical processes such as heat and/or pressure. This conservation methodology is minimally invasive and aims to inhibit metabolic activities, and therefore the proliferation of microorganisms by subtracting the free and bound water within the food. Dehydrated foods are characterized by a low moisture content (<14%), corresponding to a low value of free water with water activity (A_w) that is less than 0.75. Compared to the native product, the reduction in A_w values therefore represent one of the main parameters of ensuring the inhibition of the growth of organisms, consequently giving stability to the food from both a microbiological, enzymatic, and chemical point of view. Foodstuffs of vegetable origin such as cereals and derivatives, seeds, dried fruit, coffee, cocoa, herbs, and spices are easily attacked by molds, which in particular conditions of temperature and humidity, can produce secondary potentially dangerous metabolites such as mycotoxins [14–18].

Although spices and aromatic herbs on the market are dehydrated products and therefore have stable chemical and physical characteristics, these foods are generally subjected to drying processes at room temperature in their places of origin, which generally are developing or tropical countries where the production technologies used are not always capable of guaranteeing the implementation of good hygiene and safety practices, thus becoming the main cause of contamination [19,20]. Nevertheless, even with low water activity levels, some microorganisms, including pathogenic and toxigenic ones, are able to survive and may proliferate when vegetal matrices are added to foods.

Considering the crucial role of safety and quality in food production, in Europe, systems to detect and neutralize contaminants in herbs and spices have been developed within the project "Securing the spices and herbs commodity chains in Europe against deliberate, accidental, or natural biological and chemical contamination" (SPICED). The EU market is one of the main world markets for spices and herbs, and the problem linked to microbiological and toxicological hazards can pose a serious risk for the consumer, as spices and herbs could potentially contaminate a wide range of products due to their widespread use. Europe is one of the most important regions in the world with reference to the importation of herbs and spices, accounting for about one quarter of the world's total imports of herbs and spices [21,22].

Different microorganisms could be potentially harmful in herbs as well as in different food matrices, i.e., *Staphylococcus aureus*, *Salmonella* spp., *Escherichia coli*, *Listeria monocytogenes*, aflatoxin-producing fungi (i.e., *Aspergillus* spp.), *Clostridium perfringens*, and *Bacillus cereus* [16,23–25].

Taking into account these possible threats to health, the problem of spice and herb product consumption has global significance.

There are many studies on microflora in agricultural products, but only a small part of them address spices and officinal and aromatic herbs, which are instead increasingly present in our daily diet, thus also acquiring an increasingly important economic role.

On the other hand, monitoring the presence of metals in spices and herbs represents a key step [25–27].

Metals have important biological functions and activities, but inorganic elements can become toxic when their intake exceeds the accepted and maximum allowed levels as suggested by the European Commission and the European Food Safety Authority (EFSA) [28–36].

Considering that Europe is among the main importer of spices from extra EU countries [21,22], monitoring the contaminant levels in spices and herbs from these countries may provide relevant toxicological data on spices and aromatic herbs commonly present in the

European Union (EU) market, improving the accuracy of dietary risk exposure/assessment, and thus enhancing the feasibility of epidemiological studies. In this context, the present study aims to monitor the safety of some spices and herbs from foreign countries by evaluating: (i) microbiological contamination, and (ii) trace mineral element content.

2. Materials and Methods

2.1. Samples

Spices and aromatic herbs of different species, and of different countries of origin, were acquired in an international market in Saudi Arabia. The spices and herbs that were the object of this study are among the main ones imported into Europe and distributed in European countries. In this context, the end points of the present study were to check conformity with existing EU regulations, and the content of trace metal elements.

Thirteen samples of spices and aromatic herbs (labelled from A to O), packed in sealed plastic bags, were subjected to microbiological and chemical analysis. Table 1 reports the description of the samples according to their scientific name, vulgaris name, and country of origin. All microbial and chemical analyses have been carried out in triplicate.

Table 1. Analyzed herbs and spices and their origin.

Label	Origin	Common Name	Scientific Name
Black Cumin Seeds (A)	India	Black cumin	*Nigella sativa*
Iranian Tokhme Sharbati (B)	Iran	Chia seeds	*Salvia hispanica*
Clove Buds (C)	Indonesia	Clove buds	*Syzygium aromaticum*
Shahjeera (D)	India	Caraway	*Carum carvi*
Abbaszadeh Saffron (E)	Iran	Saffron	*Crocus sativus*
Organic Fenugreek (F)	India	Fenugreek	*Trigonella foenum-graecum*
Whole Black Pepper (G)	Vietnam	Black pepper	*Piper nigrum*
Cinnamon (H)	Indonesia	Cinnamon	*Cinnamomum verum*
Abthul Ahmar (Asario) (I)	India	Cress Sprouting Seeds	*Lapidium sativum*
Ajwan Seeds (L)	India	Thymol seeds	*Trachyspermum ammi*
Whole Coriander Seeds (M)	India	Coriander	*Coriandrum sativum*
Black Sesame Seeds (N)	India	Black sesame seeds	*Sesamum indicum*
Sabja Seeds (O)	India	Chia seeds	*Salvia hispanica*

2.2. Microbiological Methodology

The counts of total mesophilic bacteria, fungi (yeasts and molds), total coliforms, *Escherichia coli*, Enterobacteria, lactic bacteria, *Staphylococcus aureus*, *Salmonella* spp., *Listeria monocytogenes*, *Bacillus cereus*, and sulphite-reducing clostrides were performed. All media were supplied by ThermoFisher Scientific, Oxoid Ltd., Basingstoke, UK. The samples were homogenized with PBS (pH = 7.4) by mixing vigorously. Then serial dilutions in PBS (pH = 7.4) were performed, and the samples were spread aseptically over the media plates. For mesophilic counts, samples were spread over plate count agar (PCA) and incubated at 30 °C for 72 h. Malt extract agar with 10% lactic acid (MEA) was utilized to verify the counts of yeasts and molds by incubating at room temperature for 3 days. Lactic acid microorganisms and Enterobacteriacee were investigated using De Man, Rogosa, and Sharpe agar and Violet Red Bile Glucose Agar (MRSA and VRBG) using an incubation period of 72 h at 30 °C and 48 h at 37 °C, respectively. The determination of *B. cereus* was carried out by using *Bacillus Cereus* Agar Base (PEMBA) after an incubation for 24 h at 37 °C. Coliforms and *E. coli* were analyzed using the standard membrane filter technique using Chromogenic Coliform Agar (CCA) and Tryptone Bile X-Gluc agar (TBX) using an incubation period of 24 h at 37 °C and of 18/24 h at 44 °C. *S. aureus* contamination was detected using the standard membrane filter technique using Baird–Parker Agar (BPA), at 37 °C for 48 h. The standard membrane filter technique using Sulfite Polymyxin Sulfadizine agar (SPS) was used for the determination of sulphite-reducing clostrides after an incubation at 37 °C for 48 h. The determination of *L. monocytogenes* was carried out according to ISO standard 11290. Briefly, after a two-stage enrichment process, the first in half Fraser broth for 24 h and then in Fraser broth, the enriched broths were plated on Oxford and BD PALCAM *Lysteria* agars.

The presence of *Salmonella* spp. was investigated according to ISO standard 6579, which includes: pre-enrichment in a non-selective liquid medium (buffered pepton water), followed by selective enrichment (Rappaport–Vessiliadis Soy Broth; ThermoFisher Scientific, Oxoid Ltd., Basingstoke, UK), and then isolation on a selective medium (Hektoen; ThermoFisher Scientific, Oxoid Ltd., Basingstoke, UK). All microbial analyses were carried out in triplicate.

2.3. Mineral Analysis

2.3.1. Sample Preparation

A total of 0.25 g dry weight (dw) of each sample was milled in a Teflon mortar and digested with 7 mL of 65% (v/v) HNO_3, and 1 mL of 30% (v/v) H_2O_2 (J.T. Baker, Mallinckrodt Baker, Milan, Italy). A total of 1 mL of 0.8 mg L^{-1} Re (Fluka, Milan, Italy) was added as an internal standard. Mineralization was performed in an Ethos 1 digestor (Milestone, Bergamo, Italy) at 1000 W and 180 °C; this temperature was reached in 10 min and held for another 10 min. The digested samples, cooled at room temperature, were diluted with ultrapure deionized water obtained (J.T. Baker, Mallinckrodt Baker, Milan, Italy) and stored at 4 °C.

2.3.2. ICP-MS Analysis

Minerals were determined by the same procedure utilized for the determination of potentially toxic inorganic species in vegetables as reported in a previous work [16]. An Agilent 7500CX ICP-MS spectrometer (Agilent, Santa Clara, CA, USA) equipped with an Octapole Reaction System (ORS) reaction/collision cell, and with an ASX 500 auto sampler, was used for analyzing the digested samples. The system was pressurized with 99.9% pure helium (Rivoira S.p.A., Milan, Italy).

The operating conditions of the ICP-MS were as follows: RF power, 1550 W; plasma gas flow rate, 15 L min^{-1}; auxiliary gas flow rate, 0.9 L min^{-1}; carrier gas flow rate, 1.1 L min^{-1}; sample introduction flow rate, 1 mL min^{-1}; sample depth, 9 mm; spray chamber temperature 2 °C; vacuum, $<1.5 \times 10^{-7}$ Pa; interface pressure, 5.3×10^{-2} Pa.

^{7}Li, ^{59}Co, ^{80}Y, and ^{205}Tl solutions (Agilent, Santa Clara, CA, USA) at a concentration of 10 μg L^{-1} were used for tuning the instrument.

^{63}Cu, ^{60}Ni, ^{75}As, ^{51}V, ^{52}Cr, and $^{208,207,206}Pb$ (Fluka, Milan, Italy), and ^{114}Cd and ^{202}Hg isotopes (Merck, Darmstadt, Germany) were selected to optimize the sensitivity, and to minimize matrix interference.

A solution of ^{115}In, ^{45}Sc, ^{103}Rh, and ^{209}Bi at a concentration of 10 μg L^{-1} was used as an online internal standard to correct any instrumental drifts and matrix effects.

Quantitative determinations were performed using the external standard method. The calibration used a multi-standard solution of Cr, V, Cu, Cd, Pb, and Ni at different concentrations ranging from 0.5 to 2000 μg L^{-1}.

Hg was analyzed separately following a previously described procedure [5]. All analyses were carried out in triplicate.

2.3.3. Statistical Analysis

All mineral data are reported as the mean and standard deviation of three independent determinations. With regards to the selected heavy metals (Cd, Pb, As, and Hg), their contents were statistically compared with the reference limit values established by the European Commission and WHO by means of Student's t test. The statistical analysis has been performed using the Past software (Version 4.09) [37].

3. Results and Discussion

The safety aspects that were explored throughout the study were: (i) microbiological aspects; and (ii) mineral profiles.

3.1. Microbiological Analysis

The microbiological analysis of 13 dried spices, labelled with letters from A to O, is reported in Table 2.

Table 2. Bacterial contamination (CFU g^{-1}) of dried spices from India (**a**) and Iran, Indonesia, and Vietnam (**b**).

(a)

Microorganisms	A	D	F	I	L	M	N	O	Reference Values [a]
Mesophilic Bacteria	2.1×10^4	1.7×10^6 *	3.7×10^2	30	1.6×10^3	2.4×10^2	3.2×10^2	8.5×10^4	$<5 \times 10^5$
Molds	1.2×10^2	2.3×10^2	1·10	ND	ND	ND	1.2×10^2	ND	$<1 \times 10^3$
Yeasts	30	7.1×10^2	ND	ND	ND	ND	ND	ND	$<1 \times 10^3$
Coliforms	7×10^3 *	2.4×10^4 *	ND	ND	ND	ND	6.2×10^3	ND	$<1 \times 10^3$
Escherichia coli	ND	ND	ND	ND	ND	ND	ND	ND	<10
Sulfite-Reducing Clostridia	ND	ND	ND	ND	ND	ND	ND	ND	$<1 \times 10^2$
Clostridial Spores	ND	ND	ND	ND	ND	ND	ND	ND	$<1 \times 10^2$
Staphilococcus aureus	ND	ND	ND	ND	ND	ND	ND	ND	$<1 \times 10^2$
Salmonella spp. [b]	ND	ND	ND	ND	ND	ND	ND	ND	0
Listeria monocytogenes	ND	ND	ND	ND	ND	ND	ND	ND	$<10^2$
Bacillus cereus	ND	ND	ND	ND	ND	ND	2×10	ND	$<1 \times 10^3$
Enterobatteriacee	1.4×10^4 *	9×10^4 *	ND	ND	ND	ND	4.4×10^3 *	ND	<10
Lactic Bacteria	ND	ND	ND	ND	ND	ND	ND	ND	$<1 \times 10^5$

(b)

Microorganisms	B	C	E	G	H	I	Reference Values [a]
Mesofilic Bacteria	6×10^4	2.2×10^2	2.5×10^2	9.3×10^4	1.1×10^2	30	$<5 \times 10^5$
Molds	2.5×10^2	ND	ND	ND	ND	ND	$<1 \times 10^3$
Yeasts	ND	ND	ND	ND	ND	ND	$<1 \times 10^3$
Coliforms	1.6×10^3 *	1.7×10^2	ND	4·10	ND	ND	$<1 \times 10^3$
Escherichia coli	ND	ND	ND	ND	ND	ND	$<1 \times 10$
Sulfite-Reducing Clostridia	ND	ND	ND	ND	ND	ND	$<1 \times 10^2$
Clostridial Spores	ND	ND	ND	ND	ND	ND	$<1 \times 10^2$
Staphilococcus aureus	ND	ND	ND	ND	ND	ND	$<1 \times 10^2$
Salmonella spp. [b]	ND	ND	ND	ND	ND	ND	0
Listeria monocytogenes	ND	ND	ND	ND	ND	ND	1×10^2
Bacillus cereus	ND	ND	ND	ND	ND	ND	$<1 \times 10^3$
Enterobatteriacee	6.6×10^2 *	2×10 *	ND	ND	ND	ND	<10
Lactic Bacteria	ND	ND	ND	ND	ND	ND	$<1 \times 10^5$

[a] Reference values for microbiological safety and quality were based on regulations from the European Union and other international guidelines [8,38,39]. [b] CFU 25 g^{-1}. * Values are higher than the reference values. ND, not determined. A, black cumin seeds; B, Iranian Tokhme Sharbati; C, clove buds; D, Shahjeera; E, Abbaszadeh saffron; F, organic fenugreek; G, whole black pepper; H, cinnamon; I, Abthul Ahmar (Asario); L, Ajwan seeds; M, whole coriander seeds; N, black sesame seeds; O, Sabja seeds.

According to the Commission Regulation (EC) No. 2073/2005 and its amendment Regulation (EC) No. 1441/2007, which sets legal microbiological criteria for several food products, these spices could be defined safe products [8,38,39].

With regards to the indicator microorganisms, which are used to provide simple, reliable, and rapid information about processing failures, post-processing contamination from the environment, the general level of hygiene, and the presence or absence of foodborne pathogens to monitor the chain of food production, low bacterial loads were detected in saffron, fenugreek, black pepper, cinnamon, cress sprouting seed, thymol seed, coriander, and chia seed [40].

The total coliform and Enterobacteriaceae amounts in black cumin were 7×10^3 CFU g^{-1} and 1.4×10^4 CFU g^{-1}, respectively, whereas in Iranian chia seeds it was 1.6×10^3 CFU g^{-1} and 6.6×10^2 CFU g^{-1}, respectively. Caraway presented a mesophilic count of 1.7×10^6 CFU g^{-1}, and an amount of total coliform and Enterobacteriaceae of 2.4×10^4 CFU g^{-1} and 9×10^4 CFU g^{-1}, respectively. Moreover, Enterobacteriaceae were also detected in clove buds (2×10 CFU g^{-1}) and black sesame seeds (4.4×10^3 CFU g^{-1}).

These contaminations may be related to the environment, inadequate hygienic handling, unsanitary conditions, and others, and occurs in samples purchased in street markets. Indeed, the adoption of good hygiene practices in all the involved steps from land growing, harvesting, and processing can be useful to reduce the health risks of spice consumption.

3.2. Mineral Analysis

3.2.1. Method Validation

The method was validated according to Eurachem criteria [41]. Commercial standards were used for the evaluation of method linearity, limits of detection (LODs), limits of quantification (LOQs), accuracy, repeatability, and intermediate precision. LODs and LOQs were calculated as 3.3 σ/S and 10 σ/S, respectively, where σ is the standard deviation of six blanks and S is the slope of the relative calibration curve. A good linearity was obtained for all elements investigated with R2 values ranging from 0.9992 (for Cu and Se) to 0.9999 (for V). The limits of detection (LODs) ranged from 0.001 to 0.051 μg kg^{-1}, and the limits of quantification (LOQs) ranged from 0.003 to 0.168 μg kg^{-1}. The lowest average recovery was observed for mercury with 92.93%, while the highest was obtained for strontium with 103.03%. Accuracy was assessed by evaluating six determinations on certified reference materials (NIST1570A spinach leaves) and was reported as the percent recovery between the value found with the calibration curve and the true value reported in the certified reference materials. If the element was not certified in the reference materials, the matrix was spiked with the known amount of analyte, and was analyzed following the procedures discussed previously. Based on these results, the analytical characteristic (linearity, sensitivity, and accuracy) can be considered to be satisfactory for the purposes of the analysis (Table 3).

Table 3. Analytical parameters for method validation.

Element	LOD (μg kg^{-1})	LOQ (μg kg^{-1})	R^2	Experimental Value (mg kg^{-1})	Expected Value (mg kg^{-1})	Recovery (%)
Be	0.001	0.003	0.9998	1.90 *	2.00 *	95.00
Sn	0.005	0.017	0.9997	1.97 *	2.00 *	98.50
Al	0.051	0.168	0.9995	308.55	310.00	99.53
V	0.003	0.010	0.9999	0.575	0.568	101.23
Cr	0.003	0.010	0.9997	1.98	2.00 *	99.00
Mn	0.015	0.050	0.9996	75.3	76.0	99.08
Fe	0.035	0.116	0.9994	2.02	2.00 *	101.00
Co	0.002	0.007	0.9995	0.389	0.393	98.98
Ni	0.002	0.007	0.9997	2.115	2.142	98.74
Cu	0.025	0.083	0.9992	11.98	12.22	98.04
Zn	0.032	0.106	0.9993	83.0	82.3	100.85
As	0.001	0.003	0.9998	0.070	0.068	102.94
Se	0.012	0.040	0.9992	0.1112	0.1152	96.53
Sr	0.09	0.030	0.9996	57.22	55.54	103.03
Cd	0.001	0.003	0.9997	2.831	2.876	98.44
Sb	0.002	0.007	0.9996	1.96 *	2.00 *	98.00
Pb	0.002	0.007	0.9996	0.19	0.20	95.00
Hg	0.001	0.003	0.9998	0.0276	0.0297	92.93

* Not present in the certified matrix. Element added for method validation.

3.2.2. Mineral Contents

The content of inorganic elements in the samples, labelled with letters from A to O, are shown in Table 4 and reported as average values and standard deviations.

Table 4. Trace elements (**a**), potentially non-toxic elements (**b**), and potentially toxic elements (**c**) present in dried spices. The contents are expressed as the mean value (mg kg $^{-1}$) and standard deviation.

(a)

	Cr	Fe	Mn	Cu	Se	Zn
A	0.04 ± 0.001	268.850 ± 12.750	15.87 ± 0.558	11.089 ± 0.380	0.072 ± 0.001	44.060 ± 1.946
B	0.059 ± 0.000	101.221 ± 7.653	11.925 ± 0.045	10.014 ± 0.059	0.244 ± 0.002	31.539 ± 0.066
C	0.119 ± 0.006	136.998 ± 9.447	627.263 ± 28.417	2.599 ± 0.104	0.011 ± 0.001	5.798 ± 0.332
D	0.193 ± 0.002	303.000 ±10.300	14.839 ± 0.520	8.366 ± 0.023	0.254 ± 0.008	20.982 ± 0.100
E	0.314 ± 0.015	251.211 ± 6.452	13.964 ± 0.538	6.837 ± 0.379	0.028 ± 0.001	30.157 ± 1.764
F	0.021 ± 0.000	314.231 ± 6.921	9.114 ± 0.026	8.631 ± 0.095	0.052 ± 0.008	30.013 ± 0.267
G	0.096 ± 0.003	13.736 ± 1.326	95.802 ± 0.455	5.879 ± 0.054	0.008 ± 0.001	6.085 ± 0.082
H	2.357 ± 0.029	10.586 ± 2.531	85.896 ± 0.936	2.376 ± 0.080	0.003 ± 0.001	15.652 ± 0.013
I	0.088 ± 0.044	86.554 ± 4.586	16.747 ± 1.330	2.871 ± 0.052	0.287 ± 0.003	35.357 ± 0.187
L	1.113 ± 0.010	289.667 ± 10.441	19.505 ± 0.024	6.874 ± 0.023	0.254 ± 0.003	48.049 ± 0.035
M	0.129 ± 0.007	192.395 ± 13.521	16.289 ± 0.607	10.266 ± 0.085	0.505 ± 0.009	26.638 ± 309
N	0.152 ± 0.001	124.569 ± 5.114	10.536 ± 0.168	11.804 ± 0.037	0.485 ± 0.006	36.698 ± 0.148
O	0.109 ± 4.813	95.142 ± 4.236	103.145 ± 4.813	6.321 ± 0.284	0.009 ± 0.001	6.843 ± 0.263

(b)

	Sr	Ni	Sn	V	Co	Sb	Be
A	17.404 ± 0.716	3.053 ± 0.123	0.052 ± 0.001	0.028 ± 0.001	0.052 ± 0.001	0.001 ± 0.000	<LOD
B	62.373 ± 0.023	0.416 ± 0.004	0.510 ± 0.019	0.084 ± 0.001	0.271 ± 0.001	0.004 ± 0.002	<LOD
C	36.705 ± 1.657	0.753 ± 0.034	0.108 ± 0.038	0.040 ± 0.001	0.048 ± 0.002	0.008 ± 0.008	<LOD
D	95.917 ± 0.197	0.892 ± 0.010	0.617 ± 0.017	0.138 ± 0.001	0.111 ± 0.001	0.006 ± 0.001	<LOD
E	7.479 ± 0.302	1.124 ± 0.069	0.126 ± 0.073	0.303 ± 0.013	0.094 ± 0.005	0.010 ± 0.001	<LOD
F	4.299 ± 0.021	0.639 ± 0.010	<LOD	0.021 ± 0.001	0.227 ± 0.001	0.003 ± 0.000	<LOD
G	8.667 ± 0.047	1.194 ± 0.009	1.174 ± 0.011	0.066 ± 0.001	0.020 ± 0.001	0.003 ± 0.002	<LOD
H	50.586 ± 0.127	2.766 ± 0.012	1.054 ± 0.028	0.052 ± 0.001	0.048 ± 0.001	0.013 ± 0.000	<LOD
I	9.414 ± 0.929	0.732 ± 0.045	0.804 ± 0.007	0.024 ± 0.001	0.053 ± 0.001	0.005 ± 0.001	<LOD
L	89.403 ± 1.563	2.087 ± 0.021	0.813 ± 0.006	1.415 ± 0.013	0.288 ± 0.004	0.016 ± 0.000	<LOD
M	15.222 ± 0.219	1.134 ± 0.011	2.392 ± 0.039	0.049 ± 0.006	0.132 ± 0.001	0.022 ± 0.016	<LOD
N	55.034 ± 0.490	0.928 ± 0.013	0.735 ± 0.013	0.062 ± 0.001	0.145 ± 0.002	0.003 ± 0.003	<LOD
O	9.547 ± 0.287	1.293 ± 0.061	1.439 ± 0.129	0.073 ± 0.001	0.022 ± 0.001	0.028 ± 0.034	<LOD

(c)

	Cd	Pb	As	Hg	Al
A	0.026 ± 0.001	0.034 ± 0.001	0.024 ± 0.002	0.01 ± 0.001	19.963 ± 0.742
B	0.010 ± 0.000	0.023 ± 0.000	0.015 ± 0.000	0.01 ± 0.000	32.355 ± 0.008
C	0.014 ± 0.000	0.034 ± 0.001	0.005 ± 0.000	0.003 ± 0.001	20.532 ± 1.097
D	0.012 ± 0.001	0.043 ± 0.000	0.078 ± 0.001	0.003 ± 0.001	87.507 ± 0.269
E	0.017 ± 0.001	0.096 ± 0.005	0.086 ± 0.002	0.003 ± 0.001	145.216 ± 6.281
F	0.011 ± 0.000	0.010 ± 0.001	0.003 ± 0.000	0.002 ± 0.001	7.112 ± 0.108
G	0.003 ± 0.000	0.008 ± 0.000	0.002 ± 0.000	0.003 ± 0.001	43.064 ± 0.253
H	0.071 ± 0.001	0.023 ± 0.000	0.005 ± 0.000	0.002 ± 0.001	27.168 ± 0.547
I	0.079 ± 0.001	0.012 ± 0.001	0.009 ± 0.000	0.002 ± 0.001	8.863 ± 0.390
L	0.065 ± 0.000	0.544 ± 0.003	0.339 ± 0.006	0.004 ± 0.001	930.198 ± 5.269
M	0.034 ± 0.002	0.026 ± 0.002	0.009 ± 0.002	0.004 ± 0.002	19.914 ± 2.904
N	0.021 ± 0.001	0.015 ± 0.001	0.026 ± 0.001	0.003 ± 0.001	32.508 ± 0.006
O	0.005 ± 0.002	0.011 ± 0.002	0.005 ± 0.002	0.002 ± 0.001	46.889 ± 1.678

A, black cumin seeds; B, Iranian Tokhme Sharbati; C, clove buds; D, Shahjeera; E, Abbaszadeh safron; F, organic fenugreek; G, whole black pepper; H, cinnamon; I, Abthul Ahmar (Asario); L, Ajwan seeds; M, whole coriander seeds; N, black sesame seeds; O, Sabja seeds.

Although various classifications for trace elements have been proposed and may be controversial, this paper uses the World Health Organization recommendations which classifies trace elements as: essential trace elements, potentially non-toxic essential elements, and potentially toxic elements [42].

3.2.3. Trace Elements

Given the increasing consumption of spices and aromatic herbs in the daily diet, it is also interesting to elucidate the content of minerals known for their nutritional roles such as Cr, Fe, Mn, Cu, Se, and Zn. Several studies on the content of mineral and trace elements in spices and herbs have remarked that they occur in a wide range of concentrations [43–46].

Data on Cr content in these products are needed. Chromium has been quantified in several foods and beverages, and currently the most comprehensive source are the Danish food composition tables [47]. The content of chromium in foods is relatively low and most foods present a content below 0.1 mg kg^{-1}. Data from the literature indicate that the presence of Cr in spices and aromatic herbs is higher than other foods and beverages, within a range from 0.01 to 3.0 mg kg^{-1} [48–51]. In our study, the chromium level found in spices was below 0.3 mg kg^{-1}, except for the high content found in cinnamon (2.35 mg kg^{-1}) and thymol seeds (1.11 mg kg^{-1}).

Selenium content varied from 0.003 to 0.505 mg kg^{-1}; in particular, selenium was reported with values: 0.254, 0.287, 0.254, 0.505, and 0.485 mg kg^{-1} for caraways, cress sprouting seeds, thymol seeds, coriander, and black sesame seeds, respectively. Selenium (Se) is an essential trace element involved in the synthesis of various selenium-containing proteins, and also has other relevant biological functions; moreover, it has a fundamental role in the human diet since it may act as a preventive agent against some health conditions [51,52].

3.2.4. Potentially Non-Toxic Elements

Eight trace elements that are not normally known for their toxic effects were identified, but it is nonetheless important to monitor them, because if very high concentrations of these metals are ingested, they can lead to physiological disorders [53,54]. Their maximum concentrations were found to be in the decreasing order as follows: Sr > Ni > Sn > V > Co > Sb > Be. All these potentially non-toxic trace elements were contained in variable amounts in the analyzed spices. In particular among these, the major trace elements were: Sr, which ranged from 95.92 ± 0.20 mg kg^{-1} (carraway) to 4.30 ± 0.02 mg kg^{-1} (fenugreek); Ni, which ranged from 3.05 ± 0.12 mg kg^{-1} (black cumin) to 0.42 ± 0.01 mg kg^{-1} (chia seeds); Sn, which ranged from 2.39 ± 0.04 mg kg^{-1} (coriander) to 0.11 ± 0.04 mg kg^{-1} (clove buds); and Co, which ranged from 0.29 ± 0.00 mg kg^{-1} (thymol seeds) to 0.02 ± 0.00 mg kg^{-1} (black pepper).

3.2.5. Potentially Toxic Elements

Concerning Cd, Pb, As, and Hg, the following ranges have been observed in the 13 analyzed spices: Cd (0.003–0.079 mg kg^{-1}), Pb (0.008–0.544 mg kg^{-1}), As (0.003–0.339 mg kg^{-1}), and Hg (0.001–0.010 mg kg^{-1}). The contamination level of the analyzed samples followed the sequence: Pb > As > Cd > Hg.

Heavy metals should be closely monitored, considering that the absorption and bioaccumulation of those compounds, with reference to their toxic and mutagenic effects, have a negative effect on consumers' health [8,44,45,55].

According to the European Commission Regulation (EC) No. 1881/2006 and its amendments, Regulation (EU) No. 1317/2021 and Regulation (EU) No. 1323/2021, the maximum levels in fresh herbs for lead and cadmium have been set at 0.1 mg kg^{-1} and 0.2 mg kg^{-1}, respectively. However, for many elements, there is a lack of shared worldwide regulation, and reference could be made to the values reported by WHO and EFSA, namely 5.0 and 0.2 mg kg^{-1} for As and Hg, respectively [28–36,56,57].

Concerning the content of potentially toxic minerals (Cd, Pb, As, and Hg), the results showed that all the spices analyzed did not present contamination concerns for cadmium, arsenic, and mercury, as the contents of these three metals were always lower than the permitted limit values established by the European Commission regulation and WHO. On the other hand, for Pb content, two samples showed two warning values with respect to the established permitted limit of 0.1 mg kg^{-1}. Ajwan seeds from India (sample L) showed a lead content (0.544 mg kg^{-1}) five times higher than the permitted limit ($p < 0.05$) while the lead content found in Abbaszadeh saffron from Iran (0.096 mg kg^{-1}, sample E) was close to the accepted limit ($p > 0.05$), indicating that for these samples, there was a possible threat to health.

Several studies have reported a potential threat to the nervous system from aluminum [58,59]. The content of aluminum ranges from 7.112 mg kg^{-1} in fenugreek to 46.889 mg kg^{-1} in chia seeds, with the exception of carraway, saffron, and thymol seeds, in which the reported values were 87.507 mg kg^{-1}, 145.216 mg kg^{-1}, and 930.198 mg kg^{-1}, respectively, showing a high capacity to accumulate aluminum.

Lopez et al. [58] showed data on the levels of Al in 72 dried samples of 17 different spices and aromatic herbs, and aluminum levels ranged from 3.74 to 56.50 mg kg^{-1}. Bratakos et al. found Al values in spices with a mean value of 157 mg kg^{-1} [58]. For foods from plant origins, high aluminum concentration could be related to its high content in the soil where the plants are grown, or to the fact that plants grow in acid soils, because its availability depends on soil pH [59,60].

4. Conclusions

This study aims at monitoring the levels of both microbial contamination and trace elements in some spices and aromatic herbs commonly used in the Mediterranean diet. The data indicate that black cumin and Iranian chia seeds presented contamination (total coliform bacteria and Enterobacteriaceae). The concentration of trace elements was variable and related to each spice. Concerning the contents of potentially toxic heavy metals (Cd, Pb, As, and Hg), they were within the above-mentioned limits, although Pb presented a higher value in two cases.

It should be considered as a final remark that all spices and herbs are susceptible to environmental (e.g., microbial and heavy metal) contamination. Microbial contamination could be prevented by adopting good standards of practice during growing, harvesting, and processing. Environmental contamination with heavy metals should be avoided and monitored to minimize contamination levels. For this reason, spices and aromatic herbs must be strictly monitored for the aspects concerning their safety in order to prevent foodborne illness due to contamination.

Author Contributions: Conceptualization, N.C., A.S., and T.G.; methodology, N.C., T.G., R.V., and A.S.; investigation, A.M., V.N., F.G., and R.T.; writing-original draft preparation, T.G., A.D., N.C., and A.S.; writing-review and editing, T.G., N.C., and A.S.; visualization, R.V., A.D., and M.L.; project administration, N.C. and A.S.; funding acquisition, N.C. and A.S.; supervision, N.C., A.S., and T.G. All authors have read and agreed to the published version of the manuscript.

Funding: This research received no external funding.

Data Availability Statement: Data is contained within the article.

Conflicts of Interest: The authors declare no conflict of interest.

References

1. Metro, D.; Tardugno, R.; Papa, M.; Bisignano, C.; Manasseri, L.; Calabrese, G.; Gervasi, T.; Dugo, G.; Cicero, N. Adherence to the Mediterranean diet in a Sicilian student population. *Nat. Prod. Res.* **2018**, *32*, 1775–1781. [CrossRef]
2. Elviss, N.C.; Little, C.L.; Hucklesby, L.; Sagoo, S.; Surman-Lee, S.; de Pinna, E.; Threlfall, E.J. Microbiological study of fresh herbs from retail premises uncovers an international outbreak of salmonellosis. *Int. J. Food Microbiol.* **2009**, *134*, 83–88. [CrossRef] [PubMed]
3. Elshafie, A.E.; Al-Rashdi, T.A.; Al-Bahry, S.N.; Bakheit, C.S. Fungi and aflatoxins associated with spices in the sultanate of Oman. *Mycopathologia* **2002**, *1*, 155–160. [CrossRef]
4. Bua, G.D.; Annuario, G.; Albergamo, A.; Cicero, N.; Dugo, G. Heavy metals in aromatic spices by inductively coupled plasma-mass Spectrometry. *Food Addit. Contam. Part B* **2016**, *9*, 210–216. [CrossRef]
5. Salvo, A.; La Torre, G.L.; Mangano, V.; Casale, K.E.; Bartolomeo, G.; Santini, A.; Granata, T.; Dugo, G. Toxic inorganic pollutants in foods from agricultural producing areas of Southern Italy: Level and risk assessment. *Ecotoxicol. Environ. Saf.* **2018**, *148*, 114–124. [CrossRef]

6. Pecoraro, R.; Marino, F.; Salvaggio, A.; Capparucci, F.; Caro, G.D.; Iaria, C.; Salvo, A.; Rotondo, A.; Tibullo, D.; Guerriero, G.; et al. Evaluation of chronic nanosilver toxicity to adult zebrafish. *Front. Physiol.* **2017**, *8*, 1011. [CrossRef]
7. European Commission. REGULATION (EC) No 1334/2008 of the European Parliament and of the Council of 16 December 2008 on flavourings and certain food ingredients with flavouring properties for use in and on foods and amending Council Regulation (EEC) No 1601/91, Regulations (EC) No 2232/96 and (EC) No 110/2008 and Directive 2000/13/EC. *Off. J. Eur. Union* **2008**, L 354/34. Available online: https://eur-lex.europa.eu/legal-content/en/TXT/?uri=CELEX%3A32008R1334 (accessed on 21 December 2021).
8. World Health Organization. *Guidelines for Assessing Quality of Herbal Medicines with Reference to Contaminants and Residues*; World Health Organization: Geneva, Switzerland, 2007.
9. Gan, Y.; Wang, L.; Yang, G.; Dai, J.; Wang, R.; Wang, W. Multiple factors impact the contents of heavy metals in vegetables in high natural background area of China. *Chemosphere* **2017**, *184*, 1388–1395. [CrossRef]
10. Balali-Mood, M.; Naseri, K.; Tahergorabi, Z.; Khazdair, M.R.; Sadeghi, M. Toxic Mechanisms of Five Heavy Metals: Mercury, Lead, Chromium, Cadmium, and Arsenic. *Front. Pharmacol.* **2021**, *12*, 643972. [CrossRef]
11. Jabeen, S.; Shah, M.; Khan, S.; Hayat, M.Q. Determination of major and trace elements in ten important folk therapeutic plants Pakistan. *J. Med. Plants Res.* **2010**, *4*, 559–566.
12. Wafaa Muaad, M.; Khalil Abdul-Rezzak, R. Investigate the presence of some heavy metals and microbes in spices. *Samarra J. Pure Appl. Sci.* **2021**, *3*, 41–48. [CrossRef]
13. Al-Dalain, S.Y.; Haddad, M.A.; Al-Hawadi, J.S.; Freihat, A.; Arabiat, S.; Al-Qudah, M.M.; Ateyyat, M.A. Determination of heavy metals and bacterial count of some frequently consumed spices in Jordan. *Syst. Rev. Pharmacol.* **2021**, *12*, 319–325.
14. Mikušová, P.; Šrobárová, A.; Sulyok, M.; Santini, A. Fusarium fungi and associated metabolites presence on grapes from Slovakia. *Mycotoxin Res.* **2013**, *29*, 97–102. [CrossRef] [PubMed]
15. Nazhand, A.; Durazzo, A.; Lucarini, M.; Souto, E.B.; Santini, A. Characteristics, Occurrence, Detection and Detoxification of Aflatoxins in Foods and Feeds. *Foods* **2020**, *9*, 644. [CrossRef]
16. Mikusova, P.; Ritieni, A.; Santini, A.; Juhasova, G.; Srobarova, A. Contamination by moulds of grape berries in Slovakia. *Food Addit. Contam. Part A Chem. Anal. Control. Expo. Risk Assess.* **2010**, *27*, 738–747. [CrossRef] [PubMed]
17. Santini, A.; Meca, G.; Ritieni, A. Fusaproliferin, Beauvericin, and Enniatins: Occurrence in food—A review. *World Mycotoxin J.* **2012**, *5*, 71–81. [CrossRef]
18. Zhao, X.; Jin, X.; Lin, Z.; Guo, Q.; Liu, B.; Yuan, Y.; Yue, T.; Zhao, X. Simultaneous rapid detection of aflatoxin B1 and ochratoxin a in spices using lateral flow immuno-chromatographic assay. *Foods* **2021**, *10*, 2738. [CrossRef]
19. Sospedra, I.; Soriano, J.M.; Mañes, J. Assessment of the microbiological safety of dried spices and herbs commercialized in Spain. *Plant Food Hum. Nutr.* **2010**, *65*, 364–368. [CrossRef]
20. Vitullo, M.; Ripabelli, G.; Fanelli, I.; Tamburro, M.; Delfin, S.; Sammarco, M.L. Microbiological and toxicological quality of dried herbs. *Lett. Appl. Microbiol.* **2011**, *52*, 573–580. [CrossRef]
21. European Commission. *Securing the Spices and Herbs Commodity Chains in Europe against Deliberate, Accidental or Natural Biological and Chemical Contamination of Haripur Basin*; SPICED Project | FP7 | CORDIS; European Commission: Bruxelles, Belgium, 2013; Project Coordinated by Germany. Available online: https://cordis.europa.eu/project/id/312631 (accessed on 3 February 2022).
22. What Is the Demand for Spices and Herbs on the European Market? Available online: https://www.cbi.eu/market-information/spices-herbs/what-demand (accessed on 3 February 2022).
23. Śmiechowska, M.; Newerli-Guz, J.; Skotnicka, M. Spices and seasoning mixes in European Union-innovations and ensuring safety. *Foods* **2021**, *10*, 2289. [CrossRef]
24. Sagoo, S.K.; Little, C.L.; Greenwood, M.; Mithani, V.; Grant, K.A.; Mc Lauchlin, J.; De Pinna, E.; Threlfall, E.J. Assessment of the microbiological safety of dried spices and herbs from production and retail premises in the United Kingdom. *Food Microbiol.* **2009**, *26*, 39–43. [CrossRef]
25. Witkowska, A.; Hickey, D.; Alonso-Gomez, M.; Wilkinson, M. The microbiological quality of commercial herb and spice preparations used in the formulation of a chicken supreme ready meal and microbial survival following a simulated industrial heating process. *Food Control.* **2011**, *22*, 616–625. [CrossRef]
26. Panebianco, F.; Nava, V.; Giarratana, F.; Gervasi, T.; Cicero, N. Assessment of heavy- and semi-metals contamination in edible seaweed and dried fish sold in ethnic food stores on the Italian market. *J. Food Compos. Anal.* **2021**, *104*, 104150. [CrossRef]
27. Salvo, A.; La Torre, G.; Rotondo, A.; Cicero, N.; Gargano, R.; Mangano, V.; Casale, E.; Dugo, G. Multiple analytical approaches for the organic and inorganic characterization of Origanum vulgare L. samples. *Nat. Prod. Res.* **2019**, *33*, 2815–2822. [CrossRef] [PubMed]
28. Commission Regulation (EC) No 1881/2006 of 19 December 2006 Setting Maximum Levels for Certain Contaminants in Foodstuffs. Available online: https://eur-lex.europa.eu/LexUriServ/LexUriServ.do?uri=OJ:L:2006:364:0005:0024:EN:PDF (accessed on 31 January 2022).
29. Commission Regulation (EU) 2021/1323 of 10 August 2021 Amending Regulation (EC) No 1881/2006 as Regards Maximum Levels of Cadmium in Certain Foodstuffs. Available online: https://eur-lex.europa.eu/eli/reg/2021/1323/oj (accessed on 31 January 2022).
30. Commission Regulation (EU) 2021/1317 of 9 August 2021 Amending Regulation (EC) No 1881/2006 as Regards Maximum Levels of Lead in Certain Foodstuffs. Available online: https://eur-lex.europa.eu/eli/reg/2021/1317/oj (accessed on 31 January 2022).

31. EFSA (European Food Safety Authority). Opinion of the scientific panel on dietetic products, nutrition and allergies on a request from the commission related to the tolerable upper intake level of sodium. *EFSA J.* **2005**, *209*, 1–26.
32. EFSA (European Food Safety Authority). Scientific opinion on arsenic in food. *EFSA J.* **2009**, *7*, 1–199.
33. EFSA (European Food Safety Authority). Scientific opinion on lead in food. *EFSA J.* **2010**, *8*, 1–151.
34. EFSA (European Food Safety Authority). Cadmium dietary exposure in the European population. *EFSA J.* **2012**, *10*, 1–37.
35. EFSA (European Food Safety Authority). Scientific opinion on the risk for public health related to the presence of mercury and methyl-mercury in food. *EFSA J.* **2012**, *10*, 1–241.
36. European Communities Commission. Commission Directive 2008/100/EC of 28 October 2008 Amending Council Directive 90/496. EEC on Nutrition Labelling for Foodstuffs as Regards Recommended Daily Allowances, Energy Conversion Factors and Definitions; European Communities Commission: 2008. Available online: https://eur-lex.europa.eu/LexUriServ/LexUriServ.do?uri=OJ:L:2008:285:0009:0012:EN:PDF (accessed on 22 November 2021).
37. Hammer, O.; Harper, D.A.T.; Ryan, P.D. PAST: Paleontological statistics software package for education and data analysis. *Palaeontol. Electron.* **2001**, *4*, 1–9.
38. European Commission. Commission Regulation (EC) No 2073/2005 of 15 November 2005 on microbiological criteria for foodstuffs. *Off. J. Eur. Union* **2005**, *50*, 1–26.
39. European Commission. Commission Regulation (EC) No 1441/2007 of 5 December 2007 amending Regulation (EC) No 2073/2005 on microbiological criteria for foodstuffs. *Off. J. Eur. Union* **2007**, *322*, 12–29.
40. Halkman, A.K. Indicator Organisms. In *Encyclopedia of Food Microbiology*, 2nd ed.; Carl, A.B., Mary, L.T., Eds.; Academic Press: Cambridge, MA, USA, 2014; pp. 358–363, ISBN 9780123847331.
41. EURACHEM. Guide. S L R Ellison and A. Williams (Eds). Eurachem/CITAC Guide: Quantifying Uncertainty in Analytical Measurement, Third edition. 2012. ISBN 978-0-948926-30-3. Available online: https://www.eurachem.org/ (accessed on 21 December 2021).
42. WHO (World Health Organization). Trace Elements in Human Nutrition and Health. Geneva, Switzerland. 1996. Available online: https://scholar.googleusercontent.com/scholar?q=cache:1ZTWbPtjCfYJ:scholar.google.com/&hl=en&as_sdt=0,5 (accessed on 22 November 2021).
43. Tokalıoğlu, Ş. Determination of trace elements in commonly consumed medicinal herbs by ICP-MS and multivariate analysis. *Food Chem.* **2012**, *134*, 2504–2508. [CrossRef] [PubMed]
44. Ibrahim, G.; Hassan, L.; Baban, S.; Fadhil, S. Effect of heavy metal content of some common spices available in local markets in Erbil City on human Consumption. *Raf. J. Sci.* **2012**, *23*, 106–114. [CrossRef]
45. Doabi, S.; Karami, M.; Afyuni, M.; Yeganeh, M. Pollution and health risk assessment of heavy metals in agricultural soil at-mospheric dust and major food crops in Kermanshah province, Iran. *Ecotox. Environ. Saf.* **2018**, *163*, 153–164. [CrossRef]
46. Nan, C.; Changhui, Z.; Tiehua, Z. Selenium transformation and selenium-rich foods. *Food Biosci.* **2021**, *40*, 100875. [CrossRef]
47. *Food Data (frida.fooddata.dk), Version 4*; National Food Institute, Technical University of Denmark: Lyngby, Denmark, 2019. Available online: https://frida.fooddata.dk/?lang=en (accessed on 22 November 2021).
48. Castro, A. Chromium in a series of Portuguese plants used in the herbal treatment of diabetes. *Biol. Trace Elem. Res.* **1998**, *62*, 101–106. [CrossRef] [PubMed]
49. Abdul Latif, A.; Bilal, M.; Asghar, W.; Azeem, M.; Muhammad, I.; Ahmad, M.I.; Abbas, A.; Ahmad, M.Z.; Toufeeq Shahzad, T. Heavy metal accumulation in vegetables and assessment of their potential health risk. *J. Environ. Anal. Chem.* **2018**, *5*, 1000234. [CrossRef]
50. Tefera, M.; Teklewold, A. Health risk assessment of heavy metals in selected Ethiopian spices. *Heliyon* **2021**, *7*, e07048. [CrossRef]
51. Garcia, E.; Cabrera, C.; Lorenzo, M.L.; López, M.C. Chromium levels in spices and aromatic herbs. *Sci. Total Environ.* **2000**, *247*, 51–56. [CrossRef]
52. Chen, J.; Zhang, Y.; Lv, Y.; Tian, M.; You, J.; Chen, F.; Zhang, S.; Guan, W. Effects of Selenomethionine on Cell Viability, Selenoprotein Expression and Antioxidant Function in Porcine Mammary Epithelial Cells. *Front. Nutr.* **2021**, *26*, 665855. [CrossRef]
53. Khan, N.; Jeong, I.S.; Hwang, I.M.; Kim, J.S.; Choi, S.H.; Nho, E.Y.; Choi, J.Y.; Kwak, B.-M.; Ahn, J.K.; Yoo, T.; et al. Method validation for simultaneous determination of chromium, molybdenum and selenium in infant formulas by ICP-OES and ICP-MS. *Food Chem.* **2013**, *141*, 3566–3570. [CrossRef] [PubMed]
54. Llorent-Martínez, E.J.; De Córdova, M.L.F.; Ruiz-Medina, A.; Ortega-Barrales, P. Analysis of 20 trace and minor elements in soy and dairy yogurts by ICPMS. *Microchem. J.* **2012**, *102*, 23–27. [CrossRef]
55. Briffa, J.; Sinagra, E.; Blundell, R. Heavy metal pollution in the environment and their toxicological effects on humans. *Heliyon* **2020**, *6*, e04691. [CrossRef] [PubMed]
56. Kowalska, G. The safety assessment of toxic metals in commonly used herbs, spices, tea, and coffee in Poland. *Int. J. Environ. Res. Public Health* **2021**, *18*, 5779. [CrossRef]
57. Cilak, G.O.; Mujdeci, G.N.; Kabak, B. Legislation on aromatic herbs in food. In *Aromatic Herbs in Food*; Charis, M.G., Ed.; Academic Press: Cambridge, MA, USA, 2021; pp. 405–438.
58. López, F.F.; Cabrera, C.; Lorenzo, M.L.; López, M.C. Aluminium levels in spices and herbs. *Sci. Total Environ.* **2000**, *257*, 191–197. [CrossRef]

59. Bojórquez-Quintal, E.; Escalante-Magaña, C.; Echevarría-Machado, I.; Martínez-Estévez, M. Aluminum, a Friend or Foe of Higher Plants in Acid Soils. *Front. Plant. Sci.* **2017**, *8*, 1767. [CrossRef]
60. Bratakos, S.M.; Lazou, A.E.; Bratakos, M.S.; Lazos, E.S. Aluminium in food and daily dietary intake estimate in Greece. *Food Addit. Contam. Part B Surveill.* **2012**, *5*, 33–44. [CrossRef]

foods

MDPI

Article

Chemical Characterization, Antibacterial Activity, and Embryo Acute Toxicity of *Rhus coriaria* L. Genotype from Sicily (Italy)

Giovanna Lo Vecchio [1], Nicola Cicero [1,*], Vincenzo Nava [1], Antonio Macrì [1], Claudio Gervasi [2], Fabiano Capparucci [2], Marzia Sciortino [3], Giuseppe Avellone [3], Qada Benameur [4], Antonello Santini [5,*] and Teresa Gervasi [1]

1 Department of Biomedical and Dental Sciences and Morphofunctional Imaging, University of Messina, 98125 Messina, Italy; giolovecchio@unime.it (G.L.V.); vincenzo.nava@unime.it (V.N.); antonio.macri@unime.it (A.M.); teresa.gervasi@unime.it (T.G.)
2 Department of Chemical, Biological, Pharmaceutical and Environmental Sciences, University of Messina, 98166 Messina, Italy; claudio.gervasi@unime.it (C.G.); fabiano.capparucci@unime.it (F.C.)
3 Department of Biological, Chemical and Pharmaceutical Sciences and Technologies, University of Palermo, 90123 Palermo, Italy; marziasciortino@gmail.com (M.S.); beppe.avellone@unipa.it (G.A.)
4 Nursing Department, Faculty of Nature and Life Sciences, University of Mostaganem, Mostaganem 27000, Algeria; qada.benameur@univ-mosta.dz
5 Department of Pharmacy, University of Napoli Federico II, Via D. Montesano 49, 80131 Napoli, Italy
* Correspondence: nicola.cicero@unime.it (N.C.); asantini@unina.it (A.S.)

Citation: Lo Vecchio, G.; Cicero, N.; Nava, V.; Macrì, A.; Gervasi, C.; Capparucci, F.; Sciortino, M.; Avellone, G.; Benameur, Q.; Santini, A.; et al. Chemical Characterization, Antibacterial Activity, and Embryo Acute Toxicity of *Rhus coriaria* L. Genotype from Sicily (Italy). *Foods* **2022**, *11*, 538. https://doi.org/10.3390/foods11040538

Academic Editor: Manuel Viuda-Martos

Received: 27 December 2021
Accepted: 8 February 2022
Published: 14 February 2022

Abstract: This study reports a full characterization of the Sicilian sumac, *Rhus coriaria* L. This fruit represents a potential source of fiber (33.21 ± 1.02%) and unsaturated fatty acids, being the contents of linoleic and α-linolenic acids, 30.82 ± 1.21% and 1.85 ± 0.07%, respectively. In addition, the content of phenolic and total anthocyanin was 71.69 ± 1.23 mg/g as gallic acid equivalents, and 6.71 ± 0.12 mg/g as cyanidin-3-O-glucoside equivalents, respectively. The high content in mineral elements, consisting mainly of potassium, calcium, magnesium, and phosphorus, followed by aluminum, iron, sodium, boron, and zinc, was detected by inductively coupled plasma mass spectrometry (ICP-MS). Moreover, its antimicrobial activity was evaluated against multidrug resistant (MDR) microorganisms, represented by *Escherichia coli* and *Klebsiella pneumoniae* strains isolated from poultry. The activity of seven different sumac fruit extracts obtained using the following solvents—ethanol (SE), methanol (SM), acetone (SA), ethanol and water (SEW), methanol and water (SMW), acetone and water (SAW), water (SW)—was evaluated. The polyphenol profile of SM extract, which showed better activity, was analyzed by ultra-high performance liquid chromatography coupled with mass spectrometry (UHPLC-MS). The major component identified was gallic acid, followed by quercetin, methyl digallate, pentagalloyl-hexoside, and kaempferol 3-O-glucoside. The non-toxicity of Sicilian *R. coriaria* was confirmed by testing the effect of the same extract on zebrafish embryos.

Keywords: *Rhus coriaria*; sumac; polyphenols; antibacterial activity

1. Introduction

Rhus coriaria L., generally known as sumac, is a typical plant native of a large area spreading from the Canary Islands over the Mediterranean coast to Iran and Afghanistan. Its name originates from "sumaga," which means red in Syriac [1].

Among the sumac species, *R. coriaria* L. has the greatest economic importance. In Sicily, this Asian species was first imported by the Arabs; on this island, it grows spontaneously and acquires surprising nutritional characteristics determined by the pedoclimatic conditions [2].

R. coriaria is a high shrub or small tree (1–3 m high) with imparipinnate leaves, villose and red fruits with one-seeded drupe, and small greenish-white flowers organized in panicles [3].

Several studies have been carried out in order to identify the major components of the *R. coriaria* plant's different parts [3], its bioactive molecules [4], and its fatty acid composition [5]. However, only few studies have been conducted on Sicilian sumac [6,7], and in particular on its drupes, which are the main endpoints of the proposed study.

Since ancient times, given its nutritional value and its phytochemical components (flavonoids, flavones, anthocyanins, tannins, organic acids, fiber, proteins, volatile oils, nitrites, and nitrates), it has been used both as a spice by mashing and mixing the dehydrated fruits with salt and as a medicinal herb [7,8]. Its ordinary state is a fruit, and, to date, it is economically attractive because of its growing use in several biotechnological applications, from the nutraceutical and food sectors, to cosmetic and pharmaceutical industries as well as in veterinary practices, and in dying leather [3,4,6,9,10].

The *R. coriaria* components, including fatty acids, minerals, fiber, and phytochemicals, are responsible for its several beneficial properties. Its nutritional value makes this plant interesting as a food fortifier or functional food [6]. Its antimicrobial and antioxidant properties make this plant a promising tool as a food preservative [3,11,12]. In addition, its coloring properties and tannins are used in dying and tanning fine leather. The bioactive compounds, which are responsible for antioxidant, antilipidemic, antimicrobial, antiviral, antifungal, and anti-inflammatory activity [11–14], also make this plant an interesting tool for the pharmaceutical sector.

To date, there is no study investigating the antimicrobial activity of Sicilian sumac, and although several studies have reported on the antimicrobial activity of *R. coriaria* [15–17], the proposed study is novel, adding information to the area of interest.

In order to investigate all the promising potential of Sicilian sumac, in the present study, a full characterization of this fruit is reported, including the proximate composition determination, the phenolic and anthocyanin content, and the mineral content. A preliminary comparative antibacterial screening of different sumac extracts was carried out, and its non-toxicity was proven by using the Zebrafish Embryo Toxicity Test (ZFET). Further studies are in progress to have a better understanding of this plant genotype and its possible biotechnological applications.

2. Materials and Methods

2.1. Reagents and Chemicals

Heptane, methanol, ethanol, and acetone were supplied by PanReac AppliChem (Barcelona, Spain) and J.T. Baker (Phillipsburg, NJ, USA). Ultrapure water (18 mΩ cm resistivity and <5 ppb TOC) was produced by a Barnstead Smart2Pure 12 Water Purification System (Thermo Scientific, Milan, Italy). Reference standards of fatty acids methyl esters (FAMEs, C4–C24) and stock standard solutions of inorganic elements (1000 mg/L in 2% HNO_3) were obtained from Supelco (Bellefonte, USA). Sugars and polyphenol standards were obtained from Extrasynthese (Genay, France) and Sigma-Aldrich (St. Louis, MO, USA). The Kjeldahl catalyst was supplied by Carlo Erba (Milan, Italy). Methanol, standard gallic acid, cyanidin-3-O-glucoside, and Folin–Ciocalteau reagent were obtained from Sigma-Aldrich (Steinheim, Germany). PTFE syringe filters (0.45 µm) were purchased from Gelman Sciences Inc. (Ann Arbor, MI, USA). High purity water with a resistivity of 10 mΩ cm, nitric acid trace metal analysis grade, and hydrogen peroxide were acquired from J.T. Baker (Milan, Italy). Stock standard solutions of B, Mg, Na, Al, K, Ti, Cr, Mn, Fe, Zn, Ni, As, Sr, Ba, Pb, and Bi (1000 mg/L in 2% nitric acid) were purchased from Fluka (Milan, Italy). The Cd solution (1000 mg/L in 2% nitric acid) and the Hg solution (1000 mg/L in 3% hydrochloric acid) were obtained from Merck (Darmstadt, Germany). MeOH HPLC grade, quercetin, and gallic acid standards were obtained from Sigma-Aldrich (St. Louis, MO, USA).

2.2. Plant Material

R. coriaria L. drupes (1 kg) were collected in Messina (38°13′31.42″ N 15°32′21.7″ E) in September 2020. The fresh material was immediately dried in the dark at a low temperature and then pulverized.

2.3. Proximate Composition Determination

The determination of the proximate composition was carried out according to the AOAC (Association of Official Analytical Chemist) methods [18]. In particular, the following methods were used: dry matter (method 925.10), crude ash (method 923.03), crude protein (method 990.03), crude fiber (method 962.09), starch (method 996.11), and crude fat by Soxhlet extraction (method 960.39 with some modifications). The sumac sample was analyzed in triplicate.

For the fatty acids profiling, 15 g of sample was extracted for 6 h with a Soxhlet apparatus using heptane as solvent. After the extraction, the solvent was eliminated with a rotating evaporator and stored at −18 °C until the chromatographic analysis. The analysis of FAMEs was performed as described by the EU Regulation n. 1833/2015 (European Commission, 2015). One μL of each extract was analyzed by a gas chromatograph (GC) (Dani Master GC1000) equipped with a split/splitless injector and a flame ionization detector (FID) (Dani Instrument, Milan, Italy). A 60 m × 0.25 mm ID 0.20 μm film thickness Supelco SLB-IL100 capillary column (Supelco, Sigma-Aldrich, St. Louis, MO, USA) was used. The chromatographic conditions used were as follows: temperature from 165 °C to 210 °C (10 min held) at 2 °C/min, injector and detector temperature was 250 °C, and helium was at a linear velocity of 30 cm/s. The injection volume was 1 μL with a split ratio of 1:100. The Clarity Chromatography Software v4.0.2 (DataApex, Prague, Czech Republic) was used for data acquisition and processing. The sample was analyzed in triplicate. FAMEs were identified by comparing the retention times of the peaks with those of the standards. The percentage of each FAMEs was calculated by comparison with the corresponding chromatogram peak area. The precision of the method was assessed in terms of RSD% analyzing five times each sample [19].

2.4. Total Phenolic and Anthocyanin Content Determination

The total phenolic content of *R. coriaria* L. was determined using the Folin–Ciocalteu method [20]. Briefly, 3 g of dried and minced sample was homogenized with 8 mL of an 80% aqueous methanol solution and placed in a vessel at −20 °C overnight. The sample was then centrifuged (10000 rpm for 15 min), and the supernatant was filtered with a 0.45 μM filter.

The total phenolic content absorbance measurements were registered using a Ultrospec 2100 Pro UV-VIS spectrophotometer (GE Healthcare Ltd., Chicago, IL, USA) at a wavelength of 760 nm. Absorbance values were converted to gallic acid equivalents and expressed as mg/g.

The total anthocyanins content was spectrophotometrically evaluated, as described by Landi et al. [21], in acidified methanol (0.1% HCl, *v*/*v*), and the absorbance was measured at 535 nm. Absorbance values were converted to cyanidin-3-O-glucoside equivalents and expressed as mg/g.

2.5. Mineral Element Content Determination

The determination of the mineral elements in the *Rhus coriaria* samples was carried out using an ICP-MS iCAP-Qc spectrometer (Thermo Fisher Scientific, Milan, Italy) equipped with a 27 MHz radiofrequency solid-state generator at 1550 W. A closed vessel microwave digestion system Ethos 1 (Milestone, Bergamo, Italy) was used for the sample digestion. Approximately 0.50 g of *R. coriaria* sample was digested with 7 mL of HNO_3 (69% *v*/*v*) and 1 mL of H_2O_2 (30% *v*/*v*) in a pre-washed PTFE vessel. The sample was then cooled down at room temperature, diluted up to 25 mL with ultrapure water, and filtered with a 0.45 μm filter [19,22]. The certified reference materials were processed using the same conditions.

The ICP-MS operating parameters were the incident radiofrequency power 1500 W, plasma gas flow argon (Ar) at 15 L/min, auxiliary gas flow rate (Ar) 0.9 L/min, and carrier gas flow rate (Ar) 1.10 L/min. Helium (He) was the collision cell gas (4 mL/min), and the spray chamber was set at T = 2 °C. The injection volume and sample introduction rate were 200 µL and 1 mL/min, respectively. A full scan mode (dwell time 0.5 s point 1) was used for spectra acquisition. All samples and the analytical blanks were analyzed in triplicate.

Data acquisition was performed using the Qtegra™ Intelligent Scientific Data Solution™ (Thermo Scientific) software. For the quantification, an external calibration procedure was used. The determination of mercury was performed using the automatic mercury analyzer DMA-80 (Milestone Srl, Bergamo, Italy). An aliquot of the sample (100 mg) was placed in a vessel, dried for 3 min at 200 °C, and decomposed at 650 °C for 2 min. The content of Hg was determined by measuring the absorbance at 253.7 nm.

2.6. Preparation of Extracts

The dried and ground fruits were extracted with different solvents obtained from Sigma-Aldrich: methanol (SM), ethanol (SE), acetone (SA), methanol and water (SMW), ethanol and water (SEW), acetone and water (SAW), and water (SW). The samples (2 g) in the respective extraction solvents (20 mL) were sonicated for 15 min, filtered with a Whatman filter, dried with a rotary evaporator (BUCHI R-210, Merck KGaA, Darmstadt, Germany), and lyophilized. The resulting dried extracts were used for further microbiological analysis.

2.7. Polyphenols Profile in SM

Amounts of 229.16 mg and 224.29 mg of sumac powder were added to 3 mL aliquots of HPLC grade MeOH. The ultrasound-assisted extraction of phenolics from each powder was performed using a Transonic 460 H ultrasonic bath (Elma Hans Schmidbauer, Singen, Germany) at room temperature operating for 15 min at 35 kHz ultrasonic frequency. Then, each sample was centrifuged (5000 rpm, 6 min) using an SL 16 centrifuge (Thermo Fisher Scientific, Waltham, MA, USA). The supernatant was recovered, and to the solid residue, another 3 mL of MeOH were added two times. The methanolic extracts were dried under reduced pressure. Samples were then added with 2 mL of MeOH, filtered, and transferred in a glass vessel. Analyses were carried out using an HPLC Alliance e2695 (Waters, Milford, MA, USA) system equipped with an autosampler, degasser, and column heater coupled with a Q-Tof Premier quadrupole time-of-flight mass spectrometer (Waters, Milford, MA, USA).

The compounds were separated using a 50 × 2.1 mm ID Hypersil GOLD HPLC column (Thermo Fisher Scientific, Waltham, MA, USA) kept at 20 °C. The injection volume was 5 µL. A thermostated auto sampler, kept at 4 °C, was used, and all samples were injected in triplicate. The HPLC eluent was a mix of 0.1% aqueous formic acid solution and 0.1% formic acid in MeOH, with a flow rate of 0.25 mL/min.

Elution started with 95% aqueous formic acid and 5% methanol formic acid, and then isocratic for 1 min. Then, in the following 14 min, the solvent became 100% MeOH, remaining isocratic for the subsequent 5 min (from min 15 to min 20). After 30 s, the eluting solvent mixture was reverted to 95% aqueous formic acid and 5% methanolic formic acid and maintained for another 30 s. Each run lasted 21 min. Every sample was injected three times. The concentration values, which are reported in Table 4, are the arithmetic mean of the values observed in each run.

For the detection of flavonoids and other phenols, the calibration curves of quercetin and gallic acid, both of HPLC purity grade, were used. The calibration curves were performed using a standard solution of quercetin in methanol (1000 ppm 10 mg/10 mL) and a standard solution of gallic acid in methanol (1000 ppm 10 mg/10 mL), respectively. Each calibration curve was obtained using 0.5 ppm, 1 ppm, 5 ppm, 10 ppm, and 20 ppm solutions.

The following compounds were investigated: quercetin, quercetin 2′O-gallate, quercetin glucuronide, quercetin-hexose malic acid, methyl-dihydroquercetin hexoside, kaempferolol, myricetin-rhamnose malic acid, quercetin 3-O-hexuronide, kaempferol 3-O-glucoside, quercetin 3-O-galactoside, myricetin, myricetin 3-O-hexoside, apigenin glucoside, myrecetin O-rhamnosylglucose, phenols, gallic acid, methyl digallate, pentagalloyl-hexoside, p-coumaric acid, peonidin 3-O-hexoside isomer, and vanillic acid.

2.8. Antimicrobial Tests

2.8.1. Bacterial Strains

Five *Enterobacteriaceae* isolates, four *Escherichia coli*, and one *Klebsiella pneumonia* were selected to carry out the study. The strains were isolated from poultry in the Regional Veterinary Laboratory of Mostaganem, Algeria, and identified using matrix-assisted laser desorption–ionization time-of-flight mass spectrometry (MALDI-TOF-MS), as previously reported [23]. *E. coli* ATCC 25922 and *Staphylococcus aureus* ATCC 6538 (American Type Culture Collection, Rockville, MD, USA) were also tested.

2.8.2. Antimicrobial Susceptibility Testing

Disk diffusion method was used to test and confirm the antimicrobial susceptibility of the Enterobacteriaceae isolates using Muller–Hinton agar (MHA, Oxoid, Milan, Italy) and an incubation time of 16–18 h at 37 °C, following the Clinical and Laboratory Standards Institute Guidelines (CLSI) [24]. The antimicrobial used were: ciprofloxacin (CIP, 5 µg), nalidixic acid (NA, 30 µg), amoxicillin/clavulanic acid (AMC, 20/10 µg), amoxicillin (AML, 25 µg), levofloxacin (LEV, 5 µg), cefotaxime (CTX, 30 µg), sulphonamides (SSS, 300 µg), tetracycline (TE, 30 µg), trimethoprim/sulphamethoxazole (SXT, 1,25/23,75 µg), trimethoprim (TMP, 5 µg), chloramphenicol (C, 30 µg), and neomycin (N, 30 µg) (Bio-Rad, Marnes la Coquette, France). The results were assessed following the CLSI guidelines [24].

2.8.3. Determination of the Antibacterial Activity of all Extracts of *Rhus Coriaria* by Disk Diffusion Assays

The antibacterial activity of the different extracts (SE, SM, SA, SEW, SMW, SAW, and SW) against the selected Enterobacteriaceae isolates was assessed by the disk diffusion method, as previously described [25]. Briefly, the bacterial colonies were suspended in 10 mL of saline water, and the turbidity of the bacterial suspension was adjusted to 0.5 McFarland standard. MHA plates were inoculated with bacteria by spreading overnight cultures on MHA using sterile cotton swabs. Filter paper disks (6 mm diameter; Thermo Fisher, Milan, Italy) containing 10 µL of each extract at a concentration of 10 mg/mL were then applied on the agar plates. Cefotaxime served as a positive control, and a disk impregnated with sterile distilled water was used as a negative control. The plates were incubated for 24 h at 37 °C, and the antibacterial activity was evaluated by measuring the diameters of the inhibition zones. Each assay was performed in triplicate.

2.8.4. Determination of the Minimum Inhibitory Concentration (MIC) of the SM Extract

In order to determine the minimum inhibitory concentration (MIC) of SM, the serial double dilution method was performed according to CLSI guidelines [26]. Briefly, overnight bacterial cultures in log phase were used to prepare the suspension of cells adjusted to 10^6 CFU in Muller-Hinton Broth (MHB). Serial dilutions were performed in the growth medium in a concentration range between 2000 and 2 µg/mL for the SM extract. Wells containing compound-free MHB with bacteria were used as the positive control. Plates were incubated at 37 °C for 24 h. The MIC value was defined as the lowest concentration of the tested compound that inhibits the growth of bacteria at the end of the 24 h incubation. MICs were determined in triplicate. The MIC was defined as the lowest concentration inhibiting the visible growth of the tested strains after incubation.

2.9. Embryo Acute Toxicity Test

The embryo acute toxicity test was carried out according to the Organisation for Economic Cooperation and Development (OECD) guidelines for the testing of chemicals [27]. The ZFET was conducted on fertilized eggs from the Centre for Experimental Fish Pathology of Sicily (CISS, Sicily, Italy). Adult zebrafish (*Danio rerio*) were kept in a standalone facility (ZebTec, Tecniplast, West Chester, PA, United States) in water-controlled conditions: temperature 28 °C, conductivity 600 µS/cm, pH 7.5, and 14/10 h dark/light regimen. Twice a day, fish were fed with *Artemia salina* at 3% of body weight and Gemma micro 300 (Skretting, Varese, Italy). Following mating, the eggs were placed in steel grids inside tanks to avoid predation by adults and to guarantee their collection. The fertilized eggs were collected using a stereomicroscope (Leica M205 C) and exposed to *R. coriaria* extract, which was previously prepared at a concentration of 9.37 µg/mL, in a sterilized embryo medium (15 mM NaCl, 0.5 mM KCl, 1 mM $CaCl_2$, 1 mM $MgSO_4$, 0.15 mM KH_2PO_4, 0.05 mM Na_2HPO_4, 0.7 mM $NaHCO_3$; pH 7.3). The control group was held in an embryo medium. The Fish Embryo Acute Toxicity (FET) was performed, as described by Pecoraro et al. [28]. Right after the fertilization, embryos were collected, bleached as reported by Westerfield [29], and distributed as one embryo per well into 24-well plates (LABSOLUTE, Th. Geyer GmbH & Co.KG, Berlin, Germany). Embryos were incubated with a 10/14 h dark/light regimen at 26 °C for 96-hours post-fertilization (hpf). The test solutions and controls were replaced daily [27]. The exposure period started from 180 min post-fertilization and ended at 96 h. The following endpoints were used to evaluate the toxicity: embryo coagulation, tail non-detachment, somite formation lack, heartbeat non-detection, and the hatched embryos number. Acute toxicity was determined at the end of the exposure period.

2.10. Statistical Analysis

The existence of significant differences in the antimicrobial activity of the extracts was assessed by one-way ANOVA using SIGMAPLOT, version 14.0 (Systat Software Inc., San Jose, California, USA). Tukey's HSD was used as a post hoc test. The level of significance was set at $p \leq 0.05$.

3. Results and Discussion

3.1. Chemical Characterization

The proximate composition and fatty acid profile for *Rhus coriaria* from Sicily are reported in Tables 1 and 2, respectively.

Table 1. Proximate composition of *Rhus coriaria* from Sicily.

g/100 g	Sumac Fruits
Crude fiber	33.21 ±1.02
Ash	4.78 ±0.29
Crude oil	9.56 ± 0.72
Moisture	6.64 ± 0.03
Crude protein	3.47 ± 0.19
Carbohydrates	ND
Crude energy (Kcal)	ND

ND, not determined.

These results show that sumac represents a source of dietary fiber, which may be useful for relieving gastrointestinal upset [30].

In line with previous studies, fats represent the second most abundant compounds [5,9]. The fatty acid composition is reported in Table 2. As it can be observed, sumac fruits contain 65.09 ± 1.67% of polyunsaturated fatty acids, and the contents of linoleic (omega 6) and α-linolenic acid (omega 3) are 30.82 ± 1.21% and 1.85 ± 0.07%, respectively. This result confirms that this fruit could represent a source of unsaturated fatty acids, as previously reported [9]. Furthermore, oleic acid is the most abundant fatty acid, representing 31.67 ± 1.29%.

Table 2. Fatty acid profile of *Rhus coriaria* from Sicily.

Fatty Acid g/100 g	Sumac Fruits
Miristic acid	0.38 ± 0.08
Palmitic acid	31.25 ± 0.47
Palmitoleic acid	0.75 ± 0.15
Stearic acid	3.28 ± 0.55
Oleic acid	31.67 ± 1.29
Linoleic acid	30.82 ± 1.21
Linolenic acid	1.85 ± 0.07
∑ TUFA	65.09 ± 1.67
∑ TSFA	34.91 ± 1.04

Total phenolic and total anthocyanin contents of *R. coriaria* extract are 71.69 ± 1.23 mg/g as gallic acid equivalent, and 6.71 ± 0.12 mg/g as cyanidin-3-O-glucoside equivalents, respectively. These two classes of compounds in sumac are the principal constituents responsible for its phytochemical activity. *R. coriaria* is, hence, a source of healthy substances and is useful in various fields and applications.

In agreement with a previous study [31], our results confirm that sumac is a good source of minerals, so it could be a suitable tool to increase dietary mineral intake. As reported in Table 3, *R. coriaria* is mainly rich in potassium, calcium, magnesium, and phosphorus, followed by aluminum, iron, sodium, boron, and zinc.

Table 3. Mineral elements (mg/kg) in *Rhus coriaria* fruits powder.

Cr	B	Na	Mg	Al	K	Ti
0.040 ± 0.00	0.770 ± 0.09	3.980 ± 0.11	41.870 ± 3.55	4.010 ± 0.24	266.91 ± 15.55	0.480± 0.56
Mn	Fe	Ni	Zn	As	Sr	Cd
0.410 ± 0.06	2.950 ± 0.13	0.020 ± 0.00	0.360± 0.06	0.001 ± 0.00	5.390 ± 2.49	0.003± 0.00
Ba	Pb	Bi	Hg	Li	V	Co
0.270 ± 0.08	0.010 ± 0.00	0.004± 0.00	0.008± 0.00	0.010 ± 0.00	0.010 ± 0.00	0.001 ± 0.00
Cu	Se	Mo	Sb	Tl	P	Ca
0.120 ± 0.00	0.065 ± 0.00	0.005 ± 0.00	0.001 ± 0.00	0.002 ± 0.00	39.70 ± 3.05	215.53 ± 16.78

The polyphenolic composition of methanolic sumac extract is reported in Table 4. The total polyphenols account for 71.69 mg/g. The flavonoids and phenols represent 18.48% and 81.52%, respectively. The first component identified was gallic acid, followed by quercetin, methyl digallate, pentagalloyl-hexoside, and kaempferol 3-O-glucoside.

3.2. Antimicrobial Activity

In order to analyze the antimicrobial activity of sumac, a comparative study was conducted by evaluating the antibacterial potential of six different extracts against foodborne pathogens.

The isolates involved one *S. aureus* isolate, one *K. pneumoniae* isolate, and five *E. coli* isolates.

Table 4. Polyphenolic compounds in *Rhus coriaria* SM extract.

Compounds	SM Extract (mg/g)
Flavonoids	
Quercetin	23.13 ± 0.02
Quercetin 2′O-gallate	5.30 ± 0.02
Quercetin glucuronide	1.71 ± 0.01
Quercetin-hexose malic acid	11.11 ± 0.01
Methyl-dihydroquercetin hexoside	18.34 ± 0.02
Kaempferolo	3.34 ± 0.01
Myricetin-rhamnose malic acid	11.58 ± 0.01
Quercetin 3-O-hexuronide	2.84 ± 0.02
Kaempferol 3-O-glucoside	99.86 ± 0.01
Quercetin 3-O-galactoside	160.53 ± 0.02
Myricetin	2.71 ± 0.02
Myricetin 3-O-hexoside	18.55 ± 0.01
Apigenin glucoside	20.86 ± 0.01
Myrecetin O-rhamnosylglucose	4.34 ± 0.02
Phenols	
Gallic acid	142.549 ± 0.02
Methyl digallate	110.96 ± 0.01
Pentagalloyl-hexoside	128.09 ± 0.01
p-Coumaric acid	10.48 ± 0.01
Peonidin 3-O-hexoside isomer	4.54 ± 0.02
Vanillic acid	5.76 ± 0.02

The antimicrobial susceptibility test showed that the Enterobacteriaceae isolates from poultry (*E coli* and *K. pneumoniae* strains) were resistant to AML and N. The Enterobacteriaceae isolates tested were found to be resistant to three or more antimicrobial agents that belong to dissimilar antibiotic classes and are multidrug resistant (MDR). The antimicrobial profile of the tested microorganisms is reported in Table 5.

Table 5. Antimicrobial resistance profile.

Strains	Antimicrobial Resistance Profile
Escherichia. coli (S12/15)	NA, CIP, AML, AUG, SXT, TE, N
Escherichia. coli (S34/16)	NA, CIP, N
Escherichia coli (S6/15)	NA, CIP, AML, AUG, TE, N
Escherichia. coli (S2/15)	NA, CIP, AML, AUG, SXT, TE, N
Klebsiella pneumoniae	AML, SXT, TE, C, N

NA, nalidixic acid; CIP, ciprofloxacin; AML, amoxicillin; AUG, amoxicillin-clavulanic acid; SXT, trimethoprim-sulfamethoxazole; TE, tetracycline; N, neomycin; C, chloramphenicol.

These results, showing the multi-resistance of the majority of the isolates, confirm previous studies that reported a high level of multidrug resistant Enterobacteriaceae isolates from poultry in Algeria [23,32]. The causes of this antibiotic resistance phenomenon are very different, and the important cause is the massive and inappropriate use of antibiotics. Currently, antimicrobial resistance represents a growing global concern, and the development of effective therapeutic options against MDR bacteria is a public health priority.

Plant extracts can be valuable alternatives to antibiotics [33,34]. Recently, sumac has gained more attention due to its high amount of polyphenols. Numerous studies have been conducted to study the in vitro antimicrobial effectiveness of *Rhus coriaria* extracts against several bacterial species [14,35,36], while limited data are available on their effect against MDR bacteria. In this study, the antibacterial activity of *Rhus coriaria* extracts was tested against five MDR Enterobacteriaceae isolates, *S. aureus* ATCC 6538, and *E. coli* ATCC 25922. The results of the antibacterial activity of *Rhus coriaria* extracts are summarized in Table 6, which indicates that almost all the extracts exerted antibacterial activity against all the tested strains.

Table 6. Antibacterial activity of extracts from *Rhus coriaria* evaluated by disk diffusion assay.

Strains	Radius of Inhibition in mm (Mean of Three Tests)						
	SA	SE	SM	SAW	SEW	SMW	SW
Staphylococcus aureus ATCC 6538	10 A,B,C,D,H,P	11 D,E,F,G,I,R	19 H,I,L,N,O	9 A,E,L,N,Q	10 B,F,M,S	17 O,P,Q,R,S	5 C,M
Escherichia coli ATCC 25922	13 A,B,C	12 A, D, E,G	15	11 D,F,H	11 B,E,F,I	12 C,G,H,I	4
Klebsiella pneumoniae	9 A,B	14	22	8 A, C	11	19	8 B,C
Escherichia coli (S12/15)	15 A	19 B	25 B	14	17 A	22	7
Escherichia coli (S34/16)	16 A,B	17 B	24	8	14 A	21	5
Escherichia coli (S6/15)	14 A	20 B	24	12 B	15 A	20	7
Escherichia coli (S2/15)	12 A,B	23	22	11 A	16 B	19	6

SA, acetone extract; SE, ethanol extract; SM, methanol extract; SAW, acetone + water extract; SEW, ethanol + water extract; SMW, methanol + water extract; SW, water extract. Means sharing the same capital letters in the raw are not significant at $p < 0.05$ according to Tukey's HSD test.

The higher antimicrobial activity was obtained by using SM and SE. In particular, SE presents growth inhibition zones ranging from 14 to 16 mm and SM ranging from 22 to 25 mm. Given the higher antimicrobial activity of the SM, its MIC was evaluated, and the results are reported in Table 7. These results are in line with a study on Syrian sumac antimicrobial activity, which reports that the antimicrobial activity of the methanol extract was the most effective [37]. In addition, further studies reported on the higher amount of phenolics and flavonoids contained in the methanolic extract compared with the ethanolic one [37,38]. Several studies have described the antibacterial activity of the fractions of sumac extract, highlighting how some compounds, due to their polarity, can act only on Gram positive bacteria, while some others, such as gallic acid, can affect both Gram positive and Gram negative bacteria; nevertheless, their antibacterial activity is not so strong [38].

Table 7. Antibacterial activity of SM from *Rhus coriaria* evaluated by serial double dilution method.

Strains	MIC (µg/mL)
Staphylococcus aureus ATCC 6538	9.37
Escherichia coli ATCC 25992	9.37
Klebsiella pneumoniae	9.37
Escherichia coli (S12/15)	9.37
Escherichia coli (S34/16)	9.37
Escherichia coli (S6/15)	9.37
Escherichia coli (S2/15)	4.68

Several studies have also reported on the efficacy of the total extract and attribute its biological activity to its content in phenolics, which are the major extract fraction [39–41].

The maximum inhibitory action was observed at a concentration of 9.37 µg/mL for all strains except for *E coli* (S2/15), which presented an MIC of 4.68 µg/mL. This finding supports the use of *Rhus coriaria* in traditional medicine as a bactericide agent. A study reported that the water extract of *R. coriaria* had an effective in vitro antibacterial power against *S. aureus*, *P. aeruginosa*, and *S. aureus* (MRSA) [35]. Another study reported that the extract showed strong antibacterial activity against Gram positive and Gram negative bacteria, with MIC < 0.78% [42]. A similar study assessed sumac's methanolic extract antibacterial power as having the highest inhibitory activity. In all sumac extracts, increasing the concentration of sumac causes an increase in antibacterial power [14]. In addition, the methanolic extract of sumac leaves revealed antibacterial activity against *E. coli* and *S. aureus*. A MIC of 312 μg/mL was reported, although the inhibitory effect was only bacteriostatic, and the bactericidal effect was observed at a concentration of 2500 μg/mL [36].

The results of our study, highlighting the higher activity of the methanol extract, are therefore in line with previous comparative studies that showed that methanolic extracts of sumac contain a higher content of flavonoids and phenolics when compared with other extracts [13,14].

3.3. Fish Embryo Acute Toxicity (FET)

The evaluation of the toxicological profile of medicinal plant extract is of utmost importance. Zebrafish (*Danio rerio*) is one of the main study models [43]. Given that the embryo develops quickly outside the mother and that this is visually evident, it is certainly usable for testing and observation. Being that the ZFET is a valid alternative method to animal tests [44,45], the non-toxicity of the *R. coriaria* genotype from Sicily was assessed and confirmed by analyzing the effect of the extract on zebrafish larvae. To date, there is no evidence of the toxicity of Sicilian sumac.

In this study, fertilized zebrafish embryos were exposed to SM extract from *R. coriaria* at a concentration of 9.37 μg/mL. According to our results, the extract was found to be non-toxic using zebrafish FET assay. During the 96 h of exposure, no visible toxic effects of this extract on the development of embryos were observed (Figure 1). The mortality was 5% (one larva) for the whole test period. During the observation under the stereomicroscope, it was observed that at 48 hpf (hours post-fertilization), the hatched larvae were 95%.

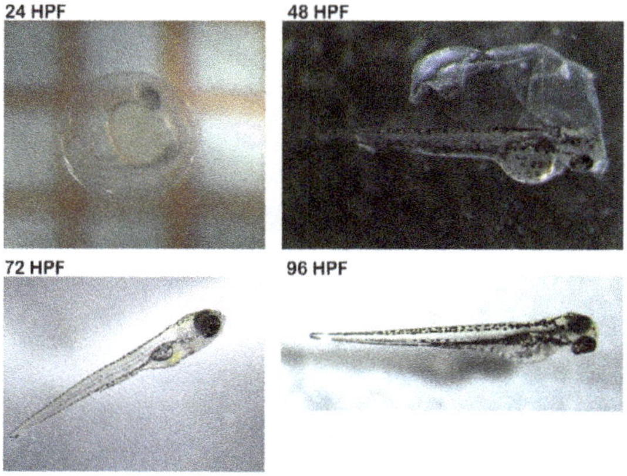

Figure 1. Development of embryos. Hpf: hours post-fertilization.

According to the OECD guidelines, the *R. coriaria* extract obtained from the Sicilian genotype did not induce any toxic effect on zebrafish embryos and larval development. These results are in line with other findings reporting the safe and even beneficial effects of the *R. coriaria* extract on both humans and animals [5]. Other studies evidenced no acute toxicity of the extract in rat model experiments and, in addition, showed beneficial cardioprotective and hepatoprotective properties under hypercholesterolemic conditions [46]. Another study using a diabetic rats model and testing 250, 500, and 1000 mg/kg of the plant extract reported good tolerance and a non-lethal oral uptake of this extract, even at 1000 mg/kg, showing not only no signs of toxicity and mortality after 3 days of daily extract administration but also a positive effect on diabetes and diabetes-related complications [47]. Taken together, these results suggest the safety of this plant.

4. Conclusions

This study provides a characterization of Sicilian sumac drupes, including the proximate composition determination, the phenolic and anthocyanin content, and the mineral content, and suggests its potential use in the food supplement area [48–51]. In addition, it offers for the first time a preliminary screening of its antimicrobial activity against MDR Enterobacteriaceae and proves the non-toxicity of this food matrix. Taken together, the results show the potential of sumac as a functional food supplement and the application of the sumac extract in the food industry, not only as a food additive but also as an efficient and natural food preservative.

Author Contributions: Conceptualization, T.G. and A.S.; methodology, G.L.V., V.N., A.M., C.G., F.C., M.S. and Q.B.; writing—original draft preparation, T.G., A.S., G.L.V. and C.G.; writing—review and editing, T.G., N.C. and A.S.; supervision, T.G., N.C., A.S. and G.A. All authors have read and agreed to the published version of the manuscript.

Funding: This work received no external funding.

Conflicts of Interest: The authors declare no conflict of interest.

References

1. Wetherilt, H.; Pala, M. Herbs and spices indigenous to Turkey. In *Spices, Herbs and Edible Fungi: Developments in Food Science*; Charalambous, G., Ed.; Elsevier Science BV: Amsterdam, The Netherlands, 1994; Volume 34, pp. 285–307.
2. Conti, F.; Abbate, G.; Alessandrini, A.; Blasi, C. *An Annotated Checklist of the Italian Vascular Flora*; Palombi Editori: Roma, Italy, 2005; ISBN 88-7621-458-5.
3. Shabbir, A. *Rhus coriaria* Linn., a Plant of Medicinal, Nutritional and Industrial Importance: A Review. *J. Anim. Plant Sci.* **2012**, *22*, 505–512.
4. Asgarpanah, J.; Saati, S. An Overview on Phytochemical and Pharmacological Properties of *Rhus coriaria* L. *Res. J. Pharmacogn.* **2014**, *1*, 47–54.
5. Dogan, M.; Akgul, A. Characteristics and fatty acid compositions of *Rhus coriaria* cultivars from southeast Turkey. *Chem. Nat. Compd.* **2005**, *41*, 724–725. [CrossRef]
6. Grassia, M.; Sarghini, F.; Bruno, M.; Cinquanta, L.; Scognamiglio, M.; Pacifico, S.; Fiorentino, A.; Geraci, A.; Schicchi, R.; Corona, O. Chemical Composition and Microencapsulation Suitability of Sumac (*Rhus coriaria* L.) Fruit Extract. *Eur. Food Res. Technol.* **2021**, *247*, 1133–1148. [CrossRef]
7. Romeo, F.V.; Ballistreri, G.; Fabroni, S.; Pangallo, S.; Li, M.G.; Nicosia, D.; Schena, L.; Rapisarda, P. Molecules Chemical Characterization of Different Sumac and Pomegranate Extracts Effective against Botrytis Cinerea Rots. *Molecules* **2015**, *20*, 11941–11958. [CrossRef]
8. Ardalani, H.; Hassanpour Moghadam, M.; Hadipanah, A.; Fotovat, F.; Azizi, A.; Soltani, J. Identification and Characterization of Chemical Composition of *Rhus coriaria* L. Fruit from Hamadan, Western Iran. *J. Herb. Drugs* **2016**, *6*, 195–198.
9. Kizil, S.; Haşimi, N.; Tolan, V.; Kilinc, E.; Yuksel, U. Mineral Content, Essential Oil Components and Biological Activity of Two Mentha Species (*M. piperita* L., *M. spicata* L.). *Turk. J. Field Crop.* **2010**, *15*, 148–153.
10. Dabas, D. Polyphenols as Colorants. *Adv. Food Technol. Nutr. Sci.-Open J.* **2018**, *SE*, S1–S6. [CrossRef]
11. Sakhr, K.; El Khatib, S. Physiochemical properties and medicinal, nutritional and industrial applications of Lebanese Sumac (Syrian Sumac -Rhus coriaria): A review. *Heliyon* **2020**, *6*, e03207. [CrossRef]
12. Gulmez, M.; Oral, N.; Vatansever, L. The Effect of Water Extract of Sumac (*Rhus coriaria*) and Lactic Acid on Decontamination and Shelf Life of Raw Broiler Wings. *Poult. Sci.* **2006**, *85*, 1466–1471. [CrossRef]
13. Fereidoonfar, H.; Salehi-Arjmand, H.; Khadivi, A.; Akramian, M.; Safdari, L. Chemical variation and antioxidant capacity of sumac (*Rhus coriaria* L.). *Ind. Crop. Prod.* **2019**, *139*, 111518. [CrossRef]
14. Abraham, A.M.; Quintero, C.; Carrillo-Hormaza, L.; Osorio, E.; Keck, C.M. Production and Characterization of Sumac Plant Crystals: Influence of High-Pressure Homogenization on Antioxidant Activity of Sumac (*Rhus coriaria* L.). *Plants* **2021**, *10*, 1051.
15. Nasar-Abbas, S.M.; Halkman, A.K. Antimicrobial effect of water extract of sumac (*Rhus coriaria* L.) on the growth of some food borne bacteria including pathogens. *Int. J. Food Microbiol.* **2004**, *97*, 63–69. [CrossRef]
16. Alsamri, H.; Athamneh, K.; Pintus, G.; Eid, A.H.; Iratni, R. Pharmacological and Antioxidant Activities of *Rhus coriaria* (Sumac). *Antioxidants* **2021**, *10*, 73. [CrossRef]
17. Shahryari, F.; Rabiei, Z.; Sadighian, S. Antibacterial activity of synthesized silver nanoparticles by sumac aqueous extract and silver-chitosan nanocomposite against Pseudomonas syringae pv. syringae. *J. Plant Pathol.* **2020**, *102*, 469–475. [CrossRef]
18. Association of Official Analytical Chemist (AOAC). *Official Method of Analysis*, 18th ed.; Association of Official Analytical Chemist (AOAC): Washington, DC, USA, 2005.
19. Nicolò, M.S.; Gugliandolo, C.; Rizzo, M.G.; Zammuto, V.; Cicero, N.; Dugo, G.; Guglielmino, S. Nutritional conditions of the novel freshwater Coccomyxa AP01 for versatile fatty acids composition. *J. Appl. Microbiol.* **2021**, *132*, 401–412. [CrossRef] [PubMed]

20. Dewanto, V.; Wu, X.; Adom, K.K.; Liu, R.H. Thermal Processing Enhances the Nutritional Value of Tomatoes by Increasing Total Antioxidant Activity. *J. Agric. Food Chem.* **2002**, *50*, 3010–3014. [CrossRef] [PubMed]
21. Landi, M.; Guidi, L.; Pardossi, A.; Tattini, M.; Gould, K.S. Photoprotection by Foliar Anthocyanins Mitigates Effects of Boron Toxicity in Sweet Basil (*Ocimum Basilicum*). *Planta* **2014**, *240*, 941–953. [CrossRef] [PubMed]
22. Panebianco, F.; Nava, V.; Giarratana, F.; Gervasi, T.; Cicero, N. Assessment of Heavy- and Semi-Metals Contamination in Edible Seaweed and Dried Fish Sold in Ethnic Food Stores on the Italian Market. *J. Food Compos. Anal.* **2021**, *104*, 104150. [CrossRef]
23. Benameur, Q.; Gervasi, T.; Giarratana, F.; Vitale, M.; Anzà, D.; La Camera, E.; Nostro, A.; Cicero, N.; Marino, A. Virulence, Antimicrobial Resistance and Biofilm Production of Escherichia coli Isolates from Healthy Broiler Chickens in Western Algeria. *Antibiotics* **2021**, *10*, 1157. [CrossRef]
24. M02. *Performance Standards for Antimicrobial Disk Susceptibility Tests*, 27th ed.; Clinical and Laboratory Standards Institute (CLSI): Wayne, PA, USA, 2017.
25. Balouiri, M.; Sadiki, M.; Ibnsouda, S.K. Methods for in vitro evaluating antimicrobial activity: A review. *J. Pharm. Anal.* **2016**, *6*, 71–79. [CrossRef] [PubMed]
26. M07-A10. *Methods for Dilution Antimicrobial Susceptibility Tests for Bacteria That Grow Aerobically: Approved Standard*, 10th ed.; Clinical and Laboratory Standards Institute (CLSI): Wayne, PA, USA, 2015.
27. Organisation for Economic Co-operation and Development (OECD). Test no. 236: Fish embryo acute toxicity (FET) test. In *OECD Guidelines for the Testing of Chemicals, Section 2*; Organisation for Economic Co-operation and Development: Paris, France, 2013; ISBN 9789264203709.
28. Pecoraro, R.; Salvaggio, A.; Marino, F.; di Caro, G.; Capparucci, F.; Lombardo, B.M.; Messina, G.; Scalisi, E.M.; Tummino, M.; Loreto, F.; et al. Metallic Nano-Composite Toxicity Evaluation by Zebrafish Embryo Toxicity Test with Identification of Specific Exposure Biomarkers. *Curr. Protoc. Toxicol.* **2017**, *74*, 1.14.1–1.14.13. [CrossRef] [PubMed]
29. Westerfield, M. *The Zebrafish Book: A Guide for the Laboratory Use of Zebrafish (Danio Rerio)*, 5th ed.; University of Oregon Press: Eugene, OR, USA, 2007.
30. Hariri, N.; Darafshi Ghahroudi, S.; Jahangiri, S.; Borumandnia, N.; Narmaki, E.; Saidpour, A. The Beneficial Effects of Sumac (*Rhus coriaria* L.) Supplementation along with Restricted Calorie Diet on Anthropometric Indices, Oxidative Stress, and Inflammation in Overweight or Obese Women with Depression: A Randomized Clinical Trial. *Phytother. Res.* **2020**, *34*, 3041–3051. [CrossRef] [PubMed]
31. Ozcan, M. The Neighboring Keidel Mineral Contents of Same Plants Used as Condiments in Turkey Food. *Chem. Nat. Compd.* **2004**, *84*, 74–84.
32. Benklaouz, M.B.; Aggad, H.; Benameur, Q. Resistance to Multiple First-Line Antibiotics among *Escherichia coli* from Poultry in Western Algeria. *Vet. World* **2020**, *13*, 290–295. [CrossRef]
33. Aghraz, A.; Benameur, Q.; Gervasi, T.; Ait Dra, L.; Ben-Mahdi, M.H.; Larhsini, M.; Markouk, M.; Cicero, N. Antibacterial activity of *Cladanthus arabicus* and *Bubonium imbricatum* essential oils alone and in combination with conventional antibiotics against Enterobacteriaceae isolates. *Lett. Appl. Microbiol.* **2018**, *67*, 175–182. [CrossRef]
34. Benameur, Q.; Gervasi, T.; Pellizzeri, V.; Pľuchtová, M.; Grul'ová, D.; Cicero, N.; Meriem-Hind, B. Comparison of Sensitivity to a Commercial Origanum Vulgare Essential Oil between Extended-Spectrum β-Lactamases (ESBL-) and Non-ESBL-Producing Enterobacteriaceae Isolates. *Nat. Prod. Res.* **2021**, 1–6. [CrossRef]
35. Gabr, S.A.; Alghadir, A.H. Evaluation of the Biological Effects of Lyophilized Hydrophilic Extract of *Rhus coriaria* on Myeloperoxidase (MPO) Activity, Wound Healing, and Microbial Infections of Skin Wound Tissues. *Evid.-Based Complement. Altern. Med.* **2019**, *2019*, 5861537. [CrossRef]
36. Nostro, A.; Guerrini, A.; Marino, A.; Tacchini, M.; di Giulio, M.; Grandini, A.; Akin, M.; Cellini, L.; Bisignano, G.; Saraçoğlu, H.T. In Vitro Activity of Plant Extracts against Biofilm-Producing Food-Related Bacteria. *Int. J. Food Microbiol.* **2016**, *238*, 33–39. [CrossRef]
37. Raodah, M.; Alia, Z.; Feleeha, H. The antioxidant and antimicrobial activity of Syrian sumac (*Rhus coriaria*) fruits extract. *J. Nat. Sci. Res.* **2014**, *4*, 36–40.
38. Almouwaly, K.; Alflayeh, K.; Ali, A. Antioxidant and free radical scavenging effects of Iraqi sumac *Rhus coriaria* L. *Baghdader Sci. J.* **2013**, *10*, 921–933.
39. Khanlarbeik, M.; Fazeli, M.R.; Jamalifar, H.; Ahmadian-Attari, M.M.; Shakiba, R.; Hajimiri, H. Synergistic effects of *Rhus coriaria* L. (sumac) and some commercial antibiotics against Meticillin-resistant Staphylococcus aureus. In Proceedings of the 71st FIP Congress, Hyderabad, India, 3–8 September 2011; Available online: https://www.fip.org/abstracts?page=abstracts&action=item&item=6260 (accessed on 25 November 2021).
40. Gabr, S.A.; El-Metwally, M.M.; Al-Ghadir, A.H. Antioxidant and antibacterial active constituents of *Rhus coriaria*. *Biotechnology* **2014**, *13*, 37–45. [CrossRef]
41. Kosar, M.; Bozan, B.; Temelli, F.; Baser, K.H.C. Antioxidant activity and phenolic composition of sumac (*Rhus coriaria* L.) extracts. *Food Chem.* **2007**, *103*, 952–959. [CrossRef]
42. Mahdavi, S. Antimicrobial and Antioxidant Activities of Iranian Sumac (*Rhus coriaria* L.) Fruit Ethanolic Extract. *J. Appl. Microbiol. Biochem.* **2018**, *2*, 2–5. [CrossRef]
43. Pecoraro, R.; Scalisi, E.M.; Iaria, C.; Capparucci, F.; Rizza, M.T.; Ignoto, S.; Salvaggio, A.; Fiorenza, R.; Impellizzeri, G.; Brundo, M.V. Toxicological assessment of CeO_2 nanoparticles on early development of zebrafish. *Toxicol. Res.* **2021**, *10*, 570–578.

44. Khan, F.R.; Alhewairini, S. Zebrafish (*Danio rerio*) as a model organism. In *Current Trends in Cancer Management*; IntechOpen: London, UK, 2018; ISBN 978-1-83880-005-5. Available online: https://www.intechopen.com/chapters/64178 (accessed on 25 November 2021).
45. Jayasinghe, C.D.; Jayawardena, U.A. Toxicity assessment of herbal medicine using zebrafish embryos: A systematic review. *Evid.-Based Complementary Altern. Med.* **2019**, *2019*, 7272808. [CrossRef]
46. Shafiei, M.; Nobakht, M.; Moazzam, A.A. Lipid-Lowering Effect of *Rhus coriaria* L. (Sumac) Fruit Extract in Hypercholesterolemic Rats. *Pharmazie* **2011**, *66*, 988–992. [CrossRef]
47. Doğan, A.; Çelik, İ. Healing Effects of Sumac (*Rhus coriaria*) in Streptozotocin-Induced Diabetic Rats. *Pharm. Biol.* **2016**, *54*, 2092–2102. [CrossRef]
48. Durazzo, D.; Lucarini, M.; Santini, A. Nutraceuticals in Human Health. *Foods* **2020**, *9*, 370. [CrossRef]
49. Santini, A.; Tenore, G.C.; Novellino, E. Nutraceuticals: A paradigm of proactive medicine. *Eur. J. Pharm. Sci.* **2017**, *96*, 53–61. [CrossRef]
50. Dziki, D.; Cacak-Pietrzak, G.; Hameed Hassoon, W.; Gawlik-Dziki, U.; Sułeke, A.; Różyło, R.; Sugier, D. The fruits of sumac (*Rhus coriaria* L.) as a functional additive and salt replacement to wheat bread. *LWT* **2021**, *136*, 110346. [CrossRef]
51. Wali, A.F.; Jabnoun, S.; Razmpoor, M.; Najeeb, F.; Shalabi, H.; Akbar, I. Account of Some Important Edible Medicinal Plants and Their Socio-Economic Importance. In *Edible Plants in Health and Diseases*; Masoodi, M.H., Rehman, M.U., Eds.; Springer: Singapore, 2022. [CrossRef]

foods

MDPI

Article

Mineral Composition and Bioaccessibility in Rocket and Purslane after Zn Biofortification Process

Massimiliano D'Imperio, Francesco Fabiano Montesano, Francesco Serio, Elisa Santovito and Angelo Parente *

Institute of Sciences of Food Production, CNR—National Research Council of Italy, Via Amendola 122/D, 70126 Bari, Italy; massimiliano.dimperio@ispa.cnr.it (M.D.); francesco.montesano@ispa.cnr.it (F.F.M.); francesco.serio@ispa.cnr.it (F.S.); elisa.santovito@ispa.cnr.it (E.S.)

* Correspondence: angelo.parente@ispa.cnr.it

Abstract: Zinc (Zn) is an essential key nutrient in different biochemical and physiological processes. The nutritional deficit of this mineral element is estimated to affect the health of over 3 billion people worldwide. Several strategies are available to reduce the negative impact of mineral malnutrition; among them, biofortification is the practice of deliberately increasing the nutrients and healthy compounds in the edible parts of vegetables. This study aims to evaluate Zn bioaccessibility in biofortified and non-biofortified rocket and purslane using an in vitro gastrointestinal digestion process and measure the concentration of other mineral elements (Al, B, Ca, Fe, K, Mg, Mn, and Sr) released during the digestion process from rocket and purslane biofortified with Zn. The bioaccessible Zn in biofortified rocket and purslane ranged from 7.43 to 16.91 mg/kg, respectively. In addition, the daily intake, the RDA coverage (%), and the hazard quotient (HQ) for the intake of Zn (resulting from the consumption of 100 g of rocket and purslane) were calculated. The calculated HQ highlights the safety of these baby leaf vegetables. The study confirms that it is possible to obtain Zn-biofortified rocket and purslane with high Zn bioaccessibility by adopting an appropriate mineral plant nutrition solution enriched in Zn.

Keywords: in vitro digestion process; floating system; baby leaf; hazard quotient; RDA

Citation: D'Imperio, M.; Montesano, F.F.; Serio, F.; Santovito, E.; Parente, A. Mineral Composition and Bioaccessibility in Rocket and Purslane after Zn Biofortification Process. *Foods* **2022**, *11*, 484. https://doi.org/10.3390/foods11030484

Academic Editors: Arun K. Bhunia and Antonello Santini

Received: 19 November 2021
Accepted: 4 February 2022
Published: 7 February 2022

1. Introduction

Zinc (Zn) is an essential key nutrient for several biochemical activities, such as human growth and development, immune system functions, and gene regulation. After iron, Zn is the second most abundant metal ion in organisms [1,2].

The Zn content in vegetables is related to various factors, such as species, genotype, type of edible portion (seed, leaf, fruit, or roots), phenological stage (microgreens, baby leaf, or mature vegetables), production method, and type of soil [3–6]. The recommended dietary allowance (RDA) of Zn for adults is 11 mg/day for men and 8 mg/day for women [7]. However, in some physiological conditions (such as pregnancy and lactation), chronic diseases (such as liver cirrhosis), diet (vegans/vegetarians), and in the elderly, it is necessary to increase the Zn intake with nutrition [7]. In humans, Zn deficiency is mostly associated with poor nutrition and poor dietary variegation and is aggravated by its poor availability in soils [5].

Zn deficiency is estimated to affect more than 3 billion of the world's population, with the vast majority occurring in underdeveloped countries [8,9].

The human and economic cost of Zn malnutrition is noteworthy, considering that about 17% of the global population suffers from this condition in developed and underdeveloped countries. More than 100,000 deaths per year in children under the age of 5 with various pathologies are attributable to the Zn deficiency [1–9]. Consequently, a series of international actions have been undertaken to improve the nutritional status of the population exposed to Zn malnutrition through the use of different approaches [10–13]. Among these, biofortification is the practice of deliberately increasing nutrients and healthy compounds and/or decreasing antinutritional factors (such as phytic and oxalate acids) in plant-based foods (cereal, vegetables, and fruit) [14,15]. Biofortified crops can be obtained through various strategies, such as genetic engineering, plant breeding, and agronomic practices [14,15].

Agronomic biofortification is generally used to increase the content of mineral nutrients (iodine, silicon, calcium, iron, zinc, magnesium, selenium, and copper) in the edible parts of various leafy vegetables and fruits, such as mizuna, tatsoi, chicory, basil, purslane, lettuce, tomato, Swiss chard, rocket, potatoes, green beans, and others [16]. This approach can be applied in different cultivation conditions, such as open field, greenhouse, and indoors; in the latter cases, also using soilless cultivation systems. Indeed, several studies have reported that the efficiency of biofortification, especially in greenhouses and indoor cultivation, can be maximized by specific management of the growing conditions [17–19]. The concentration of the nutrient solutions (NS) is an important characteristic for the quality of vegetables production [18]; therefore, changes in the composition of the NS can have a considerable impact on the nutritional quality of products, in particular, on the content of mineral elements [17,19] and bioactive organic compounds [20]. Furthermore, the choice of the plant species for biofortification represents an important aspect of the mineral biofortification process due to the effect of the phylogenetic heritage that inevitably affects plants' ability to accumulate essential mineral elements [21]. As an example, among leafy vegetables, purslane is considered a "new crop" for ready-to-eat products [22] and is characterized by a high oxalate content (2000 mg/kg of fresh weight). Rocket, on the other hand, is one of the most popular species grown in Mediterranean areas as a "ready-to-eat fresh-cut salads" product and is generally considered oxalate-free [23].

A crucial step after the biofortification process is the assessment of the bioaccessibility of the target nutrient. Ideally, in a successful biofortification protocol, the increase of a target nutrient in the edible parts parallels an increase in its bioaccessibility. The amount of nutrient that is released from the plant matrix during the gastrointestinal digestion process and its evaluation are independent of the approach and the method used to produce the biofortified crop. Furthermore, not all parts of a nutrient in the edible parts of biofortified vegetables can perform a biological activity. The release of nutrients in the intestinal tract (during the gastrointestinal digestion process) depends on different factors, such as species and type, and is subject to various influences, for example the concentration of nutrients, the activity of antinutritional compounds, texture, food processing, and the interaction of some nutrients with others [24–26]. During the gastrointestinal digestion process, the interaction of different mineral elements with similar electronic configurations (Zn^{2+}, Ca^{2+}, Fe^{2+}, Mg^{2+}, Mn^{2+}, and Sr^{2+}) can often lead to changes in the bioaccessibility and bioavailability of mineral nutrients [27,28]. Several methods are available to assess bioaccessibility using the in vitro digestion protocol. In these methods, the chemical, physical, and dynamic conditions of the gastrointestinal tract (mouth, stomach, and gut) are artificially reproduced in vitro [29].

Overall, the assessment of bioaccessibility provides information on the number of nutrients released from the food matrix, on nutrient–nutrient and nutrient–antinutrient interactions, on biochemical transformations, on chemical degradations, and on the effect of the matrix [30,31]. Furthermore, the assessment of bioaccessibility represents the starting point for the estimation of the beneficial effects of biofortified products on human health and can be used as a method to improve the food design process.

With all the above taken into account, the objectives of this study were:(i) to evaluate the overall mineral profile of rocket and purslane subjected to a process of Zn biofortification; (ii) to assess the quantity of mineral elements released by biofortified vegetables during the digestion process (bioaccessible fraction); and (iii) to calculate the RDA coverage and the hazard quotient (HQ) in relation to Zn bioaccessibility.

Two baby leaf vegetables (rocket and purslane) were produced and biofortified with Zn, the consumption of which allows an increase of zinc intake in the human diet without causing harm to the consumer. A workflow was proposed that was based on the evaluation of the efficiency of the biofortification process from a nutritional point of view, taking into account the overall bioaccessibility of the mineral nutrients.

2. Materials and Methods

2.1. Production of Zn-Biofortified Purslane and Rocket

Zn-biofortified rocket and purslane were produced in the experimental greenhouse "La Noria" located in Mola di Bari (BA), southern Italy (41°03′ N, 17°04′ E; 24 m a.s.l.) by using the floating hydroponic system. Rocket and purslane were grown in a complete NS with macro- and micro-nutrients [32]. Zn levels in the NS were 0.13 and 5.2 mg/L for growing non-biofortified and biofortified plants, respectively. The plants were harvested at the commercial stage of "baby leaf" (24 January 2020 and 30 July 2020, respectively, for rocket and purslane), as defined by Di Gioia et al. [33].

2.2. Mineral Profile of Rocket and Purslane

Al, B, Ca, Fe, K, Mg, Mn, Na, Sr, and Zn content was measured in dry samples by inductively coupled plasma optical emission spectrometry (ICP-OES) after mineralization of the dry samples with an acid microwave-assisted digestion system (MARS 6, CEM Corporation, Matthews, North Carolina) performed as reported by D'Imperio et al. [34]. To confirm the accuracy of the measurements, certified reference vegetable material (CRM, NIST tomato leaf 1535a) was analyzed using the same procedure as the rocket and purslane samples.

2.3. In Vitro Gastrointestinal Digestion Process

The assessment of mineral bioaccessibility (Al, B, Ca, Fe, K, Mg, Mn, Sr, and Zn) from plant samples (biofortified and not) during the digestion process was performed as reported by Ferruzzi et al. [35]. After the digestion process, samples were centrifuged at 10,000× *g* for 1 h at 4 °C to separate the aqueous intestinal digesta, called 'bioaccessible fraction' (BF), from the residual solids. The BFs were collected, filtered (0.2 µm PTFE filter), and dried at 50 °C for 48 h before the minerals content was measured. For the CRM sample only, the residual solids were washed with Milli-Q H_2O (18 MΩ/cm) and dried (50 °C for 48 h) until use. To evaluate the accuracy of the measurement, CRM (NIST tomato leaf 1535a) was analyzed using the same procedure adopted for the rocket and purslane samples.

2.4. Analysis of Mineral Content in Digested Sample

After the digestion process, the BF and the residual solid were mineralized with HNO_3 65% using the same protocol used for rocket and purslane (see Section 2.2). Blank correction was performed in all analyses. The protocol applied did not allow the estimation of Na bioaccessibility, because the blank correction was not performed for this mineral element. The amount of Na released from the food matrix during the digestion process was lower than the amount of Na in the blank sample (3.81 g/L). This is related to the reagents used, as also reported by another study [36]. The bioaccessibility fraction percentage ($BF_{\%}$), defined as the percentage of nutrient(s) released from the digested matrix in the gastrointestinal digestion process, was calculated as $BF_{\%}$ = (total nutrient released during digestion/total nutrient in food) × 100.

2.5. Percentage of Recommended Daily Allowance and Hazard Quotient for Zn Intake

The recommended daily allowance of Zn (RDA-Zn) is equal to 11 and 8 mg, respectively, for male and female adults [7]. The daily intake of Zn and the percentage of coverage of RDA for Zn (% RDA-Zn) were calculated in relation to the quantity of Zn released from the vegetables during the gastrointestinal digestion process. Risk assessment was also performed by using HQ, considered as the risk to consumer health resulting from the consumption of Zn-biofortified, fresh baby leaf vegetables, based on a 70 kg adult. The HQ is the ratio of the potential exposure to an organic and/or inorganic substance and the level at which no negative effects are expected. HQ allows the estimation of the potential negative effects on health related to chronic consumption of food (in our case, biofortified rocket and purslane). A HQ lower or equal to 1 indicates that adverse effects are unlikely to occur, and, thus, the product can be considered to have negligible hazard. For a HQ greater than 1, the potential for adverse effects increases [37]. The contribution of Zn from other nutritional sources was not examined. The HQ was calculated according to the protocol described by the Environmental Protection Agency [37], using the following equation: HQ = ADD/RFD, where ADD is the average daily dose of Zn (mg of Zn/kg body weight/day), and RFD is the recommended dietary tolerable upper intake level of Zn (mg of Zn/kg body weight/day). The I RFD value for a 70 kg adult is 3×10^{-1} mg Zn/kg/day [38]. The ADD for 100 g portions of rocket or purslane was computed as follows: ADD = (MI × CF × DI)/BW. MI is the Zn concentration released during the gastrointestinal digestion process after the consumption of the two vegetables (mg/kg DW); CF is the fresh-to-DW conversion factor for vegetable samples (calculated as the ratio of FW to DW; rocket: 0.093 on average; purslane: 0.054 on average); DI is the daily intake of baby leaf vegetables (kg, taken as 100 g); BW is the body weight (kg) of humans, assumed as 70 kg.

2.6. Statistical Analysis

The effects of the biofortification process were evaluated using one-way analysis of variance (ANOVA) followed by means separation with Fisher's protected least significant difference (LSD) at $p \leq 0.05$. In the bioaccessibility parameter analysis, the effects of treatments and species were estimated using a two-way analysis of variance (ANOVA) followed by means separation with Fisher's protected least significant difference (LSD) at $p \leq 0.05$. The software Statistica 10.0 (StatSoft, Tulsa, OK, USA) was used.

3. Results and Discussion

3.1. Mineral Analysis

Analysis of the accuracy of the analytical measurements of macro and trace elements in the edible parts and in digested samples, from biofortified and non-biofortified baby leaf vegetables, was performed. The mineral elements Al, B, Ca, Fe, K, Mg, Mn, Sr, and Zn were detected and measured. The limits of detection (LOD) and the limit of quantification (LOQ) of the methods were calculated as suggested by D'Imperio et al. [34]. Tomato leaves (NIST-1535a) were used as CRM to evaluate the accuracy of the measurements in the plants and in the digested samples, as reported in Tables 1 and 2. The recovery of mineral elements in the vegetable samples ranged from 90 to 107%. After the in vitro digestion of the CRM, some trace elements, such as Al, Fe, and K, showed the lowest recovery values (%), whereas B, Ca, Mg, Mn, Sr, and Zn showed higher recovery values, as reported in Table 2.

Table 1. Mineral content recovered from certified reference materials (NIST tomato leaf 1535a), LOD, and LOQ of methods.

Element	LOD	LOQ	Found	Certified	Recovery
	μg/L		mg/kg DW		(%)
Al	0.8904	2.6982	624 ± 33.65	598 ± 7.1	104
B	0.0451	0.1365	29.7 ± 0.29	33.0 ± 0.42	90
Ca	0.0698	0.2116	49,437 ± 113.4	50,450 ± 550	98
Fe	0.2923	0.8853	358.3 ± 0.92	367 ± 4.3	98
K	0.7344	2.2255	30,443 ± 99	26,760 ± 480	113
Mg	0.1458	0.4420	11,649 ± 35.03	12,000	97
Mn	0.1898	0.5752	264.1 ± 1.24	246 ± 7.1	107
Sr	0.2068	0.6267	88.0 ± 0.401	85.0	104
Zn	0.1763	0.5343	30.7 ± 0.205	30.9 ± 0.55	99

Results are reported as mean ± standard error. Magnesium and strontium: non-certified value. Insufficient information is available to assess the uncertainty associated with the value, and, therefore, no uncertainty is provided (NIST).

Table 2. Mineral content recovered from bioaccessibility assays of certified reference materials (NIST tomato leaf 1535a).

Element	BF	Residue	MB	Certified	$BF_{\%}$	Recovery
		mg/kg			(%)	
Al	25.0 ± 0.335	309 ± 18.78	334 ± 18.45	598 ± 7.1	4.19 ± 0.05	56 ± 3.08
B	21.8 ± 0.833	10.1 ± 0.099	32 ± 0.82	33 ± 0.42	66.1 ± 2.52	97 ± 2.49
Ca	31411 ± 149	16,310 ± 718	47,720 ± 867	50,450 ± 550	62.3 ± 0.29	95 ± 1.71
Fe	18.6 ± 0.03	228 ± 11.98	247 ± 12.03	367 ± 4.3	5.1 ± 0.009	67 ± 3.27
K	19,653 ± 161	589 ± 19.38	20,242 ± 182	26,760 ± 480	73.4 ± 0.61	76 ± 0.79
Mg	12,807 ± 82.3	545 ± 30.58	13,351 ± 113	12,000	107 ± 0.69	111 ± 0.94
Mn	213.6 ± 0.156	50.2 ± 2.103	264 ± 2.259	246 ± 7.1	86.8 ± 0.06	107 ± 0.92
Sr	58.9 ± 0.668	30.8 ± 1.408	90 ± 2.147	85.0	69.2 ± 0.79	105 ± 2.52
Zn	18.9 ± 0.664	10.2 ± 0.66	29 ± 1.332	30.9 ± 0.55	61.2 ± 2.15	94 ± 4.31

Results are reported as mean ± standard error. Magnesium and strontium: non-certified value. Information available is not sufficient to assess the uncertainty associated with the value, and, therefore, no uncertainty is provided (NIST). BF: bioaccessible fraction = concentration of element release from plant material during in vitro digestion process. Residue: residual concentration of the element in digested samples. MB: mass balance = BF + Residue. Certified: the certified value from the National Institute of Standards and Technology (NIST). $BF_{\%}$: bioaccessibility = (BF/Certified) × 100. Recovery = (MB/certified) × 100.

3.2. Mineral Profile of Biofortified and Non-Biofortified Rocket and Purslane

The biofortification process aims to improve the nutritional value of crops without altering the performance of the crops. In both species, the agronomic protocol applied in this study did not cause any toxic effect in the vegetables nor alteration of the crop performances (data not shown).

Using 5.2 mg/L of Zn in the NS, the tissue content of Zn in the edible parts of rocket and purslane increased, respectively, by 1.76 and 3.97-fold compared with the non-biofortified counterpart (0.13 mg/L of Zn), as reported in Figure 1. According to our results, the level of Zn used in the biofortification treatment favored its absorption. In fact, zinc is absorbed by plants from the soil as an ionic element or bound to an organic acid and transported through the xylem to the aerial parts (shoots and leaves) [39]. Similar increases in Zn content were found in lettuce [40], cabbage [41], soybean sprouts [42], and in three different types of microgreens that were produced in soilless systems using different levels of Zn in the NS [43].

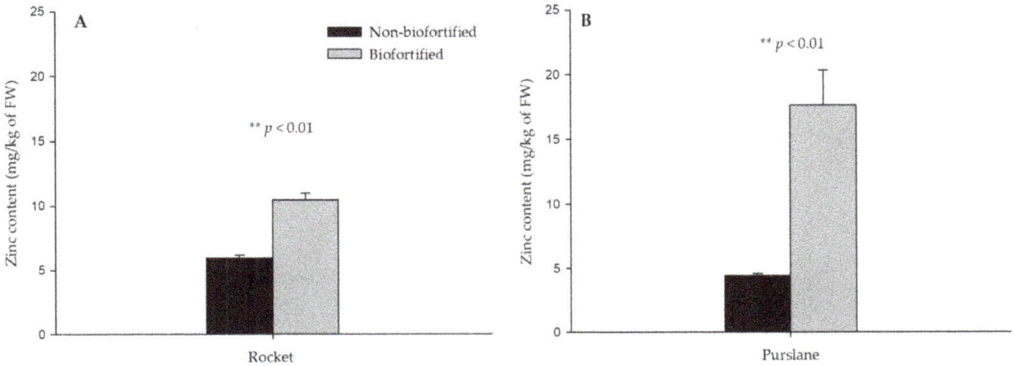

Figure 1. Zinc content in non-biofortified and biofortified rocket (**A**) and purslane (**B**), harvested at the phenological stage of "baby leaf vegetables". Results are reported as mean ± standard error of treatment (n = 3). Means separation within columns by LSD (α = 0.05). Significance: ** $p < 0.01$. Non-biofortified (0.13 mg/L of Zn in nutrient solution), Biofortified (5.2 mg/L of Zn in nutrient solution).

The content of Al, B, Ca, Fe, K, Mg, Mn, and Sr measured in rocket and purslane did not reveal significant differences imputable to biofortification (Table 3). The overall mean contents (mg/kg of FW) were 3.32 (Al), 2.69 (B), 3364 (Ca), 6.68 (Fe), 7371 (K), 520 (Mg), 2.24 (Mn), and 6.17 (Sr) in rocket and 0.89 (Al), 2.36 (B), 940 (Ca), 4.07 (Fe), 4279 (K), 856 (Mg), 8.55 (Mn), and 2.77 (Sr) in purslane. In our study, no antagonistic effects were found between Zn and other mineral elements, such as K, Ca, and Fe, although this kind of antagonism has been reported in other studies and is related to the fact that these mineral elements share the same transporters on the plasma membrane [44]. However, our result could be related to the low Zn level used in this study (5.2 mg/L of Zn in NS). Di Gioia et al. [43] reported antagonistic effects between Zn and the other mineral elements using higher levels of Zn in the NS (10 and 20 mg/L) than the level used in this study.

Table 3. Mineral content in non-biofortified and biofortified rocket and purslane harvested at the phenological stage of "baby leaf vegetables".

		Al	B	Ca	Fe	K	Mg	Mn	Sr
Species	**Treatment**	**mg/kg of Fresh Weight**							
Rocket	Non-biofortified	3.27 ± 0.13	2.66 ± 0.05	3155 ± 288	6.56 ± 0.28	7084 ± 545	482 ± 36.7	2.02 ± 0.02	6.23 ± 0.12
	Biofortified	3.37 ± 0.25	2.72 ± 0.09	3572 ± 80.9	6.81 ± 0.15	7657 ± 627	556 ± 20.9	2.46 ± 0.16	6.10 ± 0.36
	Significance	ns	ns	ns	ns	ns	ns	ns	ns
Purslane	Non-biofortified	0.89 ± 0.13	2.32 ± 0.06	875 ± 26.5	4.21 ± 0.17	4373 ± 199	818 ± 18.3	7.66 ± 0.34	2.54 ± 0.07
	Biofortified	0.89 ± 0.07	2.41 ± 0.23	1004 ± 67.8	3.93 ± 0.31	4184 ± 212	894 ± 55.4	9.44 ± 0.91	3.00 ± 0.155
	Significance	ns	ns	ns	ns	ns	ns	ns	ns

Results are reported as mean ± standard error of treatment (n = 3). Significance: ns = not significant. Means separation within columns by LSD (α = 0.05). Non-biofortified (0.13 mg/L of Zn in nutrient solution), Biofortified (5.2 mg/L of Zn in nutrient solution).

3.3. Mineral Bioaccessibility in Rocket and Purslane after the Biofortification Process

The BF is the concentration of a nutrient or a bioactive compound (mineral or organic) that is extracted from the plant matrix during the digestion process and which, potentially, becomes bioavailable in the intestinal tract. The number of mineral elements released by plant materials is related to various factors such as species, food processing (raw or cooked food), texture, nutrient concentration, and interaction with other nutrients or antinutrients [17,32,42,45,46]. In our study, after in vitro gastrointestinal digestion, Zn $BF_{\%}$ was 98% in biofortified plants and 73% in non-biofortified plants compared to the non-digested control plants. Similar results were reported for Si-biofortified green bean pods [17]. Conversely, no differences in $BF_{\%}$ values (72%) were found in rocket (biofortified and non-biofortified), although an increase in Zn was found in the edible parts (Figure 1). Therefore, the in vitro digestion protocol allows similar $BF_{\%}$ values to be obtained in both biofortified and non-biofortified plants. This result was also reported in our previous study [32,47], showing that increasing the concentration of mineral elements in the edible parts of biofortified plants does not always give an increase in $BF_{\%}$, as reported for calcium and silicon [32,47]. However, in both rocket and purslane, after the in vitro gastrointestinal digestion (bioaccessible fraction), we measured a significant release of Zn (mg/kg) in biofortified plants compared to non-biofortified ones (76% and 298%, respectively, for rocket and purslane), as shown in Figure 2. Biofortified purslane was found to be the species with the highest amount of bioaccessible Zn released during the digestion process (16.9 mg/kg). The quantity of Zn released by biofortified rocket was 7.43 mg/kg. The quantity of bioaccessible Zn released by non-biofortified purslane and rocket was 3.75 mg/kg (on average).

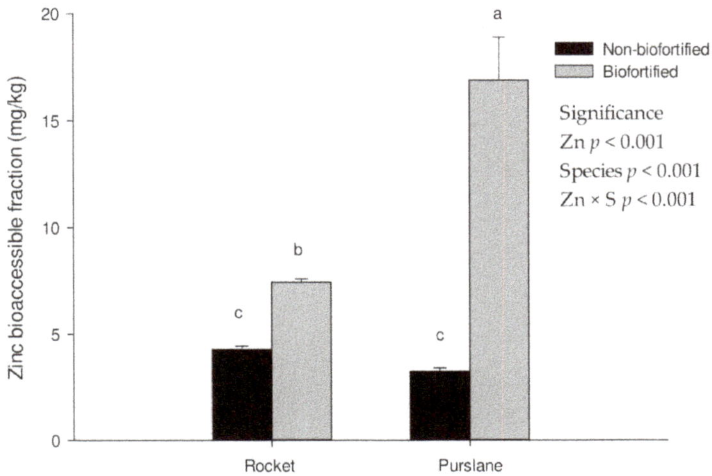

Figure 2. Bioaccessible fraction (mg/kg) of Zn in non-biofortified and biofortified rocket and purslane after in vitro digestion process. Results are reported as mean ± standard error of treatment ($n = 3$). Different letters indicate that mean values are significantly different (means separation by LSD; $\alpha = 0.05$). Non-biofortified (0.13 mg/L of Zn in nutrient solution), Biofortified (5.2 mg/L of Zn in nutrient solution).

As previously reported also in soybean sprouts [42], the BF of Zn, measured after in vitro gastrointestinal digestion, is affected by the initial content of Zn in the edible parts of the plants. The increase in the amount of Zn released during the digestion process and found in this study is a significant result, considering that this is the amount of Zn that could be potentially absorbed in the intestinal tract [48].

The BF of the mineral elements is correlated to the different compositions of the tested species and to the interaction of the plants with the intestinal juices (pancreatic enzymes and bile salts). As reported in Table 4, all mineral elements analyzed showed significant differences ($p < 0.001$) in relation to the plant species, but they were not affected by the Zn biofortification protocol used. The influence of the plant species on BF values has also been found in other studies analyzing various mineral elements, such as Si [47], Ca [32,49], K [45,49], Fe [6], Mg [49], and other trace elements [49]. In our study, the average quantities of mineral elements released in the digestion process were 0.53 mg/kg for Al, 2.36 mg/kg for B, and 7522 mg/kg for K, and these quantities were higher in rocket than in purslane. Conversely, the measured mean amounts of Fe (2.12 mg/kg) and Mg (880 mg/kg) were higher in purslane than in rocket (Table 4).

Several compounds, such as some antinutritional factors (carbonate, phytic and oxalic acids) and some healthy food components (proteins, fibers, and polyphenols), can modify the release of nutrients from the food matrix [50]. The interaction of mineral elements with these compounds generates insoluble salts and determines the reduction of BF and a reduced absorption of minerals [30,31]. Egea-Gilabert et al. [22] reported that purslane is a vegetable with a high oxalate content (2000 mg/kg of fresh weight). On the contrary, rocket is generally considered to be free of oxalate [23]. This difference in oxalate content could influence the BF of all mineral elements evaluated: in particular, Ca and Sr. Oxalate forms an insoluble salt with Ca [51] and probably also with Sr, considering the similar chemical and biological properties of these mineral elements [52]. The effects of plant species on Ca bioaccessibility and the high amount of Ca released during the digestion process were reported in our previous study [32].

Table 4. Bioaccessible fractions of Al, B, Ca, Fe, K, Mg, Mn, and Sr in non-biofortified and biofortified rocket and purslane after in vitro digestion process.

		Al	B	Ca	Fe	K	Mg	Mn	Sr
Species	**Treatment**	**mg/kg of Fresh Weight**							
Rocket	Non-biofortified	0.57 ± 0.02	2.28 ± 0.22	2352 ± 50.0 a	1.24 ± 0.12	7428 ± 342	437 ± 13.4	1.56 ± 0.05 c	6.15 ± 0.16 a
	Biofortified	0.49 ± 0.05	2.44 ± 0.22	2232 ± 1.47 b	1.4 ± 0.07	7617 ± 254	466 ± 2.99	1.61 ± 0.06 c	5.57 ± 0.04 b
Purslane	Non-biofortified	0.09 ± 0.01	1.77 ± 0.08	59.6 ± 2.39 c	2.33 ± 0.08	4422 ± 75.2	818 ± 25.1	7.33 ± 0.40 b	1.14 ± 0.02 c
	Biofortified	0.09 ± 0.01	1.96 ± 0.06	63.8 ± 3.96 c	1.91 ± 0.27	4104 ± 41.8	880 ± 83.5	8.53 ± 0.29 a	1.29 ± 0.06 c
Significance									
Zn		ns	ns	ns	ns	ns	ns	ns	ns
Species (S)		***	**	***	***	***	***	***	***
Zn × S		ns	ns	*	ns	ns	ns	*	**

Results are reported as mean ± standard error of treatment (n = 3). FW: fresh weight. Significance: ns = not significant; * $p \leq 0.05$; ** $p < 0.01$; *** $p \leq 0.001$. Different letters within column indicate that mean values are significantly different (means separation by LSD; $\alpha = 0.05$). Non-biofortified (0.13 mg/L of Zn in nutrient solution), Biofortified (5.2 mg/L of Zn in nutrient solution).

The highest amounts of Ca and Sr in the digested liquid were found in the non-biofortified rocket, followed by the biofortified rocket, whereas the purslane released lower amounts of Ca in the gastrointestinal digestion, and this result was not affected by the biofortification treatment with Zn. The high amounts of Ca observed in rocket could lead to the formation of low-solubility complexes that reduce the BF of Mn. Furthermore, mineral elements with similar electronic configurations (Zn^{2+}, Ca^{2+}, Fe^{2+}, Mg^{2+}, Mn^{2+}, and Sr^{2+}) are involved in mechanisms of mutual competition to bind antinutrient compounds [27,28,46]. Therefore, different values of BF and $BF_{\%}$ can be attributable to different factors, including the mechanisms of competition at different levels in a plant-based food system.

3.4. Daily Intake, Coverage of RDA-Zn (Male and Female), and Hazard Quotient

The DI, the RDA-Zn coverage (for men and women), and the HQ for Zn intake through digesting 100 g of baby leaf vegetables (average servings for this type of products) are shown in Table 5. The Zn biofortification significantly increased those parameters ($p < 0.001$), and differences between the two vegetables were found (Table 5). The highest values of DI, RDA-Zn coverage, and HQ were obtained for biofortified purslane, whereas the lowest values were found for non-biofortified rocket and purslane (Table 5). After digestion of 100 g of biofortified purslane, an increase in DI (3.9-fold) and RDA-Zn coverage was found in males and females, compared to non-biofortified vegetables (Table 5).

Table 5. Daily intake, coverage of RDA for Zn, and HQ for Zn intake through consumption of 100 g portions of baby leaf vegetables, biofortified and non-biofortified, by adult male and female humans (70 kg body weight).

		Daily Zn Intake (mg Zn/Day)	RDA-Zn Coverage (%)		HQ
Species	**Treatment**		**Male**	**Female**	
Rocket	Non-biofortified	0.43 ± 0.02 c	3.88 ± 0.14 c	5.34 ± 0.21 c	0.278 ± 0.011 c
	Biofortified	0.74 ± 0.02 b	6.79 ± 0.14 b	9.29 ± 0.20 b	0.534 ± 0.012 b
Purslane	Non-biofortified	0.32 ± 0.02 c	2.94 ± 0.15 c	4.05 ± 0.20 c	0.233 ± 0.017 c
	Biofortified	1.69 ± 0.19 a	15.4 ± 1.83 a	21.14 ± 2.52 a	1.086 ± 0.129 a
Significance					
Zn		***	***	***	***
Species (S)		***	***	***	***
Zn × S		***	***	***	***

Results are reported as mean ± standard error of treatment (n = 3). Significance: *** $p \leq 0.001$. Different letters within columns indicate that mean values are significantly different (means separation by LSD; $\alpha = 0.05$). Daily intake, coverage of RDA for Zn, and HQ were calculated in relation to the quantity of Zn released from vegetables during the gastrointestinal digestion process. Major details are reported in Section 2.5 of Materials and Methods.

The increase of DI and RDA-Zn coverage accentuates the efficiency of the applied biofortification protocol, suggesting its use to produce Zn-biofortified baby leaf vegetables for different target consumers groups for which the increase of the DI is advisable, such as pregnant and breastfeeding women, vegetarians/vegans, people with various diseases, and the elderly [7].

The HQ values found in rocket (biofortified and not) and in non-biofortified purslane were less than 1. However, an excessive increase of Zn in the edible portions of purslane can result in an increase of this parameter. When the HQ is higher than 1, adverse health effects are likely to occur. According to our findings, the consumption of 100 g of our biofortified products does not pose any health risk to consumers. This aspect must be taken into due consideration when approaching a biofortification process; an excessive content of Zn in the edible parts of vegetables would represent a risk for consumers (the maximum tolerable intake level is 40 mg Zn/day) since vegetables are only a relative portion of the diet and other foods and water intake can significantly contribute to the daily intake of Zn [39].

4. Conclusions

The general purpose of this study was to produce Zn-biofortified rocket and purslane and to propose a workflow for studying their nutritional qualities based on the analysis of the bioaccessible fraction of the overall mineral elements.

The agronomic biofortification protocol used in this study was based on increasing the concentration of Zn in the NS used for the cultivation of rocket and purslane in soilless conditions. This protocol allowed Zn-biofortified plants with a higher nutritional quality to be obtained. The amount of bioaccessible Zn released by the plants during the digestion process was influenced by the species (rocket and purslane) and by the initial Zn content accumulated in the edible parts of the plants in soilless cultivation using Zn-enriched NS.

The use of the in vitro gastrointestinal digestion protocol allowed the evaluation of the bioaccessible fraction of Zn and other mineral elements. Antinutritional factors (carbonate, phytic and oxalic acids) and some healthy food components (proteins, fibers, and polyphenols) can modify the release of nutrients from the food matrix, generating insoluble salts and determining the reduction of bioaccessibility and absorption of the mineral elements. Hence, it is important to quantify the bioaccessible fraction of the target mineral and also of the other mineral elements.

Our results confirmed that in vitro digestion is a valuable method for assessing the nutritional efficiency of the biofortification process. This approach can be efficiently used to improve the design process for biofortified products. Furthermore, the calculated hazard quotient demonstrates the safety of biofortified rocket and purslane.

Overall, the consumption of biofortified rocket and purslane would provide greater intake of Zn in the human diet without causing harm to the consumer, thus, providing benefits for different classes of consumers, such as the elderly, vegetarians, vegans, and people with gastrointestinal and other diseases. However, more research is needed to further explore and validate the applicability of the proposed workflow to biofortification processes for other mineral elements and in other plant species.

Author Contributions: Conceptualization, M.D. and A.P.; methodology, M.D. and A.P.; crop performance measurements, F.F.M. and F.S.; chemical analysis, M.D. and A.P.; in vitro digestion process, M.D.; statistical analysis, A.P.; original draft preparation, M.D. and A.P.; writing—review and editing, M.D., F.F.M., F.S., E.S., and A.P.; supervision of the study, M.D., F.F.M., F.S., and A.P. All authors have read and agreed to the published version of the manuscript.

Funding: This research was founded by the National Research Council (CNR) project NUTR-AGE (Ordinary fund for research organizations and institutions FOE-2019), DSB.AD004.271, and SOILLESS GO project (project code (CUP) B97H20000990009), funded by the Rural Development Program of the Apulia Region (Italy) 2014–2020, sub-measure 16.2 (Support for pilot projects and development of new products, practices, processes, and technologies and transfer and dissemination of results obtained by operational groups) (Paper n. 14).

Acknowledgments: The authors thank Nicola Gentile for the technical support.

Conflicts of Interest: The authors declare no conflict of interest.

References

1. Andreini, C.; Banci, L.; Bertini, I.; Rosato, A. Counting the zinc-proteins encoded in the human genome. *J. Proteome Res.* **2006**, *5*, 196–201. [CrossRef] [PubMed]
2. Sharma, P.; Reddy, P.K.; Kumar, B. Trace Element Zinc, a Nature's Gift to Fight Unprecedented Global Pandemic COVID-19. *Biol. Trace Elem. Res.* **2020**, *199*, 3213–3221. [CrossRef]
3. Scherz, H.; Kirchhoff, E. Trace elements in foods: Zinc contents of raw foods—a comparison of data originating from different geographical regions of the world. *J. Food Comp. Anal.* **2006**, *19*, 420–433. [CrossRef]
4. Nemzer, B.; Al-Taher, F.; Abshiru, N. Phytochemical composition and nutritional value of different plant parts in two cultivated and wild purslane (*Portulaca oleracea* L.) genotypes. *Food Chem.* **2020**, *320*, 126621. [CrossRef]
5. De Almeida, H.J.; Vergara Carmona, V.M.; Ferreira Inocêncio, M.; Furtini Neto, A.E.; Cecílio Filho, A.B.; Mauad, M. Soil Type and Zinc Doses in Agronomic Biofortification of Lettuce Genotypes. *Agronomy* **2020**, *10*, 124. [CrossRef]
6. Khoja, K.; Buckley, A.F.; Aslam, M.; Sharp, A.P.; Latunde-Dada, G.O. In vitro bioaccessibility and bioavailability of iron from mature and microgreen fenugreek, rocket and broccoli. *Nutrients* **2020**, *12*, 1057. [CrossRef]
7. National Institute of Health (NIH). Available online: https://ods.od.nih.gov/factsheets/Zinc-HealthProfessional/#h2 (accessed on 10 November 2021).
8. Wessells, K.R.; Brown, K.H. Estimating the global prevalence of zinc deficiency: Results based on zinc availability in national food supplies and the prevalence of stunting. *PLoS ONE* **2012**, *7*, e50568.
9. Kumssa, D.B.; Joy, E.J.; Ander, E.L.; Watts, M.J.; Young, S.D.; Walker, S.; Broadley, M.R. Dietary calcium and zinc deficiency risks are decreasing but remain prevalent. *Sci. Rep.* **2015**, *5*, 10974. [CrossRef]
10. Shah, D.; Sachdev, H.S.; Gera, T.; De-Regil, L.M.; Peña-Rosas, J.P. Fortification of staple foods with zinc for improving zinc status and other health outcomes in the general population. *Cochrane Database Syst. Rev.* **2016**, *6*, CD010697. [CrossRef]
11. International Zinc Nutrition Consultative Group. BIZIFED2. Available online: https://www.izincg.org/ (accessed on 2 November 2021).
12. VIRCA Plus. Available online: https://cassavaplus.org/ (accessed on 29 October 2021).
13. Harvest Plus. Available online: https://www.harvestplus.org/ (accessed on 29 October 2021).
14. Carvalho, S.M.; Vasconcelos, M.W. Producing more with less: Strategies and novel technologies for plant-based food biofortification. *Food Res. Int* **2013**, *54*, 961–971. [CrossRef]
15. Van Der Straeten, D.; Bhullar, N.K.; De Steur, H.; Gruissem, W.; MacKenzie, D.; Pfeiffer, W.; Bouis, H. Multiplying the efficiency and impact of biofortification through metabolic engineering. *Nat. Commun.* **2020**, *11*, 5203. [CrossRef]
16. Buturi, C.V.; Mauro, R.P.; Fogliano, V.; Leonardi, C.; Giuffrida, F. Mineral Biofortification of Vegetables as a Tool to Improve Human Diet. *Foods* **2021**, *10*, 223. [CrossRef]
17. Montesano, F.F.; D'Imperio, M.; Parente, A.; Cardinali, A.; Renna, M.; Serio, F. Green bean biofortification for Si through soilless cultivation: Plant response and Si bioaccessibility in pods. *Sci. Rep.* **2016**, *6*, 31662. [CrossRef]
18. Rouphael, Y.; Kyriacou, M.C. Enhancing quality of fresh vegetables through salinity eustress and biofortification applications facilitated by soilless cultivation. *Front. Plant Sci.* **2018**, *9*, 1254. [CrossRef] [PubMed]
19. Gonnella, M.; Renna, M.; D'Imperio, M.; Santamaria, P.; Serio, F. Iodine biofortification of four brassica genotypes is effective already at low rates of potassium iodate. *Nutrients* **2019**, *11*, 451. [CrossRef] [PubMed]
20. Fanasca, S.; Colla, G.; Maiani, G.; Venneria, E.; Rouphael, Y.; Azzini, E.; Saccardo, F. Changes in antioxidant content of tomato fruits in response to cultivar and nutrient solution composition. *J. Agric. Food Chem.* **2006**, *54*, 4319–4325. [CrossRef] [PubMed]
21. White, P.J.; Broadley, M.R. Biofortification of crops with seven mineral elements often lacking in human diets–iron, zinc, copper, calcium, magnesium, selenium and iodine. *New Phytol.* **2009**, *182*, 49–84. [CrossRef]
22. Egea-Gilabert, C.; Ruiz-Hernández, M.V.; Parra, M.Á.; Fernández, J.A. Characterization of purslane (*Portulaca oleracea* L.) accessions: Suitability as ready-to-eat product. *Sci. Hortic.* **2014**, *172*, 73–81. [CrossRef]
23. Santamaria, P.; Elia, A.; Serio, F.; Todaro, E. A survey of nitrate and oxalate content in fresh vegetables. *J. Sci. Food Agric.* **1999**, *79*, 1882–1888. [CrossRef]
24. Cuervo, A.; Valdés, L.; Salazar, N.; de los Reyes-Gavilán, C.G.; Ruas-Madiedo, P.; Gueimonde, M.; González, S. Pilot study of diet and microbiota: Interactive associations of fibers and polyphenols with human intestinal bacteria. *J. Agric. Food Chem.* **2014**, *62*, 5330–5336. [CrossRef]
25. Aires, A.; Marques, E.; Carvalho, R.; Rosa, E.A.; Saavedra, M.J. Evaluation of biological value and appraisal of polyphenols and glucosinolates from organic baby-leaf salads as antioxidants and antimicrobials against important human pathogenic bacteria. *Molecules* **2013**, *18*, 4651–4668. [CrossRef]
26. Fernández-Navarro, T.; Salazar, N.; Gutiérrez-Díaz, I.; Sánchez, B.; Rúas-Madiedo, P.; Clara, G.; González, S. Bioactive compounds from regular diet and faecal microbial metabolites. *Eur. J. Nutr.* **2018**, *57*, 487–497. [CrossRef] [PubMed]
27. Sandstroèm, B. Micronutrient interactions: Effects on absorption and bioavailability. *Br. J. Nutr.* **2001**, *85*, S181–S185. [CrossRef]

28. Kyomugasho, C.; Gwala, S.; Christiaens, S.; Kermani, Z.J.; Van Loey, A.M.; Grauwet, T.; Hendrickx, M.E. Pectin nanostructure influences pectin-cation interactions and in vitro-bioaccessibility of Ca^{2+}, Zn^{2+}, Fe^{2+} and Mg^{2+}-ions in model systems. *Food Hydrocoll.* **2017**, *62*, 299–310. [CrossRef]
29. D'Antuono, I.; Garbetta, A.; Linsalata, V.; Minervini, F.; Cardinali, A. Polyphenols from artichoke heads (*Cynara cardunculus* (L.) subsp. scolymus Hayek): In vitro bio-accessibility, intestinal uptake and bioavailability. *Food Funct.* **2015**, *6*, 1268–1277. [CrossRef]
30. Hemalatha, S.; Platel, K.; Srinivasan, K. Zinc and iron contents and their bioaccessibility in cereals and pulses consumed in India. *Food Chem.* **2007**, *102*, 1328–1336. [CrossRef]
31. Camara, F.; Amaro, M.A.; Barbera, R.; Clemente, G. Bioaccessibility of minerals in school meals: Comparison between dialysis and solubility methods. *Food Chem.* **2005**, *92*, 481–489. [CrossRef]
32. D'Imperio, M.; Renna, M.; Cardinali, A.; Buttaro, D.; Serio, F.; Santamaria, P. Calcium biofortification and bioaccessibility in soilless "baby leaf" vegetable production. *Food Chem.* **2016**, *213*, 149–156. [CrossRef] [PubMed]
33. Di Gioia, F.; Renna, M.; Santamaria, P. Sprouts, Microgreens and "Baby Leaf" Vegetables. In *Minimally Processed Refrigerated Fruits and Vegetables*; Springer: Boston, MA, USA, 2017; pp. 403–432.
34. D'Imperio, M.; Montesano, F.F.; Montemurro, N.; Parente, A. Posidonia natural residues as growing substrate component: An ecofriendly method to improve nutritional profile of brassica microgreens. *Front. Plant Sci.* **2021**, *12*, 580596. [CrossRef]
35. Ferruzzi, M.G.; Failla, M.L.; Schwartz, S.J. Assessment of degradation and intestinal cell uptake of carotenoids and chlorophyll derivatives from spinach puree using an in vitro digestion and Caco-2 human cell model. *J. Agric. Food Chem.* **2001**, *49*, 2082–2089. [CrossRef]
36. Hamilton, E.M.; Barlow, T.S.; Gowing, C.J.B.; Watts, M.J. Bioaccessibility performance data for fifty-seven elements in guidance materials BGS 102. *Microchem. J.* **2015**, *123*, 131–138. [CrossRef]
37. Integrated Risk Information System (Iris). *Integrated Risk Information System—Database 2011*; United States Environmental Protection Agency: Washington, DC, USA, 2011.
38. Bączek-Kwinta, R.; Baran, A.; Simlat, M.; Lang, J.; Bieniek, M.; Florek, B. Enrichment of Different Plant Seeds with Zinc and Assessment of Health Risk of Zn-Fortified Sprouts Consumption. *Agronomy* **2020**, *10*, 937. [CrossRef]
39. Noulas, C.; Tziouvalekas, M.; Karyotis, T. Zinc in soils, water and food crops. *J. Trace Elem. Med. Biol.* **2018**, *49*, 252–260. [CrossRef]
40. Barrameda-Medina, Y.; Blasco, B.; Lentini, M.; Esposito, S.; Baenas, N.; Moreno, D.A.; Ruiz, J.M. Zinc biofortification improves phytochemicals and amino-acidic profile in Brassica oleracea cv. Bronco. *Plant Sci.* **2017**, *258*, 45–51. [CrossRef] [PubMed]
41. Barrameda-Medina, Y.; Lentini, M.; Esposito, S.; Ruiz, J.M.; Blasco, B. Zn-biofortification enhanced nitrogen metabolism and photorespiration process in green leafy vegetable Lactuca sativa L. *J. Sci. Food Agric.* **2017**, *97*, 1828–1836. [CrossRef]
42. Zou, T.; Xu, N.; Hu, G.; Pang, J.; Xu, H. Biofortification of soybean sprouts with zinc and bioaccessibility of zinc in the sprouts. *J. Sci. Food Agric.* **2014**, *94*, 3053–3060. [CrossRef] [PubMed]
43. Di Gioia, F.; Petropoulos, S.A.; Ozores-Hampton, M.; Morgan, K.; Rosskopf, E.N. Zinc and Iron Agronomic Biofortification of Brassicaceae Microgreens. *Agronomy* **2019**, *9*, 677. [CrossRef]
44. Rietra, R.P.J.J.; Heinen, M.; Dimkpa, C.O.; Bindraban, P.S. Effects of nutrient antagonism and synergism on yield and fertilizer use efficiency. *Commun. Soil Sci. Plant Anal.* **2017**, *48*, 1895–1920. [CrossRef]
45. D'Imperio, M.; Montesano, F.F.; Renna, M.; Parente, A.; Logrieco, A.F.; Serio, F. Hydroponic Production of Reduced-Potassium Swiss Chard and Spinach: A Feasible Agronomic Approach to Tailoring Vegetables for Chronic Kidney Disease Patients. *Agronomy* **2019**, *9*, 627. [CrossRef]
46. Rousseau, S.; Kyomugasho, C.; Celus, M.; Hendrickx, M.E.; Grauwet, T. Barriers impairing mineral bioaccessibility and bioavailability in plant-based foods and the perspectives for food processing. *Crit. Rev. Food Sci. Nutr.* **2020**, *60*, 826–843. [CrossRef]
47. D'Imperio, M.; Renna, M.; Cardinali, A.; Buttaro, D.; Santamaria, P.; Serio, F. Silicon biofortification of leafy vegetables and its bioaccessibility in the edible parts. *J. Sci. Food Agric.* **2016**, *96*, 751–756. [CrossRef] [PubMed]
48. Fernández-García, E.; Carvajal-Lérida, I.; Pérez-Gálvez, A. In vitro bioaccessibility assessment as a prediction tool of nutritional efficiency. *Nutr. Res.* **2009**, *29*, 751–760. [CrossRef] [PubMed]
49. Bertin, R.L.; Maltez, H.F.; de Gois, J.S.; Borges, D.L.; Borges, G.D.S.C.; Gonzaga, L.V.; Fett, R. Mineral composition and bioaccessibility in Sarcocornia ambigua using ICP-MS. *J. Food Compost. Anal.* **2016**, *47*, 45–51. [CrossRef]
50. Khouzam, R.B.; Pohl, P.; Lobinski, R. Bioaccessibility of essential elements from white cheese, bread, fruit and vegetables. *Talanta* **2011**, *86*, 425–428. [CrossRef] [PubMed]
51. Holmes, R.P.; Goodman, H.O.; Assimos, D.G. Dietary oxalate and its intestinal absorption. *Scanning Microsc.* **1995**, *9*, 16.
52. Cabrera, W.E.; Schrooten, I.; De Broe, M.E.; d'Haese, P.C. Strontium and bone. *J. Bone Miner. Res.* **1999**, *14*, 661–668. [CrossRef] [PubMed]

foods

MDPI

Article

Long-Aged Parmigiano Reggiano PDO: Trace Element Determination Targeted to Health

Cristina Santarcangelo [1], Alessandra Baldi [2], Roberto Ciampaglia [1], Marco Dacrema [1], Alessandro Di Minno [1,3], Valentina Pizzamiglio [4], Gian Carlo Tenore [1] and Maria Daglia [1,5,*]

1 Department of Pharmacy, University of Naples Federico II, 80131 Naples, Italy; cristina.santarcangelo@unina.it (C.S.); roberto.ciampaglia@unina.it (R.C.); marco.dacrema@unina.it (M.D.); alessandro.diminno@unina.it (A.D.M.); giancarlo.tenore@unina.it (G.C.T.)
2 Tefarco Innova, National Inter-University Consortium of Innovative Pharmaceutical Technologies, 43124 Parma, Italy; alessandra.baldi.alimenti@gmail.com
3 CEINGE-Biotecnologie Avanzate, University of Naples Federico II, 80131 Naples, Italy
4 Consorzio del Formaggio Parmigiano Reggiano, 42124 Reggio Emilia, Italy; pizzamiglio@parmigianoreggiano.it
5 International Research Center for Food Nutrition and Safety, Jiangsu University, Zhenjiang 212013, China
* Correspondence: maria.daglia@unina.it

Abstract: The concentrations of four health-related trace elements were measured using Atomic Absorption Spectroscopy in long-ripened (24- and 40-months) Parmigiano Reggiano (PR) PDO cheese, obtained from both summer and winter milk. To date, there are limited data on PR trace element concentrations, and no data about long-ripened cheese, especially when ripened for 40 months. Thus, the aim of this investigation is to determine chromium, manganese, selenium, and zinc concentrations, improving the available data on these trace elements and increasing knowledge of the biological properties of PR linked to their content in this cheese. The results show that 40-month ripened PR is a source of selenium and chromium, according to definitions under the European Regulation 1924/2006, as a 30 g cheese portion contains 11 ± 2 μg (summer milk) and 10 ± 1 μg (winter milk) of selenium and 8 ± 1 μg (summer and winter milk) of chromium, providing in excess of 8.25 and 6 μg per portion, respectively. This represents 15% of nutrient reference intake values for adults. These findings allow for the claim to be made that PR possesses the health properties ascribed to food sources of selenium and chromium according to European Regulation 432/2012.

Keywords: Parmigiano Reggiano PDO cheese; Atomic Absorption Spectroscopy; selenium; chromium; long-ripened cheese; European health claim

Citation: Santarcangelo, C.; Baldi, A.; Ciampaglia, R.; Dacrema, M.; Di Minno, A.; Pizzamiglio, V.; Tenore, G.C.; Daglia, M. Long-Aged Parmigiano Reggiano PDO: Trace Element Determination Targeted to Health. *Foods* **2022**, *11*, 172. https://doi.org/10.3390/foods11020172

Academic Editor: Fernando Barbosa Júnior

Received: 29 November 2021
Accepted: 23 December 2021
Published: 10 January 2022

1. Introduction

Parmigiano Reggiano cheese (PR), an Italian Protected Designation of Origin (PDO) product, is known worldwide due to its sensorial and nutritional characteristics. Furthermore, despite the product category to which it belongs, it is often associated with health, primarily due to its high protein (32.4 g/100 g), calcium (1155.0 mg/100 g), and phosphorus (691.0 mg/100 g) content, with a lower fat content (29.7 g/100 g) compared to other aged cheese, and a natural absence of lactose (less than 1.0 mg/100 g) [1].

According to the specifications currently in vigor, PR is a hard, cooked, slow-ripening cheese produced with raw partially skimmed milk, coming from cows whose diet is mainly composed of feed from its area of origin. The production area includes the territories of the provinces of Bologna to the left of the Reno River, Mantova to the right of the Po River, Modena, Parma, and Reggio Emilia. Moreover, the milk cannot be thermally treated, the use of additives is not allowed, and all milk introduced into the dairy must comply with the product specifications of PR. The ripening time must last for at least 12 months, starting from the molding of the cheese, with an average ripening of 24 months, potentially lasting up to 60 months and beyond [2].

Typically, the composition of PR varies according to the microbiological and chemical composition of the milk, the cheesemaking technology applied, including the natural whey starter, ripening time, and the environmental conditions that arise during these processes [3]. The milk composition in turn depends mainly on species, breed, season, and animal diet [4].

The specific PR cheesemaking procedures and ripening processes, along with the characteristics of the milk due to its territory of origin, strongly determine the physical properties, the chemical composition, and, in turn, the health profile of the final product. In fact, these practices lead to a selective concentration of nutritional and bioactive components, which increase the health value of this product. Due to natural dehydration occurring during the ripening process, the protein and aminoacidic content increases, mineral concentrations change (e.g., the potassium and magnesium concentrations decrease and selenium increases), and lactose decreases in the early hours following the cheese making process and is no longer detectable at 12 months of ripening [3].

There are several research articles that address PR as a source of protein, vitamins, and minerals, especially calcium, with valuable nutritional properties [5–8]. Moreover, PR, as with other fully ripened cheeses, contains other health-related nutrients, such as fats, and minor components including bioactive peptides. The fat fraction of PR contains butyric acid (123.9 mg per 100 g of fat in the outer part of the wheel at 24 months of ripening) [9], which exerts beneficial effects in obesity, inflammation, and neurological disorders, and conjugated linoleic acid (CLA—0.26 g/100 g), which has shown several beneficial activities on cardiocirculatory and immune systems [4]. As far as bioactive peptides are concerned, they are generated by proteolysis, which takes place during the ripening process and in the digestion process in humans. Among the health properties of the bioactive peptides, the inhibition of the angiotensin converting enzyme (ACE) is the most studied, along with its subsequent anti-hypertensive activity. In this regard, some studies have been published which, starting from the in vitro simulated gastrointestinal digestion of PR, demonstrate that different bioactive peptides (i.e., ACE-inhibitors and antimicrobial), are released and are absorbed in the intestine [10,11]. A recently published in silico study found an inhibitory activity of some PR cheese bioactive peptides against enzymes mainly involved in glucose metabolism, suggesting a potential effect on glycemic parameters [12].

As described above, there are investigations present in the literature that identify compounds with high healthy values in PR, but limited data are published on the concentrations of trace elements [4,13]. These elements are present in living tissues in small amounts and are known to solve essential functions for biological performance, primarily acting as cofactor catalysts in enzyme systems, as well as acting as centers for stabilizing structures of enzymes and proteins or binding molecules on the receptor sites of the cell membrane. Among these trace elements, selenium, zinc, and manganese are directly involved in the antioxidant enzymatic systems as cofactors for a number of enzymes [14]. In particular, selenium is a cofactor of 25 selenoproteins (including glutathione peroxidases, thioredoxin reductases, thioredoxin-glutathione reductase, iodothyronine deiodinases, and selenophosphate synthetase) [15,16], and zinc and manganese are cofactors of superoxide [17,18]. Chromium, as a trivalent ion, is an essential trace element, although no symptoms of chromium deficiency have been reported [19,20]. The low-molecular-weight of chromium-binding substance (LMWCr) has been proposed to be the biologically active form of chromium, being able to activate the kinase activity of insulin receptors in a dose dependent manner and increase insulin sensitivity [21].

According to the current European regulations, at present it is only possible to claim health properties of PR based on proteins, calcium, and phosphorus content, due to the limited available data on its chemical composition. Regulation (EC) 1924/2006 provides harmonized legal standards across Member States concerning nutrition and health claims, to guarantee the effective functioning of the market and a high level of consumer protection. It applies to all foods, including cheese. The term "nutrition claim" means "any indication that states, suggests, or implies that a food possesses beneficial nutritional properties due to the caloric value it provides or does not provide, or to the nutrients or other substances it contains or does not contain". In addition to nutrition claims, this regulation also allows for health claims, as defined by Article 13 (general function claim) and Article 14 (reduction of disease risk claim). Specifically, a "health claim" is "any indication that affirms, suggests, or implies the existence of a relationship between a category of food, a food or one of its components and health, while, a "reduction of disease risk claim" defines claims relating to the reduction of a disease risk as any health claim that states, suggests or implies that the consumption of a food category, food or one of its constituents significantly reduces a risk factor of development of a human disease".

The Regulation (EU) 432/2012 contains a list of health claims permitted for food products, which includes all 222 functional claims currently approved for description of the health properties of a food.

Considering the widespread consumption of PR, especially in Europe, the aim of this investigation is to determine the chromium, selenium, zinc, and manganese concentrations of 24-month and 40-month ripened PR, to improve the knowledge of the concentrations of these compounds and thus of the biological properties associated with the trace element content, allowing the communication of PR healthy properties to European consumers, according to Regulations 1924/2006 and 432/2012.

2. Materials and Methods

2.1. Sampling and Treatment

Cheese samples were provided by Consorzio del Formaggio Parmigiano Reggiano. A total of 100 samples of cheese (1 kg each) were randomly taken from dairies, with a proportional and representative number of samples taken from each of the 5 provinces, reflecting their proportion of the total Parmigiano Reggiano cheese wheel production. Samples were distributed according to milk production season: 25 samples of cheese aged for 24 months obtained from summer milk, 25 samples of cheese aged for 24 months obtained from winter milk, 25 samples of cheese aged for 40 months obtained from summer milk and 25 samples of cheese aged for 40 months obtained from winter milk (see detailed samples description in Supplementary Materials n. 1). Samples were chopped with a mixer (DJ3001 Moulinette Compact, Moulinex, Milano, Italy), collected in plastic tubes, and stored at −20 °C until analysis.

2.2. Sample Preparation

Before digestion, the cheese samples were incinerated at 500 °C for 24 h. For each cheese sample, the ash was weighed (0.5–1.5 g) and rinsed with nitric acid and three times with deionised water to remove acidic debris. For digestion, samples were transferred to TFM®PTFE vessels with 6 mL of concentrated 65% HNO_3 (14.33 mol/L) and 1 mL of 30% H_2O_2. The samples were submitted to digestion in a microwave digestion apparatus (MW-AD, Ethos EZ microwave digester, Mileston, Shelton, CT, USA). The heating program for digestion consists of 4 steps: step 1 (90 °C for 7 min), step 2 (170 °C for 5 min), step 3 (210 °C for 5 min), and step 4 (210 °C for 20 min). In all steps the power was set to 1000 W. The final solutions were diluted up to 25 mL with doubly distilled water for analysis by Graphite Furnace Atomic Absorption Spectrometer (GFAAS).

2.3. Graphite Furnace Atomic Absorption Spectrometer Analysis

Analyses for Cr, Mn, Se, and Zn were performed by Atomic Absorption Spectroscopy (AAS) according to the AOAC International method, 1995. An AA-6300 atomic absorption spectrophotometer (Shimadzu, Columbia, MD, USA) was used, equipped with an ASC-6100 autosampler (Shimadzu, Columbia, MD, USA) and GFA-EX7i graphite furnace atomizer (Shimadzu, Columbia, MD, USA). The control of instruments and analysis of data were performed using Multi-Element Program Software (WizAArd software, Shimadzu, Columbia, MD, USA). Argon was used as the internal and external gas. The AAS instrument was equipped with a hollow cathode lamp for Cr, Se, Zn and Mn line sources. A deuterium lamp was used as a background corrector. Graphite pyrolytically coated tubes with a L'vov platform were employed. To optimize the analytical signal, various tests were performed with different lamp intensities, sample injection volumes and temperature ranges (1600–1800 °C for atomization).

2.4. Reagent and Calibration Curves

The water used was 18 megohm water, purified with a Milli-Q® Integral 10 system (Merck, Darmstadt, Germany).

A selenium standard solution of 1000 mg/L, and a multi-element standard solution IV (at the concentration of 1000 mg/L): Ag, Al, Ba, Bi, Ca, Cd, Co, Cr, Cu, Fe, Ga, In, K, Mg, Mn, Na, Ni, Pb, Sr, Tl, Zn, were purchased from Merck KGaA (Darmstadt, Germany).

Calibration blank (Cal Blk) and calibration standard (Cal Std) solutions were prepared at concentrations of 0 μg/L (Cal Blk), 4 μg/L (Cal Std 1), 12 μg/L (Cal Std 2), and 20 μg/L (Cal Std 3), using the selenium standard stock solution, or multi-element standard stock solution, into separate 50 mL DigiTUBE® tubes (SCP SCIENCE, Baie-D'Urfe, QC, Canada) with the addition of 0.5 mL internal standard stock solution, 0.5 mL methanol and 5 mL concentrated nitric acid, and then diluted to the final volume with water.

2.5. Statistical Analysis

Statistical analysis was performed to determine the differences in the obtained values between different aged PR cheeses (i.e., 24-month and 40-month ripened PR) and in different seasons (i.e., summer or winter milk). It was conducted using GraphPad Prism version 6.0.0 for Windows (GraphPad Software, San Diego, CA, USA). The results have been given in the form of mean ± SD, with $p < 0.05$ taken as statistically significant. Statistical significance of data was assessed through a one-way variance analysis (ANOVA) using Prism Graphpad 8 (San Diego, CA, USA). When significant differences were found, Tukey's multiple comparisons test was used to determine the difference between the groups involved.

3. Results

Five calibration curves were prepared using standard references of the selected elements. Table 1 reports the wavelength, slit width, limit of detection (LOD), limit of quantification (LOQ), the equation of the calibration curve, the linear range, and the regression coefficient for each element. The regression coefficients obtained were greater than 0.99, except for that calculated for determination of Zn, which was found to be higher than 0.94 (Table 1). The limits of detection (LOD) and the limits of quantification (LOQ) were respectively calculated to be three times and 10 times the signal of the blank.

Table 1. Analytical parameters of atomic absorption spectrometer analysis.

Element	Wavelength (nm)	Slit Width (nm)	LOD (mg/L) *	LOQ (mg/L) *	Calibration Curve	LR *	R2 *
Cr	357.9	0.7	0.0001	0.0003	y = 0.0089x + 0.0008	0–20	0.9911
Se	196	0.7	0.0001	0.0003	y = 0.0097x − 0.0008	0–20	0.9981
Zn	213.9	0.2	0.001	0.003	y = 0.0447x + 0.0949	0–20	0.9443
Mn	279.5	0.2	0.005	0.015	y = 0.0809x + 0.0104	0–20	0.9991

* LOD, limit of detection; LOQ, limit of quantification; LR, linear range; R2 regression coefficient.

According to Italian LARN (Livelli di Assunzione di Riferimento di Nutrienti ed energia—Reference Intake Levels of Nutrients and Energy) a standard portion of ripened cheese is 50 g. Considering the high nutritional value of 40-month ripened PR, a portion could be quantified as 30 g. Therefore, Table 2 reports the results expressed as average value and standard deviation (SD) of the selected trace element concentrations in 100 g and in a portion of 50 g for 24-month ripened PR, and in a portion of 30 g for 40-month ripened PR.

Table 2. Concentration of trace elements in Parmigiano Reggiano cheese at different months of ripening, expressed per 100 g and per consuming portion (30 g and 50 g).

Metal	Cheese Aged 24 Months, Summer Milk *		Cheese Aged 24 Months, Winter Milk *		Cheese Aged 40 Months, Summer Milk *		Cheese Aged 40 Months, Winter Milk *	
	(μg/100 g)	(μg/50 g)	(μg/100 g)	(μg/50 g)	(μg/100 g)	(μg/30 g)	(μg/100 g)	(μg/30 g)
Cr	13.93 ± 4.80(a)	7.00 ± 2.40	14.61 ± 3.44(b)	7.30 ± 1.72	25.34 ± 2.59(a)	7.60 ± 0.77	26.19 ± 4.12(b)	7.86 ± 1.23
Mn	15.96 ± 4.60(a)	7.98 ± 2.30	7.64 ± 7.11(b)	3.82 ± 3.55	40.24 ±16.09(c)	12.07 ± 4.83	24.96 ± 11.95(d)	7.48 ± 3.58
Se	25.27 ± 5.82(a)	12.63 ± 2.91	25.82 ± 2.97(b)	12.91 ± 1.48	37.34 ± 6.86(a)	11.20 ± 2.06	34.22 ± 4.26(b)	10.27 ± 1.28
Zn	3196 ± 1095(a)	1598 ± 547.50	1752 ± 627(b)	876 ± 313.50	3952 ± 1330(c)	1186 ± 399	2478 ± 772(a,c,d)	743 ± 231

* (n = 25, ±SD); different letters indicate statistically significant differences ($p < 0.05$) between four groups (24 summer: cheese aged 24 months obtained from summer milk; 40 summer: cheese aged 40 months, summer milk; 24 winter: cheese aged 24 months, winter milk; 40 winter: cheese aged 40 months, winter milk).

Average chromium concentration in cheese samples obtained with summer milk and aged 24 months was found to be 13.93 ± 4.80 μg/100 g, while in cheese samples obtained with winter milk aged 24 months, the concentration was 14.61 ± 3.44 μg/100 g. For samples aged 40 months, Cr was present at higher concentrations equal to 25.34 ± 2.59 μg/100 g for summer milk samples and 26.19 ± 4.12 μg/100 g for winter milk samples. Tukey's multiple comparisons test (Table 2) showed a statistically significant difference between samples aged 24 months and samples aged 40 months.

Average manganese concentration in cheese samples obtained with summer milk and aged 24 months resulted to be 15.96 ± 4.60 μg/100 g, while in cheese samples obtained with winter milk aged 24 months, the concentration was 7.64 ± 7.11 μg/100 g. For samples aged 40 months, Mn was present at higher concentrations equal to 40.24 ± 16.09 μg/100 g for summer milk samples and 24.96 ± 11.95 μg/100g for winter milk samples. Comparison tests showed a statistically significant difference between all groups (Table 2).

Average selenium concentration found in cheese aged 24 months from summer milk was 25.27 ± 5.82 μg/100 g, and from winter milk was 25.82 ± 2.97 μg/100 g. In cheese aged 40 months the Se concentration was equal to 37.34 ± 6.86 μg/100 g from summer milk samples, and 34.22 ± 4.26 μg/100 g from winter milk samples. Here too, the statistical analysis determined statistically significant differences between the samples aged at 24 months and samples aged at 40 months.

Average zinc concentration found in cheese aged 24 months made from summer milk was 3196 ± 1095 μg/100 g, and from winter milk it was 1752 ± 627 mg/100 g. Contrastingly, in cheese aged 40 months the Zn concentration was equal to 3952 ± 1330 μg/100 g in the summer milk samples, while it came to 2478 ± 772 μg/100 g from winter milk samples. Statistically significant differences were found in Zn concentration between groups: cheese aged 24 months, summer milk vs. cheese aged 40 months, summer milk; cheese aged 24 months, summer milk vs. cheese aged 24 months, winter milk; cheese aged 40 months, summer milk vs. cheese aged 24 months, winter milk; cheese aged 40 months, summer milk vs. cheese aged 40 months, winter milk. No statistical differences were found between the groups: cheese aged 24 months, summer milk vs. 40 months, winter milk; cheese aged 24 months, winter milk vs. cheese aged 40 months, winter milk. (Table 2).

4. Discussion

In this study, the concentrations of four healthy trace elements were determined in a large number of PR PDO samples with two different degrees of ripening, 24 and 40 months, taking into account different milk production seasons (summer and winter) and the percentage distribution of wheels produced among provinces of the geographical area of origin.

Trace and major elements are included in milk and cheese in a colloidal or aqueous phase depending on their type. The colloidal phase consists of proteins, caseins, organized as micelles, which during the coagulation process include the fat globules forming the curd. The aqueous phase consists of whey, which includes the soluble protein fraction, as well as monomers, small polymers, and the majority of the sugars. In the colloidal phase of milk, the casein micelles are made up of casein submicelles, cross-linked together by calcium and phosphorus, in the form of colloidal calcium phosphate (CCP). Depending on the dominant phase, colloidal or soluble, in which major minerals and trace elements can be found in milk, these can wind up in either whey or curd during the cheesemaking process. During the cheesemaking process, in the case of coagulation via lactic acid, the micelles are destabilized by acidic conditions (pH values near the isoelectric point of casein) which cause the loss of saline components including Ca and P, leading to dissociation in submicelles, which are then regrouped due to hydrophobic interaction, but not in micellar form, with Ca and P solubilized in the aqueous phase. In the case of rennet coagulation, the micelles are destabilized by rennet enzymes that cut k-casein, leading to the aggregation of many different micelles thanks to hydrophobic bonds. In this case the micelle structure is maintained, with Ca, P, and many other minerals and trace elements present in the colloidal phase (CCP). The minerals not associated with CCP (i.e., Na, and K), are almost completely lost in the whey phase, unlike the CCP minerals, such as S, Mg, and Zn, that are largely retained in curd as structural components [22]. Different factors (pH, temperature, and salinity) characterize the different cheesemaking processes, and thus play a key role in the colloidal-soluble form equilibrium of minerals and trace elements. Lactic acid fermentation leads to a decrease in milk pH, which throws off the equilibrium of the soluble form, causing a progressive solubilization of the casein-bounded minerals, losing them into the whey [23].

On the other hand, the higher temperatures promote casein-bound mineral forms. The different conditions of different cheesemaking processes lead to a variation in the concentrations of minerals and trace elements across the different kinds of cheese, despite having the same ripening period [22]. According to the literature data, the concentration of minerals with a predominantly colloidal form decreases in semi-cooked cheeses, compared to uncooked cheeses, and to lactic acid coagulation cheeses [22].

PR coagulation involves both lactic acid and rennet coagulation, with the latter being predominant. Thus, PR should be characterized by a higher concentration of minerals and trace elements that are mainly found in the colloidal casein-bound form, such as Ca, P, Zn, Se, and Cu, compared to other types of cheese, due to the predominant rennet coagulation and the long ripening process.

In addition to the cheesemaking process, cheese ripening is an important factor that influences the chemical composition of cheese, being a complex process that involves physical, chemical, and microbiological modifications, including the diffusion of salt from external to inner parts and consequent aqueous phase loss, gradual lactose loss mainly through lactic bacteria fermentation, and lactic acid neutralization leading to a pH increase. In the case of PR cheese, lactose is fully processed within 12 h post-production [24]. During the cheese ripening, minerals are progressively concentrated into the colloidal phase, due to the loss of the aqueous phase and to pH changes [25]. In addition, the basic pH environment promotes the retention of minerals in the colloidal phase. The water content of a 12-month ripened PR is 30%, decreasing to 28% in a 40-month cheese [26].

The results of our investigation show statistically significant differences between the trace element mean concentrations in 40 month-ripened PR and in 24 month-ripened PR, registered for all elements. The milk's season of origin, however, does not influence the concentrations of the studied trace elements, with the exception of the concentration of Mn and Zn in 40 and 24 month-ripened PR, which were found to have lower concentrations in cheese obtained from winter milk. This result is due to several concurrent causes, especially humidity loss and pH changes.

At the end of 1990, Gambelli et al. [14] published a paper on minerals and trace minerals in Italian dairy products, giving a healthy connotation to trace elements for the first time. These were previously considered to be toxicological components, such as heavy metals. The research group used ion exchange liquid chromatography with suppressed conductivity for the determination of the major minerals (Na, K, Mg, and Ca) and instrumental neutron activation analysis for the determination of the trace elements (Co, Cr, Fe, Rb, Se, and Zn). The results identified two subgroups within cheeses, the stirred curd group and the hard group, both being foods with high levels of nutritionally important trace elements (i.e., Se, Zn, Fe, and Co). In particular, the hard cheese analyzed, Grana Padano (1–2 years ripened cheese samples), yielded the following concentrations: Zn 4.50 + 0.00 mg/100 g, Se 10.00 + 1.03 μg/100 g, Cr 9.90 + 2.00 μg/100 g [13], in which the concentration of Zn is similar to that found in PR, while the concentrations of Se and Cr are lower in Grana Padano than in PR according to the results of the present study.

More recently, Manuelian et al. [4] published a paper regarding major and trace elements, fatty acid composition and cholesterol content of different types of PDO cheese, including PR. Mineral concentration was measured by inductively coupled plasma optical emission spectrometry. In PR ripened for a period ranging from 12 to 24 months, Zn concentration was 33.93 $\pm$ 2.30 mg/g, and Se concentration was 0.91 $\pm$ 0.13 μg/g [4]. This concentration of Zn is similar to that found in the present investigation, and so, our results are in line with those of Manuelian and colleagues. On the contrary, the concentration of Se found by Manuelian et al. is about four times higher than the one we found, and is reported by CREA Italian Council (Consiglio per la Ricerca in agricoltura e L'analisi dell' Economia Agraria—Council for Research in Agriculture and the Analysis of Agricultural Economics) [27].

The selenium concentration reported by Manuelian actually seems to be very high, as a portion of PR alone (50 g) would provide about 45 μg, satisfying more than 82% of the Dietary Reference Values for selenium. In any case, the concentration of Se found in PR in this investigation, compared to other long-ripened cheeses, could be ascribed to the specific product PDO regulation. It requires a rationing of dairy cows based on the use of fodder from the production area of PR, and at least 50% of the dry matter of said fodder must consist of hays. Interestingly, an agronomical and geological study published in 2007 has shown that the cheese production area corresponds to a one of the three geographically separate soils richer in selenium within the Italian peninsula (Figure 1) [28]. It is well established that the selenium content of soil affects the amounts of selenium in the plants that animals eat. Nevertheless, the selenium concentration in soil has a smaller effect on trace element levels in animal products than in plant-based foods, owing to the homeostatic mechanisms present in animals and their effect on the maintenance of selenium tissue concentrations.

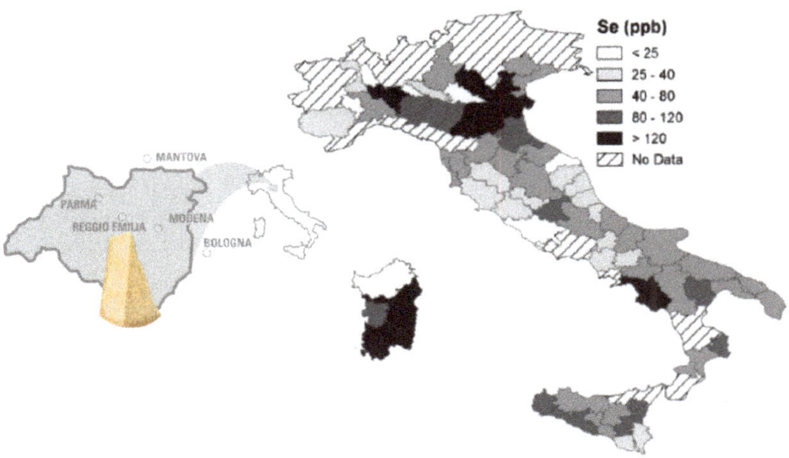

Figure 1. PR production area and selenium content data in Italian soil (modified from Spadoni et al., 2007) [28].

At the level of the European Union, as previously reported, the Regulation (EC) 1924/2006 harmonizes the laws of the Member States concerning nutrition and health claims. Generally speaking, a detailed chemical characterization of a food might be a useful approach for producers and consumers to communicate and understand the evidence-based health properties of a food.

Regarding the daily nutrient reference intakes values (NRVs) for adults, reported in annex 13 of the Regulation (EU) 1169/2011, a 30 g portion of 24-month PR could satisfy about 14% of the NRV for Se, and 11% for Cr. A 30 g portion of 40-month PR could satisfy more than 19% for both Se and Cr, making PR ripened for 40 months a confirmable source of Se and Cr, as it contains much more than the required 15% of the nutrient reference values per portion and thus easily per 100 g of food product. As far as Mn and Zn are concerned, long-aged PR contributes only a very small of the NRV for Mn. However, regarding Zn, our results showed a significant contribution towards NRV levels but only at the 100 g level, too much for a single portion consumption.

These findings allow claims to be made for PR for the health properties ascribed to the food sources of selenium and chromium, according to the Regulation (EU) 432/2012 concerning health claims (Figure 2).

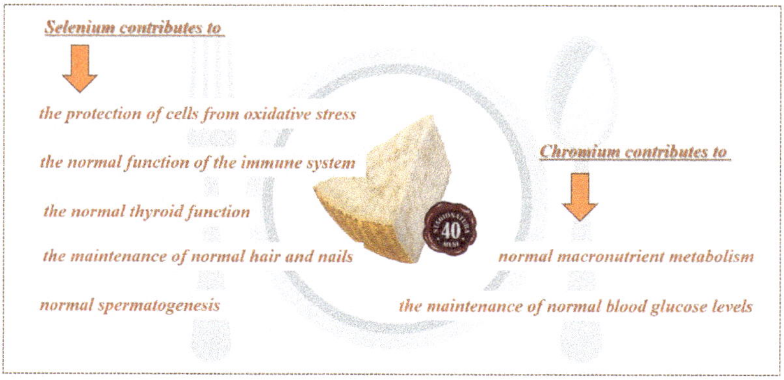

Figure 2. Health claim for selenium and chromium, according to the Regulation (EU) 432/2012.

The most studied property of selenium is its ability to protect DNA, proteins, and lipids from oxidative damage, through its role as a cofactor for antioxidant enzymes, such as gluthation peroxidase [29]. Selenium plays a key role in both thyroid function, through the selenoproteins involved in deiodination of thyroid hormones, and in the immune system, being able to stimulate the proliferation of T cells, increase the activity of natural killer cells and the response to antigen stimulation. In addition, selenium is an important factor in spermatogenesis, as the selenoproteins of the sperm mithocondrial capsule exert structural and enzymatic functions being responsible for the motility and structural integrity of the sperm tail [30]. Finally, a deficiency in selenium, shown in patients receiving total parenteral nutrition lacking selenium, results in impairment of hair and nails, with the clinical manifestation of white nail beds, pseudoalbinism, alopecia and thin hair, which disappeared after the administration of selenium [31]. A correct intake of selenium is considered important for the maintenance of normal hair and nails.

If a correct selenium intake leads to the above reported health benefits, a selenium intake much higher than the NRV induces acute selenium toxicity, which can cause severe gastrointestinal and neurological symptoms, acute respiratory distress syndrome, myocardial infarction and, in rare cases, death, as has been observed in cases of misformulated over-the-counter products containing excessive amounts of selenium. In 2008, the US National Institute of Health (NHI) reported that 201 people experienced severe adverse reactions from taking a dietary supplement containing 200 times the labelled amount [32].

The fixed selenium upper intake level (UL) from food and supplements in the adult population is 400 µg. However, a recently published observational cohort study in a diabetes-free Italian population found that a daily quantity of Selenium equal or higher to 80 μg/day is positively associated with hospitalization for type 2 diabetes [33]. In this regard, as a selenium source, long-aged PR ripened for 40 months is a useful option for nutritionists looking to give the correct daily amount of this trace element to the general healthy population, without resorting to food supplements which remain the best option for selenium deficient subjects.

As far as chromium is concerned, this occurs in nature in the forms of trivalent and hexavalent chromium. This former form is found in various foods, including cheese, reaching higher concentrations in bivalve mollusks and Brazil nuts [19]. The latter form of chromium may be found in foods as a toxic contaminant released by the tools used in the production process (e.g., certain types of stainless steel) [34]. Trivalent chromium, taken with the diet, has positive health effects in the body [35]. Adverse reactions related to chromium deficiency have been highlighted in patients undergoing parenteral nutrition for extended periods without supplementation of this element, resulting in impaired glucose tolerance and an altered metabolism [36]. In particular, trivalent chromium is linked with an increase in insulin action and an improvement in glucose tolerance in type 2 diabetes [37]. In addition to the glucose metabolism, chromium also influences nitrogen and lipid metabolism, as it inhibits the hepatic enzyme HMG-CoA reductase and lowers LDL cholesterol levels [38]. Numerous investigations have demonstrated a relationship between the consumption of dairy products and a reduction in the occurrence of diabetes. This property of dairy products, especially PR, is generally ascribed to the protein content and the type of fats. Further studies should be conducted to identify the role of chromium in the protective effects of PR against diabetes and metabolic syndrome.

Regarding the safety of trivalent chromium, it is safe to the point that the Food and Nutrition Board of the USA National Academies of Sciences, Engineering, and Medicine has stated that no adverse effects have been linked to high intakes of chromium from food or supplements, and so it did not establish a UL for chromium [34]. In fact, no case of toxicity is recorded by the intake of trivalent chromium with food, and the only negative effects recorded are in isolated case reports of misformulated chromium supplements, which might cause weight loss, anemia, thrombocytopenia, liver dysfunction, renal failure, rhabdomyolysis, dermatitis, and hypoglycemia [39,40].

5. Conclusions

In conclusion, the present research work established that a portion of long-aged PR at 40-months of ripening is a source of selenium and chromium, according to the definition given by the Regulation (EC) 1924/2006, thus allowing the communication through labelling of properties typical for a food source of these trace elements to European consumers.

Supplementary Materials: The following supporting information can be downloaded at: https://www.mdpi.com/article/10.3390/foods11020172/s1, Table S1: Selenium were determined in different cheese samples for aged and collection season milk; Table S2: Chromium were determined in different cheese samples for aged and collection season milk; Table S3: Zinc were determined in different cheese samples for aged and collection season milk; Table S4: Manganese were determined in different cheese samples for aged and collection season milk.

Author Contributions: Conceptualization, M.D. (Maria Daglia), V.P., G.C.T. and A.D.M.; methodology, C.S. and R.C.; software, M.D. (Marco Dacrema); formal analysis, C.S. and R.C.; writing—original draft preparation, M.D. (Maria Daglia), C.S., G.C.T. and A.B. All authors have read and agreed to the published version of the manuscript.

Funding: This research was founded by Consorzio del formaggio Parmigiano Reggiano.

Data Availability Statement: The data presented in this study are available on request from the corresponding author.

Conflicts of Interest: The authors declare no conflict of interest. The funder had no role in data collection, analysis, decision to publish, writing the draft. V.P. is employed by the Consorzio del

Formaggio Parmigiano Reggiano. This does not alter the author's adherence to all the journal policies on sharing data and materials.

References

1. Reggiano, C.d.F.P. Parmigiano Reggiano Nutritional Characteristics. Available online: https://www.parmigianoreggiano.com/product-guide-nutritional-characteristics (accessed on 25 September 2021).
2. Reggiano, C.d.F.P. Specifications and Legistlation of the PDO Parmigiano Reggiano. Available online: https://www.parmigianoreggiano.com/consortium-specifications-and-legislation (accessed on 25 September 2021).
3. Summer, A.; Formaggioni, P.; Franceschi, P.; Di Frangia, F.; Righi, F.; Malacarne, M. Cheese as Functional Food: The Example of Parmigiano Reggiano and Grana Padano. *Food Technol. Biotechnol.* **2017**, *55*, 277–289. [CrossRef]
4. Manuelian, C.L.; Currò, S.; Penasa, M.; Cassandro, M.; De Marchi, M. Characterization of major and trace minerals, fatty acid composition, and cholesterol content of Protected Designation of Origin cheeses. *J. Dairy Sci.* **2017**, *100*, 3384–3395. [CrossRef]
5. Franceschi, P.; Malacarne, M.; Formaggioni, P.; Faccia, M.; Summer, A. Effects of Milk Storage Temperature at the Farm on the Characteristics of Parmigiano Reggiano Cheese: Chemical Composition and Proteolysis. *Animals* **2021**, *11*, 879. [CrossRef] [PubMed]
6. Popping, B.; De Dominicis, E.; Dante, M.; Nocetti, M. Identification of the Geographic Origin of Parmigiano Reggiano (P.D.O.) Cheeses Deploying Non-Targeted Mass Spectrometry and Chemometrics. *Foods* **2017**, *6*, 13. [CrossRef] [PubMed]
7. Alessandri, C.; Sforza, S.; Palazzo, P.; Lambertini, F.; Paolella, S.; Zennaro, D.; Rafaiani, C.; Ferrara, R.; Bernardi, M.L.; Santoro, M.; et al. Tolerability of a Fully Maturated Cheese in Cow's Milk Allergic Children: Biochemical, Immunochemical, and Clinical Aspects. *PLoS ONE* **2012**, *7*, e40945. [CrossRef] [PubMed]
8. Pampaloni, B.; Bartolini, E.; Brandi, M.L. Parmigiano Reggiano cheese and bone health. *Clin. Cases Miner. Bone Metab.* **2011**, *8*, 33–36.
9. Malacarne, M.; Summer, A.; Franceschi, P.; Formaggioni, P.; Pecorari, M.; Panari, G.; Mariani, P. Free fatty acid profile of Parmigiano–Reggiano cheese throughout ripening: Comparison between the inner and outer regions of the wheel. *Int. Dairy J.* **2009**, *19*, 637–641. [CrossRef]
10. Basiricò, L.; Catalani, E.; Morera, P.; Cattaneo, S.; Stuknytė, M.; Bernabucci, U.; De Noni, I.; Nardone, A. Release of angiotensin converting enzyme-inhibitor peptides during in vitro gastrointestinal digestion of Parmigiano Reggiano PDO cheese and their absorption through an in vitro model of intestinal epithelium. *J. Dairy Sci.* **2015**, *98*, 7595–7601. [CrossRef]
11. Solieri, L.; Baldaccini, A.; Martini, S.; Bianchi, A.; Pizzamiglio, V.; Tagliazucchi, D. Peptide Profiling and Biological Activities of 12-Month Ripened Parmigiano Reggiano Cheese. *Biology* **2020**, *9*, 170. [CrossRef] [PubMed]
12. Martini, S.; Solieri, L.; Cattivelli, A.; Pizzamiglio, V.; Tagliazucchi, D. An Integrated Peptidomics and In Silico Approach to Identify Novel Anti-Diabetic Peptides in Parmigiano-Reggiano Cheese. *Biology* **2021**, *10*, 563. [CrossRef] [PubMed]
13. Gambelli, L.; Belloni, P.; Ingrao, G.; Pizzoferrato, L.; Santaroni, G. Minerals and Trace Elements in Some Italian Dairy Products. *J. Food Compos. Anal.* **1999**, *12*, 27–35. [CrossRef]
14. Wołonciej, M.; Milewska, E.; Roszkowska-Jakimiec, W. Trace elements as an activator of antioxidant enzymes. *Postępy Higieny i Medycyny Doświadczalnej* **2016**, *70*, 1483–1498. [CrossRef]
15. Qian, F.; Misra, S.; Prabhu, K.S. Selenium and selenoproteins in prostanoid metabolism and immunity. *Crit. Rev. Biochem. Mol. Biol.* **2019**, *54*, 484–516. [CrossRef]
16. Rayman, M.P. Selenium and human health. *Lancet* **2012**, *379*, 1256–1268. [CrossRef]
17. Damiano, S.; Sozio, C.; La Rosa, G.; Guida, B.; Faraonio, R.; Santillo, M.; Mondola, P. Metabolism Regulation and Redox State: Insight into the Role of Superoxide Dismutase 1. *Int. J. Mol. Sci.* **2020**, *21*, 6606. [CrossRef]
18. Yan, Z.; Spaulding, H.R. Extracellular superoxide dismutase, a molecular transducer of health benefits of exercise. *Redox Biol.* **2020**, *32*, 101508. [CrossRef] [PubMed]
19. Vincent, J.B.; Lukaski, H.C. Chromium. *Adv. Nutr.* **2018**, *9*, 505–506. [CrossRef] [PubMed]
20. Vincent, J.B. Chromium: Is It Essential, Pharmacologically Relevant, or Toxic? *Metal Ions Life Sci.* **2013**, *13*, 171–198. [CrossRef]
21. Arakawa, H.; Kandadi, M.R.; Panzhinskiy, E.; Belmore, K.; Deng, G.; Love, E.; Robertson, P.M.; Commodore, J.J.; Cassady, C.J.; Nair, S.; et al. Spectroscopic and biological activity studies of the chromium-binding peptide EEEEGDD. *JBIC J. Biol. Inorg. Chem.* **2016**, *21*, 369–381. [CrossRef]
22. Lee, M.-R. Changes in the Mineral Components in Cheese Juice (Aqueous Phase of Cheese) during Ripening. *J. Milk Sci. Biotechnol.* **2017**, *35*, 85–92. [CrossRef]
23. Lucas, A.; Rock, E.; Chamba, J.-F.; Verdier-Metz, I.; Brachet, P.; Coulon, J.-B. Respective effects of milk composition and the cheese-making process on cheese compositional variability in components of nutritional interest. *Le Lait* **2005**, *86*, 21–41. [CrossRef]
24. Pecorari, M.; Gambini, G.; Reverberi, P.; Caroli, A.; Panari, G. Andamento della glicolisi nelle prime fasi di maturazione del Parmigiano-Reggiano. *Sci. Tecn. Latt. Casearia* **2003**, *54*, 149–162.
25. McSweeney, P.L.H. Biochemistry of cheese ripening. *Int. J. Dairy Technol.* **2004**, *57*, 127–144. [CrossRef]
26. D'Incecco, P.; Limbo, S.; Hogenboom, J.; Rosi, V.; Gobbi, S.; Pellegrino, L. Impact of Extending Hard-Cheese Ripening: A Multiparameter Characterization of Parmigiano Reggiano Cheese Ripened up to 50 Months. *Foods* **2020**, *9*, 268. [CrossRef]
27. Crea. Tabella di Composizione degli Alimenti Parmigiano Reggiano DOP. Available online: https://www.alimentinutrizione.it/tabelle-nutrizionali/166000 (accessed on 16 July 2021).

28. Spadoni, M.; Voltaggio, M.; Carcea, M.; Coni, E.; Raggi, A.; Cubadda, F. Bioaccessible selenium in Italian agricultural soils: Comparison of the biogeochemical approach with a regression model based on geochemical and pedoclimatic variables. *Sci. Total Environ.* **2007**, *376*, 160–177. [CrossRef]
29. Tinggi, U. Selenium: Its role as antioxidant in human health. *Environ. Health Prev. Med.* **2008**, *13*, 102–108. [CrossRef]
30. Tapiero, H.; Townsend, D.; Tew, K. The antioxidant role of selenium and seleno-compounds. *Biomed. Pharmacother.* **2003**, *57*, 134–144. [CrossRef]
31. Masumoto, K.; Nagata, K.; Higashi, M.; Nakatsuji, T.; Uesugi, T.; Takahashi, Y.; Nishimoto, Y.; Kitajima, J.; Hikino, S.; Hara, T.; et al. Clinical features of selenium deficiency in infants receiving long-term nutritional support. *Nutrition* **2007**, *23*, 782–787. [CrossRef] [PubMed]
32. ODS. Selenium USA: USA Government. 2020. Available online: https://ods.od.nih.gov/factsheets/Selenium-HealthProfessional/ (accessed on 25 September 2021).
33. Vinceti, M.; Bonaccio, M.; Filippini, T.; Costanzo, S.; Wise, L.A.; Di Castelnuovo, A.; Ruggiero, E.; Persichillo, M.; Cerletti, C.; Donati, M.B.; et al. Dietary selenium intake and risk of hospitalization for type 2 diabetes in the Moli-sani study cohort. *Nutr. Metab. Cardiovasc. Dis.* **2021**, *31*, 1738–1746. [CrossRef]
34. Institute of Medicine (US). *Panel on Micronutrients—Dietary Reference Intakes for Vitamin A, Vitamin K, Arsenic, Boron, Chromium, Copper, Iodine, Iron, Manganese, Molybdenum, Nickel, Silicon, Vanadium, and Zinc*; National Academies Press: Washington, DC, USA, 2001. [CrossRef]
35. Anderson, R.A.; Cheng, N.; Bryden, N.A.; Polansky, M.M.; Chi, J.; Feng, J. Elevated Intakes of Supplemental Chromium Improve Glucose and Insulin Variables in Individuals with Type 2 Diabetes. *Diabetes* **1997**, *46*, 1786–1791. [CrossRef]
36. Nielsen, F.H. Summary: The metabolism, nutritional essentiality, and clinical importance of chromium—Clarity emerging after 60 years of research. In *Nutritional Biochemistry of Chromium(Iii)*; Vincent, J.B., Ed.; Elsevier: Amsterdam, The Netherlands, 2019; pp. XV–XVII.
37. Wang, Z.Q.; Cefalu, W.T. Current Concepts About Chromium Supplementation in Type 2 Diabetes and Insulin Resistance. *Curr. Diabetes Rep.* **2010**, *10*, 145–151. [CrossRef] [PubMed]
38. Maret, W. Subject Index. In *Essential Metals in Medicine: Therapeutic Use and Toxicity of Metal Ions in the Clinic*; De Gruyter: Berlin, Germany, 2019; Volume 19, pp. 393–412. [CrossRef]
39. Fowler, J.F. Systemic contact dermatitis caused by oral chromium picolinate. *Cutis* **2000**, *65*, 116. [PubMed]
40. Vincent, J.B. The potential value and toxicity of chromium picolinate as a nutritional supplement, weight loss agent and muscle development agent. *Sports Med.* **2003**, *33*, 213–230. [CrossRef] [PubMed]

foods

MDPI

Article

Physicochemical Characteristics, Microstructure and Health Promoting Properties of Green Banana Flour

Minenhle Khoza [1,*], Eugenie Kayitesi [2] and Bhekisisa C. Dlamini [1]

[1] Department of Biotechnology and Food Technology, Faculty of Science, DFC Campus, University of Johannesburg, Doornfontein, Johannesburg 2028, South Africa; bcdlamini@uj.ac.za
[2] Department of Consumer and Food Sciences, University of Pretoria, Pretoria 0028, South Africa; eugenie.kayitesi@up.ac.za
* Correspondence: minenhlek@uj.ac.za; Tel.: +27-(0)11-559-6252

Abstract: This study aimed to investigate the proximate composition, mineral content, functional properties, molecular structure, in vitro starch digestibility, total phenolic content (TPC), total flavonoid content (TFC) and antioxidant activity (DPPH, FRAP) of green banana flour (GBF) cultivars grown in South Africa. With proximate composition, Finger Rose and Pisang Awak had the highest protein (4.33 g/100 g) and fat (0.85 g/100 g) content, respectively. The highest ash content (3.50 g/100 g) occurred with both Grand Naine and FHIA-01 cultivars. Potassium and copper were the most abundant and least minerals, respectively. Pisang Awak cultivar had the highest water absorption capacity (67.11%), while Du Roi had the highest swelling power (0.83 g/g) at 90 °C. Scanning electron microscopy (SEM) images revealed that starch granules from all GBF cultivars were irregular in shape and they had dense surfaces with debris. All the GBF cultivars had similar diffraction patterns with prominent peaks from 15°–24° diffraction angles. The resistant starch (RS) and amylose content of the FHIA-01 cultivar indicates that the GBF has the potential to lower risks of type 2 diabetes and obesity. The highest TPC, TFC and antioxidant activity occurred with the Grande Naine cultivar. Based on their functional characteristics, the Grand Naine and FHIA-01 GBF cultivars could potentially be used as raw materials for bakery products as well as for the fortification of snacks.

Keywords: green banana flour; functional characteristics; in vitro starch digestibility; antioxidant activity

Citation: Khoza, M.; Kayitesi, E.; Dlamini, B.C. Physicochemical Characteristics, Microstructure and Health Promoting Properties of Green Banana Flour. *Foods* **2021**, *10*, 2894. https://doi.org/10.3390/foods10122894

Academic Editor: Antonello Santini

Received: 20 September 2021
Accepted: 15 October 2021
Published: 23 November 2021

1. Introduction

The diabetes endemic continues to increase associated with obesity, inactive lifestyles and high-energy diets [1]. These diseases are the leading cause of demise and disability worldwide. Globally 366 million people suffer from diabetes. The International Diabetes Federation predicts this number to rise to 552 million by 2030 [2]. The prevalence of diabetes in South Africa has remarkably increased over the past two decades, making South Africa one of the countries with the highest predictable upsurge in diabetes for the next twenty-five years [3,4]. Epidemiological studies show that frequent consumption of high glycemic index (GI) food may lead to a high risk of obesity and type II diabetes. As one of the solutions, studies have indicated that obesity and type II diabetes can be prevented by eating low GI foods such as green banana flour (GBF) [5]. Further, studies have revealed that adequate consumption of fruits and vegetables is vital for reducing the burden of heart diseases and diabetes, possibly due to their relatively high dietary fiber, resistant starch, antioxidants and bioactive compounds contained in these foods [1,6,7]. As a result, there has been an intensive development of secondary food products made from fruits and vegetables as sources of dietary fiber and indigestible starches, with more focus on developing new products. The development of such food products allows the consumer to have permanent access to the nutritional benefits of fruit and vegetable products, in spite of their seasonality, and thus healthy food can be made available throughout the year [8].

The use of green banana flour is important as an alternative raw material for the processing of healthy functional products. According to Kumar et al. [9], green banana flour has the following composition: 52.7–54.2% Resistant starch (RS); 1.81% Total Soluble Solids (TSS), 76.77% Total Starch and 14–17% non-starch polysaccharides. Similar to dietary fiber, starch from green banana flour is not digestible in the small intestine; hence, it is fermented in the colon by gut flora [10]. When RS reaches the colon, it is used as a substrate for microbial fermentation, and it may lead to the production of short-chain fatty acids (butyrate, propionate and acetate), carbon dioxide, hydrogen and methane [11]. Each type of short-chain fatty acid has impacts on health. Butyrate is well-known for health enhancement as it plays a vital role in human gut health, including: decreasing inflammation, reducing the risk for colon cancer and enhancing gut barrier functions [12]. The lower digestibility of RS leads to a reduced release of blood glucose. This has been shown to have a reduction in leptin and post-prandial glucose reactions in people after the consumption of food products with a high RS content [13]. Recently, food products with a low glycemic index are highly favored by consumers due to their resistance to glucoamylase and α-amylase. Further, their digestion rate in the gut is relatively low due to the presence of resistance starch, thus causing reduced energy intake by gastral cells [14,15]. The aim of the study was to investigate the proximate composition, mineral content, functional, molecular, microstructure, TPC, TFC and antioxidant activity of green banana flour cultivars grown in South Africa.

2. Materials and Methods

2.1. Materials

Flour from five green banana cultivars, namely Grande Naine, Pisang Awak, Finger Rose, FHIA-01 and Du Roi, was kindly provided by the Agricultural Research Counsel (ARC) Tropical and Subtropical Crops, Nelspruit, Mbombela in South African. All reagents were analytical grade, Trolox, Folin–Ciocalteu reagent, gallic acid and quercetin were purchased from Sigma-Aldrich Pty. Ltd. (Johannesburg, South Africa). The resistant starch assay kit and the amylose/amylopectin kit were purchased from Megazyme Ltd. (Johannesburg, South Africa).

2.2. Preparation of Banana Starch

A water-alkaline extraction method was used to prepare banana starch, as described by Jiang et al. [16], with a few changes. Briefly, GBF (100 g) was macerated in distilled water (1 L) for 20 min at a low speed, then sieved through 100-mesh screens. The collected milk was centrifuged at 4000× g for 10 min to remove soluble fiber, and then 1 L NaOH solution (0.2%, w/v) was added to the sediment. The starch sediment was mixed with water and stirred for 5 min before resting for 2 h. Thereafter, the sediment was again suspended in water and allowed to settle. This was repeated until the wash water reached a neutral pH. The resultant material was then dried at 45 °C for 24 h. The desiccated starch was pulverized and passed through a 100 μm sieve.

2.3. Proximate Composition of Green Banana Flour

The moisture content was assayed using a vacuum oven dryer at 60 °C for 16 h using 2–3 g of sample, according to Rodriguez-Jimenez et al. [17]. A furnace was used to measure ash content using a method described by [9]. Soxhlet extraction was used for total fat content [14]. For protein content, the Kjeldahl method was followed, as demonstrated by Kumar et al. [9]. The carbohydrate percentage was calculated using the formula below.

$$\text{Carbohydrate (g)} = 100 - (\text{protein (g)} + \text{moisture content (g)} + \text{lipid (g)} + \text{ash (g)})$$

2.4. Mineral Composition of Green Banana Flour

Mineral analysis was conducted following a method by Jakavula et al. [18]. Briefly, the sample was digested using ultra-pure HNO3 on a microwave-accelerated reaction system

(CEM, Matthews, NC, USA). This was conducted at high temperature and pressure for the extraction of acid-extractable elements with the sample material. After that, deionized water was added (50 mL), followed by analyses of the sample by ICP-OES (Thermo Scientific, Basingstoke, UK).

2.5. Characterization of Functional Properties of Green Banana Flour

2.5.1. Water Absorption Capacity (WAC)

The WAC of GBF was determined using the method described by Kumar et al. [9], with some modifications. Precisely, 0.5 g flour sample was weighed into 50 mL centrifuge tubes followed by the addition of 5 mL distilled water. The suspensions were vortexed and rested for 1 h at room temperature (26 $\pm$ 2 °C). Thereafter, they were centrifuged at 3000 rpm for 30 min at 25 °C. The WAC was expressed as mL of water absorbed per gram of flour.

2.5.2. Water Solubility Index and Swelling Power

The water solubility index and swelling power were determined following the method detailed by Kumar et al. [9]. Green banana flour (0.2 g) was mixed with distilled water (5 mL) for 30 s using a vortex. After that, the mixture was heated at 50 °C, 70 °C and 90 °C for 20 min, followed by cooling and centrifugation at 3000 rpm for 10 min. The supernatant was evaporated at 105 °C for 16 h in an oven. The solubility index was calculated as the ratio of the mass of dried supernatant to the mass of the flour expressed in percentage (g/100 g DW). After centrifugation, the filtrate was also weighed to obtain the swelling power.

2.6. Microstructure Analysis of Green Banana Flour

2.6.1. Scanning Electron Microscopy (SEM)

A scanning electron microscope (SEM-EDX) (JEOL, JSM 7500F) was used to study the microstructures of the banana starch granules. The GBF starch samples were placed on aluminum cylinders that had a double-sided tape followed by coating with carbon. The acceleration voltage was 10.00 kV, as previously described by Maziya et al. [19]. An electron beam with the resolution set at a particle size of 20–200 µm was used to view the microstructure of the samples.

2.6.2. X-Ray Diffraction (XRD)

The XRD analysis of the GBF samples was determined using Philips X'Pert XRD equipment (Malvern PANalytical, Almelo, The Netherlands). The power source was set at 40 kV and 40 mA power with a scanning interval of 5°/min. The scanning range was $2\theta = 5°$ to 90° [19].

2.7. Molecular Structure Analysis of Green Banana Cultivars

2.7.1. Fourier Transform Infrared (ATR-FTIR) Spectroscopy

The ATR-FTIR spectra of GBF samples were measured using a 4000 FTIR spectrophotometer (JASCO, South Africa). The functional groups of the isolated compound were detected by ATR (JASCO, South Africa) with a diamond crystal plate with a scan rate of 16 runs per scan at a resolution of 4 cm^{-1} in wavenumbers from 500 to 4000 cm^{-1} [20].

2.7.2. Determination of Rapidly Digestible, Slow Digestible, Resistant, and Total Starch Contents of Green Banana Flour

The determination of rapidly digestible, slow digestible, resistant, and total starch contents of green banana starch was carried out using a Megazyme Resistant Starch Assay Kit (Megazyme Ltd., Johannesburg, SA). Briefly, the method involved incubating the GBF sample (80 mg) in a mixture of enzymes (pancreatic α-amylase and amyloglucosidase) in maleate buffer (pH 6.0) (K-RNTDF; AOAC Method 2017.16) [21].

2.7.3. Amylose and Amylopectin

A commercial amylopectin/amylose kit (Megazyme Ltd., Johannesburg, South Africa) was used to quantify amylose content. The principle of the method involves the separation of amylopectin and amylose. Thereafter, amylopectin is precipitated with concanavalin-A (Con A), followed by centrifugation to eliminate it Jiang et al. [16].

2.8. Total Phenols, Flavonoids Content and Antioxidant Properties

2.8.1. Total Phenolic Content (TPC)

The TPC was determined using the Folin–Ciocalteu assay method according to the procedure outlined by Blainski et al. [22]. Briefly, one gram of GBF was mixed with a 25 mL mixture of methanol and water (v/v, 20:5, respectively) followed by incubation at 37 °C for 4 h with shaking. The mixture was then centrifuged (4000× g for 10 min). Thereafter, the supernatant was mixed with 500 μL deionized water in a test tube with 30 μL standard/extracts and 50 μL Folin–Ciocalteu reagent (Sigma-Aldrich, Johannesburg, South Africa). This was followed by the addition of 245 μL deionized water and 200 μL of Na_2CO_3. The sample mixture was then incubated at 27 °C for 30 min, and a microplate reader was used to measure the absorbance (750 nm). Gallic acid was used as standard, and the results were expressed as mg gallic acid equivalent (GAE)/g dry weight using the standard curve (R^2 = 0.9982).

2.8.2. Total Flavonoid Content (TFC)

The TFC was determined following the method described by Jabri-Karoui et al. [23]. Briefly, where Quercetin (Sigma-Aldrich, Johannesburg, South Africa) was used as a standard. An aliquot of 30 μL of each extract or standard {Quercetin (Sigma-Aldrich, Johannesburg, South Africa)} was mixed with 20 μL of 10% $AlCl_3$ and 20 μL of 2.5% $NaNO_3$. After 5 min, 100 μL of NaOH solution was added to the mixture. From the mixture using a micropipette, 200 μL was pipetted into a microplate (96 well). A microplate reader was used to measure the absorbance (450 nm). The TPC was expressed as mg QE/mg dry weight using the standard curve (R^2 = 0.9991).

2.8.3. Antioxidant Activity

The antioxidant activity of GBF samples was determined through the use of the ferric-reducing-ability-plasma (FRAP) with slight adjustment as described by Hofmann et al. [24]. The method involved the use of 2,2-diphenyl-1-picrylhydrazyl radical (DPPH) to assay the scavenging activity [25] of green banana flour. Trolox stock was used as standard, and the absorbance was read at 517 nm. The results were expressed as μM Trolox equivalent per 100 g of green banana flour (d.w.).

2.9. Statistical Analysis

One-way analysis of variance (ANOVA) was performed using the Statistica statistical software (Version 13.0/September 2015) for data analysis. The significant difference between the samples was determined at 95% ($p \leq 0.05$). The results were shown as means ± standard deviation. The contrast of mean values was analyzed by Fisher Least Significant Difference (LSD) tests. All the experiments were performed in triplicates.

3. Results and Discussion

3.1. Proximate Composition of Green Banana Flour

The proximate composition of green banana flour (GBF) grown in South Africa is shown in Table 1. The GBF cultivars varied significantly in moisture content, with the Grand Naine and Finger Rose cultivars showing the same and highest (10.50 g/100 g d.w.) moisture content, while the FHIA-01 cultivar showed the lowest (9.40 g/100 g d.w.) moisture content. The moisture content of GBF cultivars reported in the current study is within the range generally reported in the literature for unripe/green banana flour. Similar to the findings of the present study, Kumar et al. [9] recorded 8.59% moisture content on green Grand Naine banana flour. Utrilla-Coello et al. [26] reported 7.03% moisture content for unripe Enano cultivar and 8.96% for unripe Valery banana cultivar. The moisture content of flour products is critical as it can have an influence on both the physical and chemical properties of foods. It can affect the shelf life and stability of foods since high moisture tends to cause changes in chemical, biochemical and textural properties as well as promoting microbial growth [9]. The relatively low moisture content of the GBF of this study suggests that it could be stable and may have an extended shelf life.

Table 1. Proximate analyses of green banana cultivars grown in South Africa.

	Parameters (g/100 g d.w.)				
Samples	**Moisture**	**Ash**	**Fat**	**Protein**	**Carbohydrate**
Grande Naine	10.50 ± 0.71 ^{b}	3.50 ± 0.11 ^{b}	0.52 ± 0.00 ^{b}	3.60 ± 0.69 ^{a}	81.88 ± 1.4 ^{a}
Pisang Awak	9.50 ± 0.33 ^{a}	2.50 ± 0.05 ^{a}	0.85 ± 0.02 ^{e}	4.12 ± 0.48 ^{c}	83.03 ± 0.81 ^{b}
Finger Rose	10.50 ± 0.84 ^{b}	3.43 ± 0.31 ^{b}	0.70 ± 0.71 ^{d}	4.33 ± 0.30 ^{d}	81.04 ± 0.76 ^{a}
FHIA-01	9.40 ± 1.34 ^{a}	3.50 ± 0.22 ^{b}	0.68 ± 0.61 ^{c}	3.63 ± 0.77 ^{a}	84.82 ± 0.90 ^{d}
Du Roi	9.50 ± 0.51 ^{a}	2.46 ± 0.32 ^{a}	0.42 ± 0.51 ^{a}	3.81 ± 0.43 ^{b}	83.81 ± 0.56 ^{c}

Values with different alphabets in a column indicate significant differences ($p < 0.05$). g/100 g—gram per hundred grams; d.w.—dry weight basis. ($n = 3$).

The abundance of minerals in GBF has made it a valuable fruit. In this study, the total ash content statistically ($p \leq 0.05$) varied from 2.46–3.50 g/100 g d.w., with the Grand Naine cultivar recording the highest ash content. Campuzano et al. [27] reported an ash content of 2.61 g/100 g d.w. in GBF (Cavendish), and this is within the range of ash content found in the present study. Elsewhere, Kumar et al. [9] reported ash content of 2.06 and 2.50% for unripe Grand Naine and Nendran flours, respectively, with the former cultivar relatively lower in ash content than that of the current study. The variation in ash content could possibly be an indication of differences in mineral contents of the GBF cultivars, which can be attributed to agricultural practices and climate change [28] Further, the variations could also be linked to the differences in the type of soil under which they were grown. In general, the ash content of food is associated with a high presence of minerals such as calcium, magnesium, potassium and phosphorus [29].

The protein content of the five GBF cultivars significantly differed ($p \leq 0.05$). High protein content occurred with the Finger Rose cultivar (4.33%), while the Grand Naine cultivar had the lowest protein content (3.60%). The Grand Naine protein content in this study was comparable with the protein content (3.53%) for the Grand Naine cultivar reported by Kumar et al. [9] on GBF from dessert and plantain banana (*Musa* spp.). In the study reported by Ferreira et al. [30] a protein Ferreira content of 1.94% was determined for green banana flour. Elsewhere, Ferreira et al. [31] reported similar protein content (1.89%) in unpeeled green banana flour. Bi et al. reported a protein content of 2.90% for Pisang Awak, which is significantly lower than the 4.12% found in the present study. The observed variation in results could be due to differences in the soil type and the stage of growth of the fruit [32].

The Pisang Awak cultivar had the highest fat content (0.85 g/100 g d.w.) with the lowest fat content recorded for the Du Roi cultivar (0.42 g/100 g d.w.) (Table 1). The fat content results reported here are within the range (0.92–0.93 g/100 g d.w.) reported by Khoozani et al. [29] in green Cavendish flour. According to Ye et al. [33] low-fat content reduces the extent of starch granule swelling. The low-fat content of banana flour creates an environment that is not suitable for oxidation reactions, resulting in extended shelf life. In fact, it reportedly decreases the risk of lipid oxidation which may result in extended shelf life. Variations in the chemical composition of banana cultivars are associated with various factors, such as regional climate, agronomic methods, harvesting conditions, among others [34]. However, the differences that occurred in the current study are attributed to cultivar variation since the cultivars were grown under the same environmental condition.

3.2. Mineral Composition of Green Banana Flour

The mineral profiles of GBF used in this study are shown in Table 2. In general, potassium (K) was the most abundant (290.95–1033.25 mg/100 g) mineral, while copper (Cu) was the least abundant (0.25–0.50 mg/100 g) mineral among the five GBF cultivars. FHIA-01 recorded the highest amount of K (1033.25 mg/100 g). The results attained from this study confirm that some banana cultivars (FHIA-01, Grande Naine and Finger Rose, respectively) cultivated in Mpumalanga Province, South Africa, appear to be an excellent source of K. The K level in this study was within the range (9117.32–14,746.73 mg/kg) reported by Tasnim et al. [35] in unripe banana flour cultivars attained from domestic and commercial farms in Limpopo, South Africa. With magnesium, FHIA-01 had the highest concentration (118.15 mg/100 g), while the Finger Rose cultivar had the lowest (82.10 mg/100 g). The research has indicated that the risk of diabetes can be reduced by consuming a high-Mg diet, and this has been associated with the role that Mg plays in glucose metabolism. Phosphorus (P) was within the range of 31.72–99.25 mg/kg for all GBF cultivars. There were significant differences ($p \leq 0.05$) between the content of sulfur (S) amongst the five GBF cultivars. With essential minerals, calcium (Ca) was the least abundant mineral with the lowest concentration observed in the Finger Rose (8.70 mg/100 g) cultivar.

Table 2. Essential minerals in green banana flour from different cultivars (mg/100 g dry weight).

Minerals	Grande Naine	FHIA-01	Finger Rose	Pisang Awak	Du Roi
			Essential macro minerals		
Ca	18.38 ± 0.23 c	10.50 ± 0.33 b	8.70± 0.19 a	28.25 ± 0.34 e	19.68 ± 0.41 d
Mg	100.10 ± 0.17 d	118.15 ± 0.14 e	82.10 ± 0.15 c	35.85 ± 0.38 b	32.4 ± 0.20 a
K	934.7 ± 0.11 d	1033.25 ± 0.15 e	878.95 ± 0.31 c	501.58 ± 0.22 b	290.95 ± 0.35 a
P	99.25 ± 0.40 e	85.43 ± 0.12 d	72.50 ± 0.41 c	38.38 ± 0.21 b	31.72 ± 0.38 a
S	77.23 ± 0.11 d	136.61 ± 0.21 e	66.69 ± 0.51 c	55.55 ± 0.21 a	58.35 ± 0.12 b
			Essential trace minerals		
Zn	0.93 ± 0.21 e	0.53 ± 0.48 b	0.57 ± 0.30 c	0.28 ± 0.33 d	0.18 ± 0.43 a
Cu	0.5 ± 1.40 d	0.43 ± 0.31 c	0.33 ± 0.22 b	0.33 ± 0.30 b	0.25 ± 0.26 a
Fe	2.88 ± 0.21 e	1.50 ± 0.27 c	1.33 ± 0.31 a	2.30 ± 0.44 d	1.48 ± 0.42 b
Mn	3.20 ± 0.41 e	1.23 ± 0.29 d	0.98 ± 0.25 c	0.60 ± 0.21 b	0.48 ± 0.11 a

Values with different alphabets in a column indicate significant differences ($p < 0.05$). d.w.—dry weight basis ($n = 3$).

As expected, the overall concentration of macro-minerals was higher than that of trace minerals. The Grande Naine cultivar recorded the highest (2.88 mg/100 g) concentration of iron (Fe), while Finger Rose recorded the lowest (1.50 mg/100 g). The concentration of Fe and Zn reported in the present study was lower than that reported by Ferreira et al. [36] who reported an Fe concentration of between 8.15–33.72 mg/kg and a Zn concentration of 3.55–7.78 mg/kg for commercial as well as uncommercial unripe banana flour. The Fe concentration of GBF cultivars in this study were higher than that reported by Pessoa et al. [37] (22.5–62.8 mg/kg) in GBF cultivars from Brazil. According to Freeland-Graves et al. [38], food such as beans, bovine liver and seafood are known to be good sources of iron; hence, they are termed "iron-rich foods". Interestingly, the concentration of Fe in GBF cultivars investigated here were similar or higher than that of the aforementioned food products. This means the GBF cultivars could potentially be used as a good source of Fe in foods. The copper (Cu) concentration significantly ($p < 0.05$) varied from 0.25–5.0 mg/100 g, with the Grande Naine cultivar showing the highest concentration. The concentration of manganese (Mn) ranged from 0.48–3.2 mg/100 g, and the highest and lowest concentrations were observed in Grande Naine and Du Rio cultivars, respectively. Interestingly, the mineral composition of the soil has been reported to influence the mineral content of food crops [38]. Further, the pH as well as the amount of organic matter in the soil may also influence the mineral content. Other studies have reported similar influence with different agricultural practices and climate change [36,39].

3.3. Characterization of Functional Properties Banana Flours

3.3.1. Water Absorption Capacity (WAC) of Green Banana Flour Cultivars

The WAC indicates the volume occupied by starch granules after swelling in excess of water [34]. It is affected by how much native starch granules have disintegrated. In addition, it is influenced by the physical state of starch, dietary fiber and proteins [40]. In this study, the WAC varied with the type of cultivar (Table 3). Pisang Awak had the highest WAC (67.11%), while the least WAC (40.00%) was observed with the Finger Rose cultivar. Campuzano et al. [27] reported an almost similar WAC range (48.50–70.00%) to that of the current study in GBF at different stages of ripeness. Pereira et al. [40] reported 80.00% WAC for green banana flour. The hydrophilic sites in starch chains allow for interaction with water through hydrogen bonding. The high WAC observed with Pisang Awak GBF here suggests that it could be suitable for baking. It must also be noted that WAC influences gelatinization through available water, and thus, a lower WAC is desirable for a thinner consistency [29].

Table 3. The water absorption capacity of green banana flour cultivars.

Banana Flour Samples	Water Absorption Capacity (%)
FHIA-01	58.01 ± 0.31^{d}
Grande Naine	43.18 ± 0.10^{b}
Pisang Awak	67.11 ± 0.00^{e}
Finger Rose	40.00 ± 0.58^{a}
Du Roi	50.12 ± 0.69^{c}

Values with different alphabets in a column indicate significant differences ($p < 0.05$). The water absorption capacity was expressed on a dry weight basis (d.w.). ($n = 3$)

3.3.2. The Water Solubility and Swelling Power of Green Banana Flour Cultivars

The solubility and swelling power are parameters used to investigate the quality of starch granules. The solubility index is linked to the soluble solid contents in flour, whereas swelling power is a measure of the retention of starch granule integrity when subjected to high cooking temperatures [35,41]. Here, swelling patterns differed amongst the GBF cultivars. The swelling power increased with an increase in temperature with all cultivars (Table 4). Flour from the Du Roi cultivar had the highest swelling power (0.83 g/g) at 90 °C, while the FHIA-01 cultivar had the lowest swelling power (0.52 g/g) compared to all the other cultivars at the same temperature. The swelling of starch granules follows different stages. First, thermal energy is attained with heating, and this helps to loosen the intra-granular links of starch granules. When the temperature exceeds 70–80 °C, more rapid swelling of starch granules occurs possibly due to intermolecular hydrogen bonds breaking in the amorphous area [42,43]. The current results suggest that swelling of starch granules and high water penetration are attained at high temperatures for the investigated GBF cultivars. The solubility index and swelling power denote the range of interaction within the crystalline (amylose) and amorphous (amylopectin) regions of the starch molecule, along with the degree of branching and the length of branches [29]. Therefore, an increase in solubility index and swelling power cause gelatinization, which is the foundation for making pre-gelatinized starch. According to Khoozani et al. [29] significant differences in swelling may be attributed to low solubility, restricted swelling, the amylose content of flour and slight retrogradation (a reaction that takes place in gelatinized starch, when disaggregated amylopectin and amylose chains reassociate to form more ordered structures). Comparably, low solubility coupled with low swelling power indicates a more well-arranged, denser and strongly bonded granule structure. Another factor that can be attributed to the differences in swelling and solubility indices in the present study could be differences in the starch granule crystallinity. Viscosity patterns, the weak internal organization of starch, can also contribute to variations in solubility and swelling power of flour. The way in which amylose and amylopectin are distributed in the starch granule is thought to be another factor that greatly impacts the solubility index [44].

Table 4. The water solubility and swelling index of banana flour (d.w.).

Banana Flour Samples	Solubility (%)			Swelling Power (g/g)		
	50 °C	70 °C	90 °C	50 °C	70 °C	90 °C
FHIA-01	6.49 ± 0.73 [b]	9.50 ± 0.71 [d]	15.01 ± 0.71 [d]	0.29 ± 0.71 [a]	0.42 ± 0.71 [a]	0.52 ± 0.95 [a]
Grande Naine	7.40 ± 0.00 [c]	9.61 ± 0.34 [d]	10.21 ± 0.59 [b]	0.50 ± 0.19 [d]	0.67 ± 0.71 [b]	0.75 ± 0.00 [b]
Pisang Awak	6.50 ± 0.32 [b]	8.47 ± 0.58 [b]	11.4 ± 0.58 [c]	0.33 ± 0.21 [b]	0.38± 0.44 [a]	0.53 ± 0.01 [a]
Finger Rose	7.0 ± 0.08 [c]	9.01 ± 0.34 [c]	10.21 ± 0.59 [b]	0.41± 0.79 [c]	0.67 ± 0.71 [b]	0.79 ± 0.04 [b]
Du Roi	5.50 ± 0.71 [a]	7.59 ± 0.06 [a]	8.03 ± 0.53 [a]	0.38 ± 0.24 [c]	0.63 ± 0.27 [b]	0.83 ± 0.54 [c]

Values with different alphabets in a column indicate significant differences ($p < 0.05$). The results are expressed on a dry weight basis (d.w.). g/g—gram per gram. ($n = 3$).

3.4. Microstructure Analysis of Green Banana Flour

3.4.1. Scanning Electron Microscopy (SEM) of Starch Isolated from Green Banana Flour

The SEM images of starch isolated from GBF are shown in Figure 1. SEM is used to study the surface morphology, structural integrity, as well as determinations of the size and shape of starch granules. In the present study, SEM revealed that starch granules from GBF were irregular in shape, and they had dense surfaces that had debris. The starch granules diameter ranged from 4.5 µm (Finger Rose) to 21.67 µm (FHIA-01). According to Reyes-Atrizco et al. [45] banana starch granules can vary from 4–35 µm in size, and this is in line with the size of GBF starch granules from this study. Finger Rose and Grand Naine exhibited longer, oval-shaped granules which had fragments on their surfaces. Du Roi granules were intact and elliptical in shape, while Pisang Awak and Finger Rose GBF had smaller and compact granules. The observed variation in GBF starch granules can affect the thermal property and swelling power. In the present study, samples with a bigger granule size had a higher water holding capacity. From the SEM images, it can be observed that the Finger Rose cultivar has the smallest starch granules compared to the other flours. Additionally, FHIA-01, Pisang Awak, Du Roi and Grande Naine had higher water holding capacities compared to Finger Rose, respectively. The SEM images of green banana flour show that there is a correlation between the flour morphology and water holding capacity.

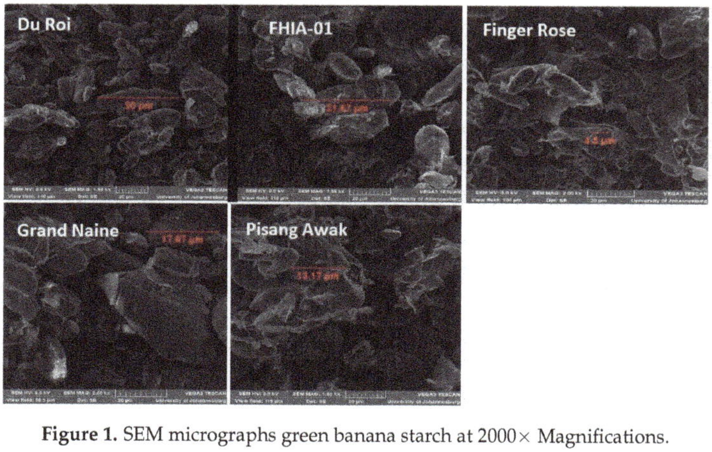

Figure 1. SEM micrographs green banana starch at 2000× Magnifications.

According to Pandey et al. [46] the fragments that can be seen on the surface of the granules are probably amyloplast membranes which enclose starch granules in the banana fruit cell. The findings of the present study are akin to those by Reyes-Atrizco et al. [45]

who reported that banana flour starch granules appear to be irregularly shaped, elongated and flattened, while the small granules are compact with spheroids and elongated forms.

3.4.2. X-Ray Diffraction of Green Banana Flour

There is a semi-crystalline nature of starch particles that can be assessed by XRD. In this study, the crystalline structure of green banana starch granules was analyzed using XRD, as shown in Figure 2. All GBF cultivars studied had similar diffraction patterns with prominent peaks at 15.00°, 18° and 24.00° diffraction angles. The GBF starch granules exhibited XRD patterns with three distinct peaks that were observed as a small peak at 15.00°, strong peak at 18.00°and a broad peak at 24. 00°. Generally, starch granules that originate from different sources appear to have varying crystallization characteristics. The three types of patterns that are displayed by starch are the A pattern (cereal starch), B (tuber, amylo-maize, and retrograded starch) pattern and C pattern (root and seed starches-pea and bean) [46–48]. In line with previous reports, the XRD pattern of green banana starch depicts the B-type crystallinity pattern irrespective of the variety and starch source [48].

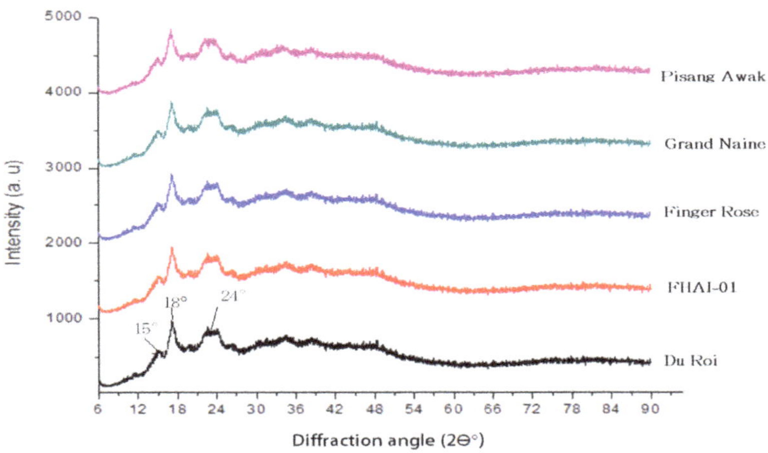

Figure 2. X-ray diffraction patterns of Pisang Awak, Grand Naine, Finger Rose, FHIA-01 and Du Roi.

3.4.3. Fourier Transform Infrared Spectroscopy (FTIR)

The FTIR analysis was performed to identify various characteristic functional groups present in the Grande Naine, Finger Rose, Du Roi and Pisang Awak green banana flours as shown in Figure 3. The identification of different functional groups in this study was conducted following the band/group assignments provided in the appendices section. For the Grand Naine, the absorption bands centered around 1643.05 cm^{-1} and 1002.80 cm^{-1} show the occurrence of hydroxyl (–OH), amine groups (–NH) and carbonyl group (=C=O) bonds, respectively [9]. The characteristic absorption bands of Grande Naine were similar to that of previous reports [9,47,49]. In general, the absorption bands between 800–1600 cm^{-1} are defined as the fingerprint region [49] There were characteristic bands of Finger Rose at 1002.80 cm^{-1}, bands between 990 cm^{-1} and 1160 cm^{-1}, attributed to carbonyl group (=C=O) bonds stretching. These compounds may contribute to the characteristic flavor and order of the banana flour. For all the analyzed banana flours, bands in similar regions were observed. Similar results have been reported elsewhere in the literature [20].

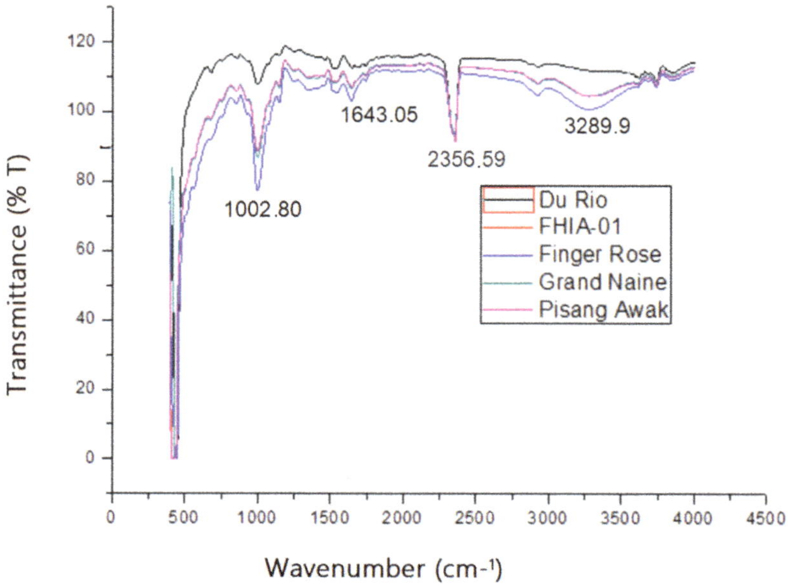

Figure 3. Comparative plots of the FTIR spectra of Du Roi, FHIA–01, Finger Rose, Grand Naine and Pisang Awak.

3.4.4. In vitro Starch Digestion and Amylose Content of Green Banana Flour

Since humans generally consume cooked starch more than raw starch, the digestion performance of cooked banana starch is more important to the food industry. The digestibility fractions of green banana starch are indicated in Table 5. The GBF varied significantly ($p \leq 0.05$) in their RDS, SDS and RS. Du Roi had the lowest RDS of 4.46%, while Grande Naine (6.02%) had the highest amount of RDS. The SDS ranged from 10.17% (FHIA-01) to 14.87% (Finger Rose). FHIA-01 had the highest amount of resistant starch (RS) (86.50%), while Grande Naine had the lowest amount of RS (80.38%). It is widely acknowledged that the GI and RS contents are two significant indicators of starch digestibility [12,50] These findings are an indication that GBF is a source of high RS, which could be linked with a lower GI. This suggests that diets that include GBF can positively influence blood glucose control and can potentially manage diabetes in patients. It is, however, worth noting that the GI of GBF may vary based on protein content, fat content, particle size and maturity and ripeness of the fruit [50]. In a study by Soto-Maldonado et al. [13] on the GI of whole banana and overripe banana pulp, it was observed that extended maturation resulted in an increase in GI, possibly due to a decrease in starch content. In the present study, the RS constituted the highest fraction in the green banana starch. These results are in agreement with the results previously reported for native banana starch (88.7%) and native plantain starch (85%) by Reyes-Atrizco et al. [45]. Recently, the health benefits of RS have been reported to be similar to those of dietary fiber when considering factors such as maintenance of gut homeostasis and promotion of the growth of beneficial gut microflora [50]. Thus, the GBF cultivars in this study may also be used in food applications as pre-biotics. Furthermore, RS is believed to control the amount of glucose released from starchy food, thus lowering the risk of obesity. Since starch is the most available carbohydrate in GBF, it must be noted that carbohydrates in food can influence processing characteristics and the development of designer foods [12].

Table 5. Rapidly digestible starch (RDS), slowly digestible starch (SDS), resistant starch (RS) and amylose content green banana flour.

Samples	RDS (%)	SDS (%)	RS (%)	Amylose (%)
Grande Naine	6.02 ± 0.11 [c]	13.30 ± 0.00 [d]	80.38 ± 1.41 [a]	18.95 ± 0.98 [b]
Pisang Awak	5.50 ± 0.05 [b]	11.73 ± 0.02 [c]	84.35 ± 1.51 [c]	23.00 ± 0.91 [d]
Finger Rose	5.43 ± 0.31 [b]	14.87 ± 0.01 [e]	81.70 ± 1.21 [b]	15.55 ± 0.90 [a]
FHIA-01	4.50 ± 0.22 [a]	10.17 ± 0.61 [a]	86.50 ± 0.21 [e]	24.82 ± 0.00 [e]
Du Roi	4.46 ± 0.82 [a]	10.42 ± 0.51 [b]	85.50 ± 0.40 [d]	21.32 ± 0.16 [c]

Values with different alphabets in a column indicate significant differences ($p < 0.05$). ($n = 3$).

There are several properties that are affected by how amylose and amylopectin are arranged in GBF. These include gelatinization, retrogradation as well as digestibility [25,51]. Here, a statistically significant ($p < 0.05$) variation in amylose content was observed, with the FHIA-01 cultivar showing the highest amylose content (24.82%) and Finger Rose recording the lowest amylose content (15.55%). It was also noted that the flours with high amylose content appeared to have high RS content. Thus, we propose that the amylose content could somewhat be positively correlated to the resistant starch. Flour with a higher amylose content is known to have a high solubility index since the amorphous region of starch granules primarily contains high amylose content [52,53]. Previous studies [54–56] suggest that a high amylose content of foods generally tends to give rise to a lower GI. The aforementioned was observed in the present study, as FHIA-01 had the highest amylose content and solubility index in comparison to the other studied GBF.

3.5. Phenolic Content and Antioxidant Activity of Green Banana Flour

3.5.1. Total Phenolic Content and Total Flavonoid Content

The TPC of GBF cultivars studies here significantly ($p < 0.05$) varied (Table 4). Grand Naine had the highest TPC (524.87 mg GAE/100 g), while Du Roi had the lowest TPC (298.73 mg GAE/100 g). Phenolic compounds are essential secondary metabolites that are relatively high in bananas when compared to other fruits [57] They have been associated with health benefits that include the prevention of several diseases, such as diabetes, obesity and cardiovascular disease. The TPC content of Grand Naine in this study was six times higher than that reported by Anyasi and Mchau [49], possibly because of differences in the stage of ripening, the growth conditions and agricultural practices. Moreover, the extent of maturity has been reported to substantially affect the total phenolic content in green banana flour [1,24,56]. Passo et al. [58] reported that over-ripened banana flour had 52% less phenolic content than GBF, while ripe banana flour had 15–45% less phenolic content than GBF. Banana flour contains phenolic compounds such as catecholamines, phenolic acids and flavonoids [57,59]. Turola et al. [60] also reported the presence of gallic acid, catechin, epicatechin and myricetin3-O-rhamnosyl-glucoside in ripe and unripe banana flour cultivars. Furthermore, phenolic compounds can be used as food additives in the food industry to prevent lipid oxidation reactions in food formulations. With its high TPC, the Grand Naine cultivar has the potential for being used as a raw material in functional foods. The variation in total phenolic content observed in the present study may be attributed to genetic differences amongst the different banana flours. Bananas contain phenolic compounds such as catecholamines, phenolic acids and flavonoids [49]. The variation in total phenolic content observed in the present results may be attributed to genetic differences amongst different banana flours. Grand Naine recorded higher TFC (407.08 mg QE/100 g) among the flours studied, while Du Roi cultivar had the lowest TFC (287.40 mg QE/100 g). According to Hofmann et al. [24], green banana is abundant in TPC and contains various flavonoids.

3.5.2. Antioxidant Activity (AOA) of Green Banana Flour

The antioxidant capacity of GBF is shown in Table 6. Among the cultivars, Grand Naine recorded the highest antioxidant activity (437. 22 and 474.23 mg TE/100 g d.w.) by both DPPH and FRAP assays, respectively. The second highest antioxidant activity was recorded with the Finger Rose cultivar and again with both assays. From the results obtained in this study, it is worth noting that cultivars with high TPC showed high antioxidant capacity. Therefore, the finding that Grand Naine was the richest in antioxidant activity was credited to its relative great quantity of phenolic compounds. Previous studies have shown that food with high antioxidants (e.g., carotenoids) can improve immunity in humans. Such a benefit has been linked to the reduction in the occurrence of diseases such as cancer, cardiovascular diseases and diabetes [9]. The health benefits associated with antioxidants are believed to be due to the vital role they play in impeding the initial stages of lipid peroxidation and scavenging singlet oxygen [48]. According to Turola Barbi et al. [61] there is a relationship between DPPH inhibition for plant materials and the TPC and TFC. This was observed through increases in DPPH that occur with an increase in the concentration of phenolic compounds or degree of hydroxylation of the phenolic compounds. Such is consistent with the fact that the antioxidant activity in plants is greatly associated with the phenolic fraction. Although not investigated here, it must be noted that different structures within the same plant contain different concentrations of phenolic compounds [45,55,62]. The FRAP assay is commonly used to study the antioxidant capacity of plant materials. In this study, all GBF samples showed a high correlation between the FRAP value and DPPH value. This can be attributed to the fact that both DPPH and FRAP assays generally follow the same mechanism [49]. High correlations between different antioxidant activity methods have also been reported by other researchers [56,57]. The above indicates that banana fruits with high antioxidant capacity could have a high value for their potential health-promoting benefits.

Table 6. Total flavonoid content (TFC), total phenolic content (TPC) of green banana flour.

Samples	TPC (mg GAE/100 g d.w.)	TFC (mg QE/100 g d.w.)	DPPH (mg TE/100 g d.w.)	FRAP (mg TE/100 g d.w.)
FHIA-01	307.03 ± 0.7 ^{b}	293.87 ± 0.91 ^{b}	359.11 ± 0.7 ^{b}	411.72 ± 0.7 ^{b}
Grande Naine	524.87 ± 1.6 ^{d}	407.08 ± 1.7 ^{d}	437. 22 ± 1.0 ^{c}	474.23 ± 0.2 ^{c}
Pisang Awak	312.00 ± 2.1 ^{b}	291.80 ± 0.9 ^{b}	363. 28 ± 0.4 ^{b}	397. 11 ± 1.1 ^{b}
Finger rose	321.87 ± 0.1 ^{c}	305.01 ± 1.0 ^{c}	421.00 ± 0.1 ^{c}	448.87 ± 1.3 ^{c}
Du Roi	298.73 ± 1.1 ^{a}	287.40 ± 2.1 ^{a}	301.34 ± 1.1 ^{a}	324.27 ± 0.1 ^{a}

Values with different alphabets in a column indicate significant differences ($p < 0.05$). d.w.–dry weight. ($n = 3$).

4. Conclusions

The varying functional and physicochemical properties of the GBF cultivars suggest that they can be utilized as raw materials in different food products. The morphological characteristics of the GBF starch and the fact that it appears to be linked to high WAC suggest their possible application in the development of edible films. The GBF were all found to contain relatively high RS, which makes them suitable for the development of low-GI food products. The high TPC, TFC and antioxidant activity in the Grand Naine cultivar suggests its possible use in health-promoting food products. Green banana flour cultivars, such as Pisang Awak, FHIA and Du Roi, indicate that they could be stable when added to food that is processed at a high temperature due to their high amylose content. While, on the other hand, Grand Naine and Finger Rose have low amylose content, which suggests a possible use in food products with low thermal characteristics.

Author Contributions: Conceptualization, M.K., E.K. and B.C.D.; methodology, M.K.; software, M.K.; validation, E.K., B.C.D. and M.K.; formal analysis, M.K.; investigation, M.K.; resources, B.C.D.; data curation, M.K.; writing—original draft preparation, M.K.; writing—review and editing,

E.K. and B.C.D.; visualization, E.K.; supervision, B.C.D. and E.K.; project administration, B.C.D.; funding acquisition, E.K. and B.C.D. All authors have read and agreed to the published version of the manuscript.

Funding: This research was funded by National Research Foundation South Africa (TTK170508230032) and Centre of Excellence–Smart Food Project.

Institutional Review Board Statement: Not applicable.

Informed Consent Statement: Not applicable.

Data Availability Statement: Not applicable.

Acknowledgments: The University of Johannesburg's postgraduate funding supported the research.

Conflicts of Interest: The authors declare no conflict of interest.

References

1. Rebello, L.P.G.; Ramos, A.M.; Pertuzatti, P.B.; Barcia, M.T.; Castillo-Muñoz, N.; Hermosín-Gutiérrez, I. Flour of banana (Musa AAA) peel as a source of antioxidant phenolic compounds. *Food Res. Int.* **2014**, *55*, 397–403. [CrossRef]
2. Wild, S.; Roglic, G.; Green, A.; Sicree, R.; King, H. Global Prevalence of Diabetes: Estimates for the Year 2000 and Projections for 2030. 2004. Available online: http://care.diabetesjournals (accessed on 9 October 2021).
3. Macaulay, S.; Ngobeni, M.; Dunger, D.B.; Norris, S.A. The prevalence of gestational diabetes mellitus amongst black South African women is a public health concern. *Diabetes Res. Clin. Pract.* **2018**, *139*, 278–287. [CrossRef]
4. Okaiyeto, K.; Oguntibeju, O.O. Trends in diabetes research outputs in South Africa over 30 years from 2010 to 2019: A biblio-metric analysis. *Saudi J. Biol. Sci.* **2021**, *28*, 2914–2924. [CrossRef]
5. Diabetes in South Africa: Assessing the Data with Fear and Trembling. Available online: http://www.statssa.gov.za/publications/P0302/P03022015.pdf (accessed on 9 October 2021).
6. Ho, L.H.; Abdul Aziz, N.A.; Azahari, B. Physico-chemical characteristics and sensory evaluation of wheat bread partially sub-stituted with banana (Musa acuminata X balbisiana cv. Awak) pseudo-stem flour. *Food Chem.* **2013**, *139*, 532–539. [CrossRef] [PubMed]
7. Naknaen, P.; Charoenthaikij, P.; Kerdsup, P. Physicochemical Properties and Nutritional Compositions of Foamed Banana Powders (Pisang Awak, *Musa sapientum* L.) Dehydrated by Various Drying Methods. *Walailak J. Sci. Tech.* **2016**, *13*, 177–191. Available online: http://wjst.wu.ac.th (accessed on 9 October 2021).
8. Martínez, P.; Peña, F.; Bello-Pérez, L.A.; Núñez-Santiago, C.; Yee-Madeira, H.; Velezmoro, C. Physicochemical, functional and morphological characterization of starches isolated from three native potatoes of the Andean region. *Food Chem.* **2019**, *2*, 100030. [CrossRef]
9. Kumar, P.S.; Saravanan, A.; Sheeba, N.; Uma, S. Structural, functional characterization and physicochemical properties of green banana flour from dessert and plantain bananas (Musa spp.). *LWT* **2019**, *116*, 108524. [CrossRef]
10. Borges, C.V.; Maraschin, M.; Coelho, D.S.; Leonel, M.; Gomez, H.A.G.; Belin, M.A.F.; Diamante, M.S.; Amorim, E.P.; Gianeti, T.; Castro, G.R.; et al. Nutritional value and antioxidant com-pounds during the ripening and after domestic cooking of bananas and plantains. *Food Res. Int.* **2020**, *132*, 109061. [CrossRef]
11. Artavia, G.; Cortés-Herrera, C.; Granados-Chinchilla, F. Total and resistant starch from foodstuff for animal and human consumption in Costa Rica. *Curr. Res. Food Sci.* **2020**, *3*, 275–283. [CrossRef]
12. Lal, M.K.; Singh, B.; Sharma, S.; Singh, M.P.; Kumar, A. Glycemic index of starchy crops and factors affecting its digestibility: A review. *Trends Food Sci. Technol.* **2021**, *111*, 741–755. [CrossRef]
13. Zhang, L.; Xing, R.; Huang, Z.; Zhang, N.; Li, X.; Wang, P. Inhibition of Synovial Macrophage Pyroptosis Alleviates Synovitis and Fibrosis in Knee Osteoarthritis. *Mediat. Inflamm.* **2019**, *2019*, 108524. [CrossRef]
14. Kumar, S.R.; Sadiq, M.B.; Anal, A.K. Comparative study of physicochemical and functional properties of pan and microwave cooked underutilized millets (proso and little). *LWT* **2020**, *128*, 109465. [CrossRef]
15. Rai, S.K.; Chaturvedi, K.; Yadav, S.K. Evaluation of structural integrity and functionality of commercial pectin based edible films incorporated with corn flour, beetroot, orange peel, muesli and rice flour. *Food Hydrocoll.* **2019**, *91*, 127–135.
16. Jiang, H.; Zhang, Y.; Hong, Y.; Bi, Y.; Gu, Z.; Cheng, L.; Li, Z.; Li, C. Digestibility and changes to structural characteristics of green banana starch during invitro digestion. *Food Hydrocoll.* **2015**, *49*, 192–199. [CrossRef]
17. Rodriguez-Jimenez, J.R.; Amaya-Guerra, C.A.; Baez-Gonzalez, J.G.; Aguilera-Gonzalez, C.; Urias-Orona, V.; Nino-Medina, G. Physi-cochemical, functional, and nutraceutical properties of eggplant flours obtained by different drying methods. *Molecules* **2018**, *23*, 3210. [CrossRef]
18. Jakavula, S.; Biata, N.R.; Dimpe, K.M.; Pakade, V.E.; Nomngongo, P.N. Multi-ion imprinted polymers (MIIPs) for simultaneous ex-traction and preconcentration of Sb(III), Te(IV), Pb(II) and Cd(II) ions from drinking water sources. *J. Hazard. Mater.* **2021**, *416*, 126175. Available online: https://linkinghub.elsevier.com/retrieve/pii/S0304389421011390 (accessed on 9 October 2021). [CrossRef] [PubMed]

19. Maziya, K.; Dlamini, B.C.; Malinga, S.P. Hyperbranched polymer nanofibrous membrane grafted with silver nanoparticles for dual antifouling and antibacterial properties against Escherichia coli, Staphylococcus aureus and Pseudomonas aeruginosa. *React. Funct. Polym.* **2020**, *148*, 104494. [CrossRef]
20. Timilsena, Y.P.; Vongsvivut, J.; Tobin, M.J.; Adhikari, R.; Barrow, C.; Adhikari, B. Investigation of oil distribution in spray-dried chia seed oil microcapsules using synchrotron-FTIR microspectroscopy. *Food Chem.* **2018**, *275*, 457–466. [CrossRef]
21. Rapid Integrated Total Dietary Fiber Assay Procedure [Internet]. 2020. Available online: www.megazyme.com (accessed on 9 October 2021).
22. Blainski, A.; Lopes, G.C.; De Mello, J.C.P. Application and Analysis of the Folin Ciocalteu Method for the Determination of the Total Phenolic Content from *Limonium Brasiliense* L. *Molecules* **2013**, *18*, 6852–6865. [CrossRef] [PubMed]
23. Jabri-Karoui, I.; Marzouk, B. Bioactive Compounds, Antioxidant Activities and Heat Stability of Corn Oil Enriched with Tunisian Citrus aurantium L. Peel Extract. *J. Am. Oil Chem. Soc.* **2014**, *91*, 1367–1375. [CrossRef]
24. Hofmann, T.; Visi-Rajczi, E.; Albert, L. Antioxidant properties assessment of the cones of conifers through the combined evalua-tion of multiple antioxidant assays. *Ind. Crop. Prod.* **2020**, *145*, 111935. [CrossRef]
25. Someya, S.; Yoshiki, Y.; Okubo, K. Antioxidant Compounds from Bananas (Musa Cavendish). Available online: www.elsevier.com/locate/foodchem (accessed on 9 October 2021).
26. Utrilla-Coello, R.; Rodríguez-Huezo, M.; Carrillo-Navas, H.; Hernández-Jaimes, C.; Vernon-Carter, E.; Alvarez-Ramirez, J. In vitro digestibility, physicochemical, thermal and rheological properties of banana starches. *Carbohydr. Polym.* **2014**, *101*, 154–162. [CrossRef]
27. Campuzano, A.; Rosell, C.M.; Cornejo, F. Physicochemical and nutritional characteristics of banana flour during ripening. *Food Chem.* **2018**, *256*, 11–17. [CrossRef] [PubMed]
28. Herrera-Agudelo, M.A.; Miró, M.; Arruda, M.A. In vitro oral bioaccessibility and total content of Cu, Fe, Mn and Zn from transgenic (through cp4 EPSPS gene) and nontransgenic precursor/successor soybean seeds. *Food Chem.* **2017**, *225*, 125–131. [CrossRef]
29. Khoozani, A.A.; Kebede, B.; Birch, J.; Bekhit, A.E.-D.A. The Effect of Bread Fortification with Whole Green Banana Flour on Its Physicochemical, Nutritional and In Vitro Digestibility. *Foods* **2020**, *9*, 152. [CrossRef]
30. Do Prado Ferreira, M.; Tarley, C.R.T. Bioaccessibility estimation of metallic macro and micronutrients Ca, Mg, Zn, Fe, Cu and Mn in flours of oat and passion fruit peel. *LWT* **2021**, *150*, 111880. Available online: https://linkinghub.elsevier.com/retrieve/pii/S0023643821010331 (accessed on 9 October 2021). [CrossRef]
31. Ferreira, D.C.; Molina, G.; Pelissari, F. Biodegradable trays based on cassava starch blended with agroindustrial residues. *Compos. Part B Eng.* **2020**, *183*. [CrossRef]
32. Bezerra, C.V.; Amante, E.R.; de Oliveira, D.C.; Rodrigues, A.M.C.; da Silva, L.H.M. Green banana (Musa cavendishii) flour obtained in spouted bed—Effect of drying on physico-chemical, functional and morphological characteristics of the starch. *Ind. Crop. Prod.* **2013**, *41*, 241–249. [CrossRef]
33. Ye, J.; Hu, X.; Luo, S.; McClements, D.J.; Liang, L.; Liu, C. Effect of endogenous proteins and lipids on starch digestibility in rice flour. *Food Res. Int.* **2018**, *106*, 404–409. [CrossRef]
34. Amini Khoozani, A.; Birch, J.; Bekhit, A.E.D.A. Production, application and health effects of banana pulp and peel flour in the food industry. *J. Food Sci. Technol.* **2019**, *56*, 548–559. [CrossRef] [PubMed]
35. Tasnim, T.; Das, P.C.; Begum, A.A.; Nupur, A.H.; Mazumder, M.d.A.R. Nutritional, textural and sensory quality of plain cake en-riched with rice rinsed water treated banana blossom flour. *J. Agric. Food Res.* **2020**, *2*, 100071. [CrossRef]
36. Do Prado Ferreira, M.; Tarley, C.R.T. Assessment of in vitro bioacessibility of macrominerals and trace elements in green banana flour. *J. Food Compos. Anal.* **2020**, *92*, 103586. [CrossRef]
37. Brito, A.L.B.; Oliveira, A.F.; Nogueira, A.R.A.; Pessoa, A.G.; Pontes, L.F. Determination of inorganic constituents and physico-chemical characterization of functional flour samples. *Microchem. J.* **2017**, *132*, 112–118. [CrossRef]
38. Freeland-Graves, J.H.; Sachdev, P.K.; Binderberger, A.Z.; Sosanya, M.E. Global diversity of dietary intakes and standards for zinc, iron, and copper. *J. Trace Elem. Med. Biol.* **2020**, *61*, 126515. [CrossRef]
39. Savlak, N.; Türker, B.; Yeşilkanat, N. Effects of particle size distribution on some physical, chemical and functional properties of unripe banana flour. *Food Chem.* **2016**, *213*, 180–186. [CrossRef] [PubMed]
40. Pereira, J.; Malairaj, S.; Brohi, S.A.; Boateng, E.F.; Zhang, W. Impact of unripe banana flour on water states, rheological behaviour and structural properties of myofibrillar protein composite gel. *LWT* **2020**, *125*, 109276. [CrossRef]
41. Ahmed, J.; Thomas, L.; Khashawi, R. Influence of hot-air drying and freeze-drying on functional, rheological, structural and dielectric properties of green banana flour and dispersions. *Food Hydrocoll.* **2019**, *99*, 105331. [CrossRef]
42. Salazar, D.; Arancibia, M.; Lalaleo, D.; Rodríguez-Maecker, R.; López-Caballero, M.E.; Montero, M.P. PHYSICO-CHEMICAL prop-erties and filmogenic aptitude for Edible packaging of Ecuadorian Discard green banana flours (Musa acuminanta AAA). *Food Hydrocoll.* **2021**, *122*, 107048. Available online: https://linkinghub.elsevier.com/retrieve/pii/S0268005X21004641 (accessed on 9 October 2021). [CrossRef]
43. Ojha, S.; Bußler, S.; Schlüter, O.K. Food waste valorisation and circular economy concepts in insect production and processing. *Waste Manag.* **2020**, *118*, 600–609. [CrossRef] [PubMed]
44. Cui, F.; Zi, H.; Liu, H.; Zhang, S.; Yuan, B. A study of starch-urea-water mixtures with a combination of molecular dynamics sim-ulation and traditional characterization methods. *Int. J. Biol. Macromol.* **2020**, *148*, 121–128. [CrossRef]

45. Reyes-Atrizco, J.N.; Agama-Acevedo, E.; Bello-Perez, L.A.; Alvarez-Ramirez, J. Morphological, molecular evolution an in vitro digestibility of filamentous granules of banana starch during fruit development. *Int. J. Biol. Macromol.* **2019**, *132*, 119–125. [CrossRef]
46. Pandey, S.; Kumar, A.; Rao, P.S. Optimization, modeling, and characterization study for the physicochemical properties of raw banana and defatted soy composite extrudates. *Food Chem.* **2020**, *339*, 127865. [CrossRef] [PubMed]
47. Rayo, L.M.; e Carvalho, L.C.; Sardá, F.A.; Dacanal, G.; Menezes, E.W.; Tadini, C.C. Production of instant green banana flour (Musa cavendischii, var. Nanicão) by a pulsed-fluidized bed agglomeration. *LWT* **2015**, *63*, 461–469. [CrossRef]
48. Mahloko, L.M.; Silungwe, H.; Mashau, M.E.; Kgatla, T.E. Bioactive compounds, antioxidant activity and physical characteristics of wheat-prickly pear and banana biscuits. *Heliyon* **2019**, *5*, e02479. [CrossRef]
49. Anyasi, T.; Jideani, A.I.O.; Mchau, G.R. Effect of organic acid pretreatment on some physical, functional and antioxidant properties of flour obtained from three unripe banana cultivars. *Food Chem.* **2014**, *172*, 515–522. [CrossRef]
50. Chakravarty, A.; Tandon, M.; Attri, S.; Sharma, D.; Raigond, P.; Goel, G. Structural characteristics and prebiotic activities of re-sistant starch from Solanum tuberosum: Kufri Bahar, a popular Indian tuber variety. *LWT* **2021**, *145*, 111445. Available online: https://linkinghub.elsevier.com/retrieve/pii/S0023643821005983 (accessed on 9 October 2021). [CrossRef]
51. Zheng, Y.; Tian, J.; Kong, X.; Yang, W.; Yin, X.; Xu, E.; Chen, S.; Liu, D.; Ye, X. Physicochemical and digestibility characterisation of maize starch–caffeic acid complexes. *LWT* **2019**, *121*, 108857. [CrossRef]
52. Lawal, O.S.; Lapasin, R.; Bellich, B.; Olayiwola, T.O.; Cesàro, A.; Yoshimura, M.; Nishinari, K. Rheology and functional properties of starches isolated from five improved rice varieties from West Africa. *Food Hydrocoll.* **2011**, *25*, 1785–1792. [CrossRef]
53. Alimi, B.A.; Workneh, T.S. Structural and physicochemical properties of heat moisture treated and citric acid modified acha and iburu starches. *Food Hydrocoll.* **2018**, *81*, 449–455. [CrossRef]
54. Navghare, V.V.; Dhawale, S.C. In vitro antioxidant, hypoglycemic and oral glucose tolerance test of banana peels. *Alex. J. Med.* **2017**, *53*, 237–243. [CrossRef]
55. Vu, H.T.; Scarlett, C.J.; Vuong, Q.V. Phenolic compounds within banana peel and their potential uses: A review. *J. Funct. Foods* **2018**, *40*, 238–248. [CrossRef]
56. Passo Tsamo, C.V.; Herent, M.F.; Tomekpe, K.; Happi Emaga, T.; Quetin-Leclercq, J.; Rogez, H.; Larondelle, Y.; Andre, C. Phenolic profiling in the pulp and peel of nine plantain cultivars (*Musa* sp.). *Food Chem.* **2015**, *167*, 197–204. [CrossRef] [PubMed]
57. Anyasi, T.A.; Jideani, A.I.; Mchau, G.R. Phenolics and essential mineral profile of organic acid pretreated unripe banana flour. *Food Res. Int.* **2017**, *104*, 100–109. [CrossRef] [PubMed]
58. Aydogdu, A.; Yildiz, E.; Aydogdu, Y.; Sumnu, G.; Sahin, S.; Ayhan, Z. Enhancing oxidative stability of walnuts by using gallic acid loaded lentil flour based electrospun nanofibers as active packaging material. *Food Hydrocoll.* **2019**, *95*, 245–255. [CrossRef]
59. Ayseli, M.T.; Yilmaz, M.T.; Cebi, N.; Sagdic, O.; Ozmen, D.; Capanoglu, E. Physicochemical, rheological, molecular, thermal and sensory evaluation of newly developed complementary infant (6–24 months old) foods prepared with quinoa (Chenopodium quinoa Willd.) flour. *Food Chem.* **2020**, *315*, 126208. [CrossRef] [PubMed]
60. Barbi, R.C.T.; Hornung, P.S.; Ávila, S.; Alves, F.E.D.S.B.; Beta, T.; Ribani, R.H. Ripe and unripe inajá (Maximilia maripa) fruit: A new high source of added value bioactive compounds. *Food Chem.* **2020**, *331*, 127333. [CrossRef]
61. Naseer, B.; Naik, H.; Hussain, S.Z.; Bhat, T.; Nazir, N. Development of instant phirni mix (a traditional dairy dessert) from high amylose rice, skim milk powder and carboxymethyl cellulose-resistant starch, predicted glycemic index and stability during storage. *Food Biosci.* **2021**, *42*, 101213. [CrossRef]
62. Tuaño, A.P.P.; Barcellano, E.C.G.; Rodriguez, M.S. Resistant starch levels and in vitro starch digestibility of selected cooked Philippine brown and milled rices varying in apparent amylose content and glycemic index. *Food Chemistry: Molecular Sciences.* **2021**, *2*, 100010.

foods

Article

Characterization and Extraction Influence Protein Profiling of Edible Bird's Nest

Nurul Nadiah Mohamad Nasir [1], Ramlah Mohamad Ibrahim [1], Md Zuki Abu Bakar [1,2], Rozi Mahmud [3] and Nor Asma Ab Razak [1,*]

[1] Natural Medicine and Products Research Laboratory, Institute of Bioscience, Universiti Putra Malaysia, Serdang 43400, Malaysia; nurul.nadiah.86@gmail.com (N.N.M.N.); ramlah86ibrahim@gmail.com (R.M.I.); zuki@upm.edu.my (M.Z.A.B.)

[2] Department of Veterinary Preclinical Sciences, Faculty of Veterinary Medicine, Universiti Putra Malaysia, Serdang 43400, Malaysia

[3] Department of Radiology, Faculty of Medicine and Health Sciences, Universiti Putra Malaysia, Serdang 43400, Malaysia; rozi@upm.edu.my

* Correspondence: norasmarazak@upm.edu.my

Citation: Mohamad Nasir, N.N.; Mohamad Ibrahim, R.; Abu Bakar, M.Z.; Mahmud, R.; Ab Razak, N.A. Characterization and Extraction Influence Protein Profiling of Edible Bird's Nest. *Foods* **2021**, *10*, 2248. https://doi.org/10.3390/foods10102248

Academic Editor: Antonello Santini

Received: 20 July 2021
Accepted: 18 September 2021
Published: 23 September 2021

Abstract: The edible bird nest (EBN) from *Aerodramus fuciphagus* has been consumed as a Chinese traditional food for health and medicinal purposes due to its elevated nutritional value. The present study focused on the influence of characterization and extraction methods on protein profiling, which could be a guideline for grading the EBN. The proposed extraction method is similar to the common food preparation methods of consumers and thus can accurately establish the bioactive protein available upon human consumption. The characterization includes physicochemical analysis (physical, morphology, elemental composition, and microbial content) and chemical analysis (crude protein and amino acid). The morphology of half-cup EBN was found to be uniformly shaped and rich in calcium as compared to rough surface of stripe-shaped EBN, and there was no significant microbial growth in both types of EBN. The crude protein and amino acid content in half-cup EBN were significantly higher than stripe-shaped EBN. The full stew (FS) and stew (SE) extraction methods produced a maximal yield of soluble protein. Sialic acid content in SE extract (8.47%, *w/w*) and FS extract (7.91%, *w/w*) were recorded. About seven parent proteins (39.15 to 181.68 kDa) were identified by LC-MS/MS Q-TOF, namely 78 kDa glucose-regulated protein, lysyl oxidase-3, Mucin-5AC-like, acidic mammalian chitinase-like, 45 kDa calcium-binding protein, nucleobindin-2, and ovoinhibitor-like. In conclusion, the characteristics and extraction methods influence the availability of bioactive protein and peptides, demonstrating the potential usage of EBN in improving its biological activities and nutritional properties.

Keywords: food; nutraceutical; edible bird's nest; *Aerodramus fuciphagus*; half-cup; stripe-shaped; physicochemical; proteomic; protein; extraction

1. Introduction

The edible bird's nest (EBN) is the nest of the swift. It is constructed with the saliva secreted from the pair of its sublingual glands as the main material. *Aerodramus fuciphagus* (white nests) and *Aerodramus maximus* (black nests) are the two main species of swiftlets that are known to produce valuable EBNs. Malaysia is the second largest exporter in the world, contributing about 20% to the total market of EBN production. Less than half of the EBN produced are consumed by locals, while a larger percentage is exported to other countries, such as Hong Kong, Singapore, and China [1]. Due to its high nutritional and medicinal therapeutic values, EBNs can cost USD 2000–10,000 per kilogram and are regarded as the most expensive animal by-product in the world [2]. White EBNs are regarded as the "Caviar of the East" due to their unique taste and smooth texture, they have also been widely used by people, especially the Chinese community, as traditional medicines. In traditional Chinese medicine, EBNs are prescribed to treat diseases such as tuberculosis, asthma, dry cough, hemoptysis, asthenia, difficulty in breathing, gastric troubles, and general bronchial ailments [3].

Interestingly, EBNs have also been the focus of modern research and technology, which have revealed their nutritional value and pharmacological activities, including anti-aging, anti-cancer, cough-suppressing, anti-tuberculosis, voice-improving, and phlegm-dissolving properties [4]. Meanwhile, in the food industry, EBN extract has been used as one of the ingredients in foods, drinks, and nutraceutical products. For example, ice cream with 0.2% EBN extract was found to be more favorable than the others due to high melting rates and lower sweetness [4]. These positive nutritional and health effects have increased the demand and supply of EBNs and have generated increasing interest among researchers. Investigations of the hidden nutritive and pharmacological properties of EBNs have become the primary focus of this research due to their potential as a therapeutic agent.

EBNs consist of protein (42.0–63.0%), carbohydrate (10.63–27.26%), moisture (7.5–12.9%), ash (2.1–7.3%), and fat (0.14–1.28%) [5]. Bioactive peptides released from EBN hydrolysate can exhibit various biological activities and nutritional properties, such as lowering blood pressure by ACE inhibitory peptides [6]. The main carbohydrate found in EBNs is sialic acid, which is a type of glycoprotein that consists of oligosaccharide chains (glycans) covalently attached to amino acid sidechains. Sialic acid has been proven to ameliorate cardiovascular disease biomarkers [7] and significantly improve the learning and memory function of the preterm infant [8]. To date, EBN containing bioactive peptides were found to improve bone loss and skin aging in post-menopausal women [9], to be effective in the treatment of neurodegenerative diseases [10], to prevent obesity-related inflammation and oxidative stress in rats [11], to lower oxidative stress and inflammatory markers [12], and to improve spatial learning performance in children [13].

Consumption of EBN-based nutritional beverages has been found to promote the improvement of human health. EBN glycopeptides were used in the formulation of ready-to-drink products and have been shown to have significantly higher antioxidant activity ($p < 0.05$) compared to non-EBN drinks [14]. In the cosmetic industry, EBN extracts have been incorporated into skincare products and have been used to promote skin cell renewal, radiant complexion, and anti-aging [4]. In some studies, EBN has been shown to improve digestive problems [5], with the prebiotic properties of EBN promoting a healthy human gut environment.

The bird's nest industry in Malaysia suffered a setback in 2011 when China banned EBN exports due to high concentrations of nitrate, lead, and arsenic detected in certain EBN products. As a result, strict regulations regarding EBN standards and specifications were implemented and enforced. The requirement listed in the EBN protocol for China specifies that EBN products must be free from avian influenza and the nitrite content must be less than 30 ppm. The quality of EBNs from Malaysia for export to China is set by [15], a Malaysian Standard that includes (1) MS 2333:2010 Good manufacturing practice (GMP) for processing raw-unclean and raw-clean edible birdnest (EBN); (2) MS 2334:2011 Edible birdnest (EBN)-specification; (3) MS 2612:2015 Raw-unclean edible birdnest (EBN)-house nest specification; and (4) MS 2509:2012 (P) Test method for edible bird nest (EBN)-determination for nitrite (NO_2^-) and nitrate (NO_3^-) content. The Malaysian Standards also provide requirements for EBN swiftlet farming—MS 2273:2012 and MS 2333:2010—as references for the EBN industry to reduce potential contamination in EBNs. The Standard MS 2273:2012 specifies guidelines for ranching practices of edible nest swiftlets, including ranch design and maintenance, hygiene of the premises, and signs of illness in swiftlets. The Standard MS 2333:2010 provides guidelines for the design of processing facilities in order to avoid cross-contamination and to control procedures that affect operations, building maintenance, personal hygiene, and animal hygiene control systems (contaminant control).

Grading of harvested EBNs depend on factors such as shape (half-cup or stripe-shaped), type (white, red, or grass), color (white, yellow, or red), and cleanliness of the nest. In addition, the dry mass, duration of nest building, and protein content of cleaned house-farmed EBNs also contribute to the EBN grading, reflecting quality [5]. The detailed characteristics and grading of EBNs, including physicochemical analysis (physical [12], morphology [16], elemental composition [17], and microbial content [18]) and chemical analysis (crude protein [2], amino acid [19] and sialic acid [7]) have been reported. However, research on the relationship between the physical and chemical characteristics of EBNs and their nutritional and bioactive properties is lacking.

Several methods have been reported for obtaining EBN extract and bioactive peptides, such as solvent extraction, heat extraction, enzymatic hydrolysis, and microbial fermentation [4]. Solvent extraction may result in low extraction efficiency, low selectivity, solvent residue, and environmental pollution. In contrast, enzymatic hydrolysis is preferable compared to solvent extraction and microbial fermentation due to high recovery, less solvent residue, and being more environmentally friendly, making this method popular among food and pharmaceutical industries [20]. However, data on the impact of characterization and extraction of different shapes of EBNs on the protein profiling and bioactive content is insufficient; this research area needs to be explored for a better understanding of EBN applications. A bioactive compound, when extracted using different extraction methods, may potentially produce a wide range of bioactivity, which may be attributed to the synergistic effects of diverse constituents of the food [21].

Proteomic research into EBNs has resulted in a bottleneck, due to its poor solubility and low extractive rate [22] and the limited number of protein sequences deposited in the database. The effects of simulated gastrointestinal digestion provide insight into the amounts of bioactive peptides that are likely to be derived from hydrolysis of EBNs in real conditions of humans. The optimum conditions implicated for the bioavailability of EBN glycopeptides is thus double boiling followed by enzymatic hydrolysis; this combination of methods uncoils the glycoprotein structure, which in turn may unlock the potential benefit of EBNs [4].

Even though both half-cup and stripe-shape EBNs originate from the same source, the comparison data from physicochemical analysis (physical, morphology, elemental composition, and microbial content) and chemical analysis (crude protein, amino acid, and sialic acid) is important to understand and correlate between protein hydrolysis and bioactive peptide studies. Four types of water extraction combined with heat treatment were selected to simulate the common traditional preparation of EBN, aimed to profile the water-soluble protein. Protein profiling portrayed by SDS-PAGE and LC-MS/MS analysis demonstrates confirmation of the good quality of protein produced by the hydrolysis technique, which is related to the functional characterization of bioactive glycopeptides. This peptide sequence clarified the specific bioactive activity with different characteristics, which is important for food, pharmaceutical, and medical applications.

2. Materials and Methods

2.1. Edible Bird's Nest (EBN)

Different types (half-cup and stripe-shaped) of raw-cleaned house Edible Bird's Nest (EBN) were purchased from Blossom View Sdn. Bhd in Terengganu (East Coast of Malaysia). Feathers and impurities identified during the screening procedure were manually removed using tweezers, ground with a mortar to form EBN powder, and stored in an air-tight container at 4 °C until further use.

2.2. Characterization

2.2.1. Physicochemical Analysis (Physical, Morphology, Elemental Composition, and Microbial Content)

EBNs with half-cup and striped-shaped physical characteristics were measured by a ruler in triplicate. The morphology characteristics were obtained by fixing EBN powder to a stub with carbon tape, coating with gold, and placing on the sample holder. Next, the sample holder was fixed on a rotatable disc inside the machine and observed using a scanning electron microscope (SEM) (FEI Quanta 400F). Meanwhile, energy dispersive

X-ray spectroscopy (EDX) was used to investigate the elemental constituents; the atomic and weight percentage of the elements were determined under low vacuum at an accelerating voltage of 15 kV and a current of 60–90 mA. The microbiological analysis for both EBN types was performed according to standard methods specified by the Food and Drug Administration/Bacteriological Analytical Manual (FDA/BAM) and Official Methods of Analysis of Association of Analytical Chemists (AOAC), 16th Edition (1995). The FDA/BAM analytical methods were used to identify aerobic plate count (APC) (Chapter 3), yeasts and molds (Chapter 18), *Staphylococcus aureus* (*S. aureus*) (Chapter 12), and *Salmonella* (Chapter 5) [23], while AOAC: 991.14 (1995) was used to identify coliforms and *Escherichia coli* (*E. coli*) in the EBNs [24].

The APC was used to indicate the level of microorganisms in products. Briefly, the EBN sample (10 mL) was mixed with diluent (90 mL) to obtain 10^{-2}, 10^{-3}, and 10^{-4} decimal dilutions. The inoculated diluents sample (1.0 mL) was mixed with plate count agar in a Petri dish and incubated aerobically at 37 °C for 48 h. Meanwhile, the yeast and mold count in the EBN samples were determined out using a media supplemented by 100 mg/L chloramphenicol to inhibit bacterial growth. Peptone water (0.1%) was added to the EBN sample (25–50 g) and homogenized for 2 min. The solution (0.1 mL) was pipetted onto solidified dichloran rose bengal chloramphenicol (DRBC) agar and subjected to incubation in a dark place at 25 °C for 5 days. Coliforms and *E. coli* in EBN samples were determined using 3 M Petri film (3M Co., Saint Paul, MN, USA). The film was placed on a flat surface and inoculated with 1 mL of test suspension onto the center of the film base. The plate was incubated at 37 °C for 24 h. The red and blue colonies referred to coliforms and *E. coli*, respectively.

The presence of *S. aureus* in EBNs was examined by transferring 1 mL of EBN sample onto a Baird-Parker agar plate. The inoculum was spread evenly using a sterile bent glass streaking rod, and the plate was then incubated at 35–37 °C for 45–48 h. Detection of *Salmonella* in EBN was carried out by mixing a 25 g sample and 225 mL sterile lactose broth in a sterile blending container. The mixture was transferred into a sterile, wide-mouth, screw-cap jar and left at room temperature for 60 ± 5 min, with the jar securely capped. Then, the mixture was mixed well, and the jar cap was slightly loosened before incubation at 35 °C for 24 ± 2 h. Rappaport-Vassiliadis (RV) medium (1:100) and tetrathionate (TT) broth (1:10) were added, and the mixture was subsequently incubated as follows: RV: 24 ± 2 h at 42 °C; TT: 24 ± 2 h at 35 °C. About 10 µL of incubated TT broth was then used to streak 3 mm loopful on bismuth sulfite (BS) agar, xylose lysine deoxycholate (XLD) agar, and Hektoen enteric (HE) agar. The loopful step was repeated using RV medium (10 µL) on the same agar plates before being incubated for 24 ± 2 h at 35 °C. The results for APC, yeasts and molds, coliforms, and E. coli were expressed as $\log_{10}$ colony-forming unit per gram (CFU/g) EBN. Meanwhile, the result for *S. aureus* was reported as the most probable number per gram (MPN/g) of EBN, and *Salmonella* species were reported as either detected or not detected in 25 g of EBN.

2.2.2. Chemical Analysis (Crude Protein and Amino Acid)

The crude protein in both EBNs was determined by the Kjeldahl method, using 6.25 as a conversion factor [25]. About 500 mg of EBN, 1 tablet of a catalytic amount of $CuSO_4/K_2SO_4$, and 12 mL of sulfuric acid were added into the digestion tube to initiate the digestion process, until clear green or blue solution was obtained. The solution was cooled for 10–20 min before the addition of 75 mL of distilled water. The analysis was continued with the distillation process through the addition of 25 mL boric acid and 10 drops of bromocresol green as an indicator. Next, the cooled digestion tubes were placed in the digestion unit, and 50 mL of sodium hydroxide solution was added to samples for 5 min in a distillation unit. The distillate was titrated with hydrochloric acid until the changed color was obtained and was calculated in accordance with [26].

The amino acid composition was analyzed based on [27] using high-performance liquid chromatography (HPLC). About 0.3 g of EBN sample was hydrolyzed with 5 mL of 6 N

HCl at 110 °C for 24 h. The samples were cooled to room temperature and filtered through filter paper into a 100 mL volumetric flask. The 400 μL of internal standard (50 μmol/mL α-aminobutyric acid (AABA) in 0.1 M HCl) was added into the same flask and topped up to 100 mL with distilled water. The aliquot was filtered by using a 0.20 mm polytetrafluoroethylene microfilter. As for derivatization, 10 μL filtered hydrolysate samples or standard and 70 μL borate buffer solution were mixed well into a 1.5 mL glass vial. Then, 20 μL AccQ Flour reagent (3 mg/mL in acetonitrile) was added to the mixture and vortexed for a few seconds. The 10 μL of samples and standards was injected into the HPLC (Waters 2475, Waters Co., Milford, MA, USA), with a 1 mL/min flow rate. Amino acid analysis was performed using AccQ Tag (3.9 × 150 mm) column with mobile phase A (Eluent A—200 mL AccQ Tag to 2 L of Milli-Q water) and mobile phase B (Eluent B—60% acetonitrile). The linear gradient condition was set according to the following times: 100% A and 0% B (start), 98% A and 2% B (0.5 min), 91% A and 9% B (15 min), 87% A and 13% B (19 min), 65% A and 35% B (32 min), 65% A and 35% B (34 min), 0% A and 100% B (35 min), 0% A and 100% B (38 min), 100% A and 0% B (39 min) and 100% A and 0% B (50 min). Amino acid content was read at 250 nm using a fluorescence detector (λ excitation and λ emission).

2.3. *EBN Extraction*

EBN extraction was performed according to a previous study [28], with some modifications. Water extraction was selected to simulate the traditional technique commonly used before the consumption of EBNs. Water extraction gives a general idea of the protein consumed by people. EBN extraction was done using four types of methods: stew (SE), full stew (FS), sonication (SO), and hot water (HW). The cleaned EBN was ground and soaked with deionized water, with a ratio of 1:100 (*w/v*), for 16 h at 4 °C. All the extracts were subsequently frozen at −80 °C, lyophilized, and stored at −20 °C until further use.

2.3.1. Stew and Full Stew Extraction

SE and FS extraction methods were carried out using a double-boiling technique, where the vapor from the boiling water in the bottom pot heated the top pot containing the EBN. The soaked EBN was double-boiled for 30 min at 100 °C and cooled to room temperature. The SE extract consisted of the double-boiled EBN, filtered using a muslin cloth, and the FS extract consisted of the whole double-boiled EBN, without filtration.

2.3.2. Sonication

SO extraction was performed by sonicating soaked EBN in a beaker for 30 min at room temperature (25 ± 1 °C) using an ultrasonic cleaner (42 kHz, 135 W; Branson Ultrasonic Corporation, Fairfield County, CT, USA). The water level in the beaker was kept at the same level of water as the ultrasonic bath, maintained at a constant room temperature (25 ± 1 °C). Next, the aliquot was filtered using a muslin cloth to obtain the extract.

2.3.3. Hot Water

Direct heat from boiling water was used to obtained the HW extract. The soaked EBN was placed in a beaker and heated for 30 min at 100 °C in a water bath. It was then cooled to room temperature and subsequently filtered through a muslin cloth to obtain the extract.

2.4. *Determination of Sialic Acid by HPLC*

The sialic acid content was evaluated based on [13] using high-performance liquid chromatography (HPLC) (Agilent1200) with a reversed-phase Agilent HC-C18 column (4.5 × 250 mm, 5 μm). The methanol, acetonitrile, and water solution (7:8:85) were used as a mobile phase, with a flow rate of 0.9 mL/min. A fluorescent detector was used, with the excitation wavelength at 373 nm and emission wavelength at 448 nm.

2.5. Enzymatic Hydrolysis of EBN

Enzymatic hydrolysis, following simulated gastrointestinal digestion, was performed to produce edible bird's nest (EBN) protein hydrolysate as previously described by [3], with slight modifications. The pH of EBN extract (5 mg/mL) was adjusted to 2.0 using HCl (20 mL) and digested with pepsin (1% (*w/w*) in 0.1 M potassium chloride). The pH-adjusted sample was put in a shaking incubator for 2 h at 37 °C and boiled for 20 min to stop the pepsin activity. The pH of the mixture was increased to 8 with 1 M NaOH and further digested with pancreatin (1% (*w/w*) in 0.1 M potassium phosphate buffer) for another 2 h at 37 °C to simulate small intestine conditions. The reaction was stopped by boiling the sample for 30 min and immediately cooling at pH 8.9. The hydrolyzed solution was centrifuged at 10,000× *g* at 4 °C for 30 min and desalted using snakeskin pleated dialysis tubing at 7.0 K MWCO (Thermo Fisher Scientific Inc., Watham, MA, USA). The hydrolysate was subsequently frozen at −80 °C, lyophilized, and stored at −20 °C until further use.

2.6. Estimation of Soluble Protein

The soluble protein content of EBN samples was determined using the Bradford protein assay kit following the manufacturer's instructions (Catalog Number: P010; Gene Copeia, Rockville, MD, USA) [29]. About 100 μL of the sample and bovine serum albumin (BSA) standard were added into 96-well plate, followed by the addition of 100 μL of Bradford reagent to each well. The mixture was mixed with a plate shaker for 30 s and incubated for 5 min at room temperature. Protein concentration was calculated according to the standard protein curve of BSA, and absorbance was read at 595 nm.

2.7. Protein Separation and Molecular Weight Determination Using SDS-PAGE

The samples for purification and determining the molecular weight of the protein were prepared using optimized half-cup EBN extracts (FS and SE) and hydrolysates (FSh and SEh). The sodium dodecyl sulfate-polyacrylamide electrophoresis (SDS-PAGE) was prepared according to [30], with some modifications. The 5% (*w/v*) of EBN was loaded in 12% resolving gel and 4% stacking gel in a 1:1 (*v/v*) ratio containing Tris buffer. The solution was heated in a 90 °C water bath for 20 min and then cooled immediately. Then, 20 μL of the sample and 20 μL of the protein standard were loaded into individual wells and run under the constant current setting of 30 mA and 150 V for 15 min before being increased to 200 V for another 45 min. Proteins were stained with 0.1% (*w/v*) Coomassie blue, with protein markers in the range of 11 to 245 kDa.

2.8. Protein Identification by LC-MS/MS Q-TOF

2.8.1. EBN Protein Digestion

In-solution digestion of the EBN sample was carried out according to the manufacturer's instructions [31]. RapiGest solution (0.2%, *w/v*) was prepared by resuspending 1 mg RapiGest^{TM} (Waters) in 500 μL of 50 mM ammonium bicarbonate. EBN extract (100 μg) was then dissolved in 50 μL of 0.2% RapiGest solution and vortexed. DTT was added to the mixture to a final concentration of 5 mM for the reduction step before boiling at 60 °C for 30 min. The mixture was cooled to room temperature, before being alkylated with iodoacetamide to a final concentration of 15 mM for 30 min in the dark environment. The proteolytic digestion step was performed by adding mass spectrometry grade Trypsin Gold (Promega, Madison, WI, USA) at a ratio of 1:50 (trypsin/protein), followed by incubation overnight at 37 °C. At the end of the digestion step, 1 μL of formic acid was added to stop trypsin activity. The digested protein samples were stored at −20 °C before protein and peptide identification.

2.8.2. Liquid Chromatography-Tandem Mass Spectrometry Coupled Quadrupole-Time of Flight (LC-MS/MS Q-TOF) Analysis

LC-MS/MS analysis was performed using 6550 iFunnel Q-TOF LC/MS from Agilent Technologies (Santa Clara, CA, USA). A total of 5 μL digested EBN was loaded into an

Agilent Large Capacity Chip consisting of a 75 μm × 150 mm analytical column and a 160 mL enrichment column, which was packed by 5 mM of Zorbax 300SB-C18 for chromatographic separation. The mobile phase consisted of solvent A (0.1% formic acid in MilliQ water) and Solvent B (9:1 ratio of 0.1% formic acid in acetonitrile: MilliQ water); a flow rate of 1.0 mL/min was used to elute the peptides. The mobile phase gradient was programmed as 3–50% of solvent B for 30 min; 50–95% of solvent B for 2 min; 95% of solvent B for 7 min; and 95–3% of solvent B for 47 min. The polarity of Q-TOF was set at positive, the voltages for capillary (2050 V) and fragment (300 V) were set accordingly, and the gas flow was set at 5 L/min and 325 °C. The peptide spectra were acquired using Agilent MassHunter Workstation Data Acquisition software (Agilent Technologies, Santa Clara, CA, USA) by monitored positive ion acquisition in the range of 110 to 3000 m/z for the MS scan, and 50 to 3000 m/z for the MS/MS scan. The chromatograms obtained were analyzed using the Agilent MassHunter Qualitative Analysis B.05.00 software (Agilent Technologies, Santa Clara, CA, USA).

2.8.3. Data Analysis

A Swiss-Prot database search for protein and peptide identification matched with *Charadrius vociferus* species was performed using the Spectrum Mill MS Proteomics Workbench (Agilent Technologies, Santa Clara, CA, USA). Carbamidomethylation was selected as the fixed modification parameter and trypsin as the digestive enzyme, with the maximum number of cleavage equal to 2. Auto-validation was set for MS searches at a false discovery rate (FDR) of 1.2% and data export to Mass Profiler Professional (MPP v 14.9.1, Agilent Technologies, Santa Clara, CA, USA) for further analysis. The software confirmed the identified protein and the post-translation modifications if present, and the data files were then processed by principal component analysis (PCA) using the MPP software.

2.9. Statistical Analysis

Statistical analysis was performed using SPSS version 24.0. All results were expressed as mean ± standard deviation (SD). The data were statistically treated by paired-sample t-test and analysis of variance (ANOVA), with $p < 0.05$ considered to be statistically significant, and the mean was compared using Duncan's multiple range tests.

3. Results and Discussion

3.1. Physicochemical Analysis (Physical, Morphology, Elemental Composition, and Microbial Content)

The physicochemical analysis included physical measurement, chemical composition, and microbial content of half-cup and stripe-shaped EBNs. The nest cups used were half-cup and stripe-shaped, as shown in Figure 1. The half-cup EBN is U-shaped, and the stripe-shaped EBN is a hard, incomplete cup shape, with some broken parts. Table 1 shows the physical measurement of half-cup and stripe-shaped EBNs. The height, length, and weight were 4.08 ± 0.71 cm, 7.06 ± 1.33 cm, and 5.19 ± 0.18 g (half-cup) and 1.28 ± 0.15 cm, 3.26 ± 0.48 cm, and 2.26 ± 0.18 g (stripe-shaped), respectively; the weight of stripe-shaped EBN was half of that of the half-cup EBN, and the height, length, and weight between half-cup and stripe-shaped EBNs were statistically significantly different ($p < 0.05$). Similar findings of average height, length, and weight of EBNs have also been reported [16], ranging from 3.5 to 5.0 cm, 7.0 to 13.0 cm, and 5.6 ± 1.3 g, respectively. The shape of an EBN is crucial for grading purposes, whereby the price is determined. The shape of the half-cup EBN is Grade A, and the stripe-shaped EBN derived from the broken part of the half-cup EBN is classified as Grade B.

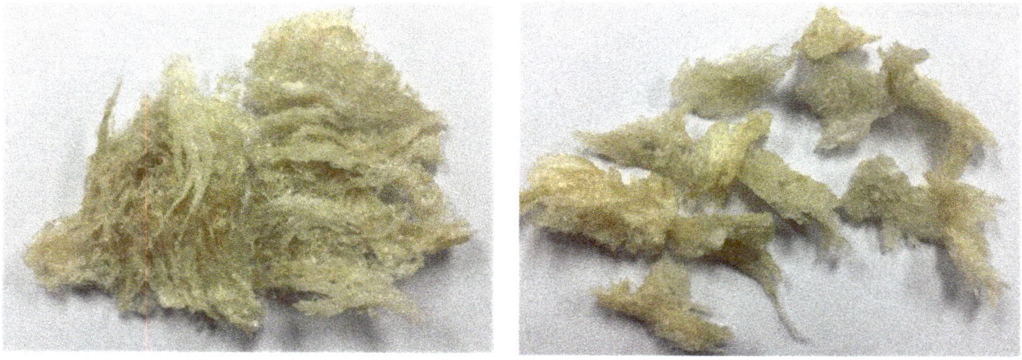

(**a**) (**b**)

Figure 1. EBN image: (**a**) half-cup and (**b**) stripe-shaped.

Table 1. Physical measurement, elemental distribution, and microbial content of half-cup and stripe-shaped EBN.

Parameter		**Edible Bird Nest**			
		Half-Cup		**Stripe-Shaped**	
		Macroscopic			
Measurement	$n = 10$	Range	Mean ± SD	Range	Mean ± SD
Height	cm	3.0–5.1	4.08 ± 0.72^{a}	1.0–1.5	1.28 ± 0.15^{b}
Length	cm	5.5–9.0	7.06 ± 1.33^{a}	2.5–4.0	3.26 ± 0.48^{b}
Weight	g	5.0–5.5	5.19 ± 0.18^{a}	2.0–2.5	2.26 ± 0.18^{b}
Element	$n = 3$	Microscopic Weight			
Carbon	%	4.78 ± 11.72^{b}		28.49 ± 22.26^{a}	
Oxygen	%	5.61 ± 13.75^{b}		27.13 ± 21.30^{a}	
Magnesium	%	ND		0.22 ± 0.53	
Calcium	%	72.94 ± 43.58^{a}		44.16 ± 43.31^{b}	
	Unit	Microbial Content			
Aerobic Plate Count	CFU/g	1.8×10^{8}		1.8×10^{8}	
Coliforms	CFU/g	1.4×10^{4}		1.4×10^{4}	
Escherichia coli	CFU/g	ND (<10)		ND (<10)	
Salmonella	in 25 g	Absent		Absent	
Staphylococcus aureus	MPN/g	ND (<3)		ND (<3)	
Yeasts and molds	CFU/g	67		ND (<10)	

Data shown are means ± standard deviation. [a b] Different superscript letters in the same row indicate a statistically significant difference between half-cup and striped-shaped EBNs ($p < 0.05$). ND: Not detected; <: Less than the minimum detection limit reported; CFU: Colony-forming unit; MPN: Most probable number.

Figure 2 shows the surface morphology of half-cup and stripe-shaped EBNs at 50× and 500× magnification by SEM. These structures are seen to be coated with a layer of a partially clear or transparent substance. Substances appearing rough and hard covered some of the surfaces of the EBNs. Micrograph images at 500× magnification showed that the EBN was unevenly structured. The surface formation of the half-cup EBN was uniformly shaped, while a rough surface can be seen in the stripe-shaped EBN. Visualization under 50× magnification for both half-cup and stripe-shaped EBNs revealed ground EBN as an irregular prism with a crystalline structure. Irregular prisms with lustrous translucent to opaque silver crystalline structures were also reported by [16]. The washing and drying process is a common step in processing EBNs, to increase the shelf-life, lower the water activity, and facilitate storage before manufacturing and packaging. Thus, no foreign or unwanted particles, such as mites, fungal structures, and feather strands, were observed.

However, drying involving heat can cause a crystalline appearance, which was observed in both half-cup and stripe-shaped EBNs using SEM.

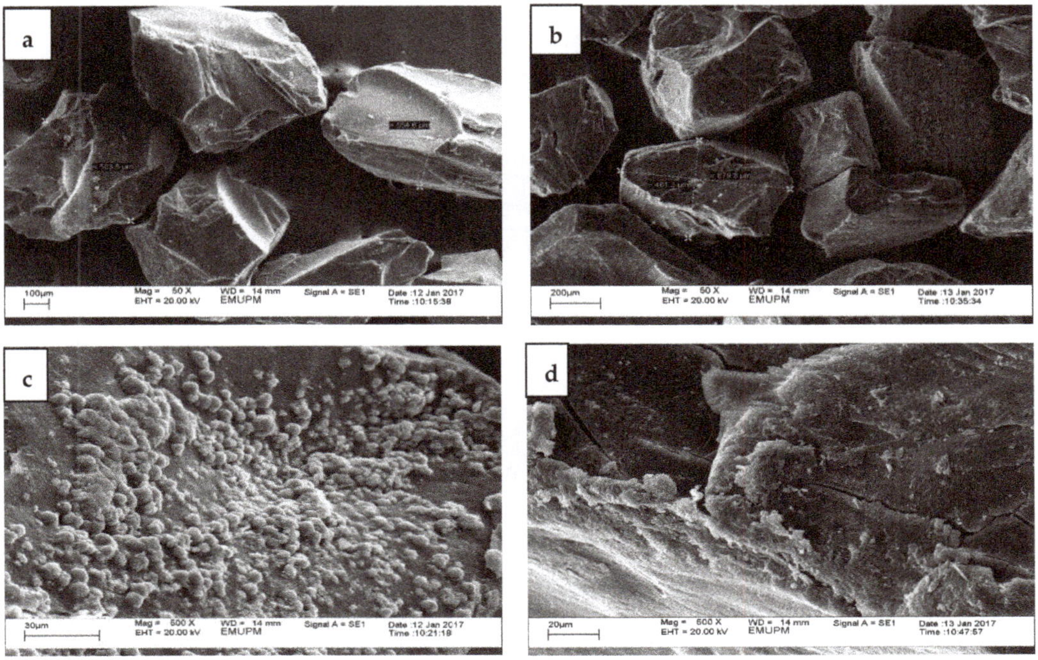

Figure 2. Scanning electron micrograph of EBNs at magnification of 50× for (**a**) half-cup and (**b**) stripe-shaped, and magnification of 500× *g* for (**c**) half-cup and (**d**) stripe-shaped.

SEM with energy-dispersive X-ray (EDX) for elemental analysis was done to characterize the composition and chemical characteristics of both EBNs, as shown in Table 1. Carbon and oxygen elements in stripe-shaped EBNs were five-fold higher than half-cup EBNs. Magnesium (0.22%) was only found in the stripe-shaped EBN sample. Oxalic acid secretion, calcium availability, hydration state, and other environmental factors may influence crystal production and morphology [17]. The high calcium content in the half-cup EBN can be assumed from the translucent crystalline structure (Figure 2a), which is glassier than the stripe-shaped EBN (Figure 2b) under the same resolution at 50×; this structure is supported by calcium, which in the half-cup EBN is two-fold higher (72.94%) than calcium element in striped-shaped EBN (44.16%).

Microorganisms such as coliforms, *Escherichia coli* (*E. coli*), *Salmonella*, *Staphylococcus aureus* (*S. aureus*), yeast, and mold were also quantified (Table 1). The results showed that both EBNs had a similar number of aerobic plate counts (1.8 × 10^8 CFU/g), coliforms (1.4 × 10^4 CFU/g), *E. Coli* (not detected at less than 10 CFU/g), *Salmonella* (absent), and *S. aureus* (not detected at less than 3 CFU/g). However, 67 CFU/g of yeast and mold were detected in the half-cup compared to stripe-shaped EBN. Bacteria present in EBNs may lead to food-borne disease if ingested by a human. *Salmonella sp.* and *E. coli* are the most common food-borne pathogens that can affect health conditions; they can cause severe diarrhea or meningitis, which can be fatal if untreated [32]. Based on the latest standards set by the Ministry of Health Malaysia, both types of EBN used in this study were in accordance with the specified standard that raw-clean EBNs should be free from *E. coli*, *Salmonella* sp., and *S. aureus* before being exported to another country [18]. The Standard and Industrial Research Institute of Malaysia (SIRIM) provides guidelines for Edible Bird Nest (EBN) (MS 2334:2010), including the permitted levels of microbial content:

total plate count (<2.5 × 106 CFU/g), coliform (<1100 most probably number (MPN)/g), *E. coli* (<100 (MPN)/g), *S. aureus* (<100 (MPN)/g), yeast and mold (<10 CFU), and no presence of *Salmonella sp.* [33].

Yeast, mold, and other contaminants such as mites and feathers can be located within the strands of EBNs. This was shown in the quantification analysis, after the EBN was ground into small pieces. Thus, it is important to educate consumers on the need for additional washing and boiling, which can assist in removing contaminants [33] and significantly reduce the number of bacteria before consumption [18].

3.2. Chemical Analysis (Crude Protein and Amino Acid)

Table 2 shows the crude protein content and amino acid profile of half-cup and stripe-shaped EBNs. A higher crude protein content in the half-cup (56.96 ± 0.09%) than stripe-shaped (54.70 ± 0.16%) EBN was obtained, which was significantly different for both EBNs ($p < 0.05$). These results were comparable with EBN samples from East Coast Peninsular Malaysia, ECM (Pahang: 55.48 ± 3.60%) [2], Northern Peninsular Malaysia, NM (Perak, Penang, Kedah: 53.8 ± 0.18%) and East Malaysia, EM (Sabah, Sarawak: 52.8 ± 1.04%) [18]. The major amino acids (AAs) found in both half-cup and stripe-shaped EBNs were glutamic acid (Glu), followed by aspartic acid (Asp), serine (Ser), and valine (V), which were also reported in EBN samples from ECM [2]. The similarity in the protein and amino acid profiles of the EBN samples studied by [2] and those of present study was due to the EBN samples being collected from the same region (ECM). However, the major amino acids in EBN from EM [18] were different due to different collection locations (NM and EM). Overall, both half-cup and stripe-shaped EBNs possess a high percentage of amino acids and nonessential amino acid composition. These variances could be due to different locations of EBN harvesting, the effects of processing methods, etc. Breeding sites, climate, and swiftlet diet may also affect the EBN nutrient composition [18].

Table 2. Crude protein content and amino acid profile of EBNs.

Parameter		This Study		[2]	[19]	
		Half-Cup	Stripe-Shaped			
Location	Compound (%)	Terengganu		Pahang	Perak, Penang, Kedah	Sabah, Sarawak
Crude protein		56.96 ± 0.09 [a]	54.70 ± 0.16 [b]	58.55 ± 0.62	53.8 ± 0.18	52.8 ± 1.04
	Arginine	6.74	6.61	3.80	4.50	4.10
Essential amino acids (EAAs)	Histidine	4.61	4.66	1.40	1.50	1.40
	Isoleucine	3.38	3.38	3.40	0.60	0.50
	Leucine	6.74	6.80	5.30	2.70	2.50
	Lysine	4.49	4.34	5.40	1.30	1.10
	Methionine	2.85	2.90	2.20	0.70	0.70
	Phenylalanine	6.23	6.33	2.70	2.30	2.20
	Threonine	6.73	6.66	2.90	2.70	2.40
	Valine	7.58	7.43	3.30	1.60	1.40
	Total EEA	49.35	49.11	30.40	17.90	16.30
	Alanine	4.36	4.43	3.90	1.30	1.20

Table 2. *Cont.*

Parameter		This Study		[2]	[19]	
		Half-Cup	Stripe-Shaped			
Location	**Compound (%)**	**Terengganu**		**Pahang**	**Perak, Penang, Kedah**	**Sabah, Sarawak**
Non-essential amino acids (nEAAs)	Aspartic acid	9.14	8.94	6.30	4.00	3.70
	Cysteine	1.37	1.29	1.70	1.10	1.40
	Glutamic acid	11.20	11.68	9.60	3.00	2.60
	Glycine	3.68	3.66	2.50	1.50	1.60
	Proline	6.95	6.84	2.90	3.20	2.90
	Serine	8.60	8.37	2.40	4.30	4.00
	Tyrosine	5.34	5.64	2.90	2.70	2.60
	Total nEAA	50.64	50.85	32.20	21.10	20.00

Values are expressed as mean (%) ± S.D. Means with different superscript letters in a row indicate a significant difference between half-cup and striped-shaped EBNs ($p < 0.05$).

The nutritional value of essential amino acids (EAAs) composition contained in protein may contribute to the functional biopeptides. The readily digestible dietary EAAs are needed at an optimum level for human body requirements, e.g., EAA Gly supplementation has been reported to promote anti-inflammatory effects during endothelial inflammation, while Arg has been found to improve endothelial function in cardiovascular or overweight patients [34]. The presence of nEAA Glu in the body is very important, especially for transamination reactions in amino acid metabolism. It is involved in the synthesis of key molecules such as glutathione, which is important in the alleviation of oxidative stress and modulation of the immune response [35]. Asp plays a role as a regulator in hormone secretion and acts as a precursor for methionine (Met), threonine (Thr), isoleucine (Ile), and lysine (Lys). Similarly, Ser is the precursor of glycine (Gly), cysteine (Cys), and tryptophan (Trp) and is involved in cell signaling.

3.3. EBN Extracts and Soluble Protein Concentration

EBN extraction is affected by its solubility behavior towards water. Since protein is the major compound in EBNs, preliminary experiments found temperature affected the extracted protein of the EBN. The extraction yield of half-cup and stripe-shaped EBNs is shown in Figure 3a. The FS method showed the highest extraction yield (half-cup = 92.29 ± 2.45%; stripe-shaped = 79.35 ± 0.91%) as compared to the SE method (half-cup = 12.50 ± 0.89%; stripe-shaped = 10.99 ± 0.11%), with significant difference ($p < 0.05$). Meanwhile, less than 5% extraction yield was obtained using the SO and HW methods for both EBNs. The temperature recorded for the double-boiling technique applied for FS and SE was 73 °C to 80 °C; SO, 25 °C; and HW, 100 °C.

The double-boiled technique with indirect heating contributes to the maximal yield of the FS and SE extraction methods. There was a significant difference in extraction yield between SE and FS methods for both EBNs ($p < 0.05$), while no significant difference was observed in both EBNs using the SO and HW methods ($p > 0.05$). Thus, the different temperatures and the characteristics of the different EBNs affected the extraction yield, with the optimum temperature between 73 °C and 80 °C. Meanwhile, low temperature (below 60 °C) or high temperature (more than 80 °C) resulted in a low extraction yield, as seen with the SO and HW extraction methods, respectively. These results agree with findings reported by [16], emphasizing that the protein concentration of the EBN extract drastically increases under extraction temperatures between 60 °C and 80 °C, but gradually reduces above 80 °C. Thus, temperature influences the extraction and solubility of proteins in EBNs.

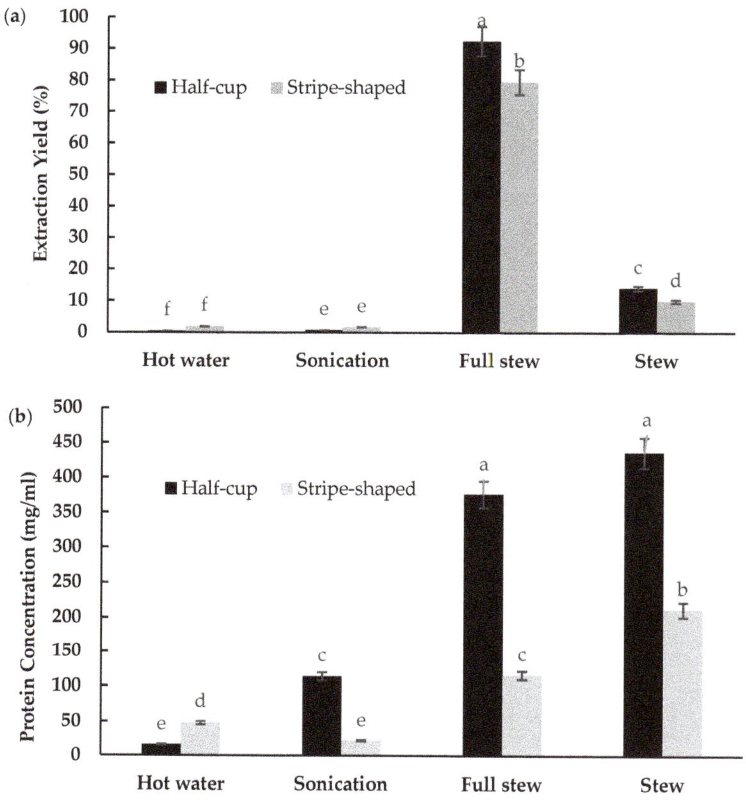

Figure 3. The (**a**) extraction yield and (**b**) protein concentration of half-cup and stripe-shaped EBNs using different extraction methods. Data shown are means ± standard deviation of triplicates. a,b,c,d,e,f: means with different letters indicate significant differences among the extraction methods and types of EBN ($p < 0.05$).

Figure 3b shows the soluble protein concentration of EBN extracts obtained from four different extraction methods. The highest protein concentration was in the half-cup EBN (SE = 435.6 mg/mL; FS = 375.6 mg/mL; SO = 113.6 mg/mL) compared to the stripe-shaped EBN (SE = 211.6 mg/mL; FS = 115.6 mg/mL; SO = 21.6 mg/mL), while the lowest soluble protein concentration was obtained in the HW extract (half-cup = 15.6 mg/mL). There were significant differences in soluble protein concentration between the two EBNs for the extraction methods applied ($p < 0.05$), but protein concentration of the half-cup FS and SE extracts were not significantly different ($p > 0.05$).

The extraction yield and soluble protein concentration in the SE method were inversely proportional with temperature. The high yield of the FS extract was a combination of soluble and insoluble fractions of the half-cup EBN, while the SE extract was mainly in soluble form. According to [36], water extraction produced the highest protein content as compared to other extraction methods, indicating that EBNs have an abundance of water-soluble proteins. Thus, the concentrated soluble protein fraction in the SE extract was contributed by its protein concentration. The protein in the insoluble fraction of the FS extract may not fully break down during the double-boiling process, resulting in the incomplete release of soluble protein. The high temperature of the HW methods (100 °C) may cause denaturation of the protein, and the mild heat of the SO methods (25 ± 1 °C) produced an incomplete breakdown of complex protein, resulting in low yield and low soluble protein concentration in both extraction methods.

The optimum extraction condition needs to be chosen to obtain the highest protein concentration and bioactivity of EBNs for further application. The FS and SE extraction methods produced a high yield of crude extract and protein concentration in the half-cup EBN compared to the stripe-shaped EBN. Thus, FS and SE extraction methods for the half-cup EBN were selected to be used for sialic determination and enzymatic hydrolysis.

3.4. Sialic Acid Content of EBN Extracts

The SE extract of the EBNs contains higher sialic acid than the FS extract, with 8.47% (*w/w*) and 7.91% (*w/w*), respectively, compared to other findings of 8.6% [4] and 11% [11]. In addition, the sialic acid content in EBNs from Malaysia are varied, with 0.70% to 1.50% from North, South, and East Peninsular Malaysia [37] and 1.17% to 3.15%, collected from North, South, and Borneo Sabah [8]. The types of habitats, environment surroundings, breeding site, nest harvesting season, and diet of the swiftlets are the factors that influence the sialic acid content of EBNs from different locations [8].

These findings agree with the protein content of half-cup EBNs in the present study, suggesting that high protein may contributed to high sialic acid content in EBNs. This is because sialic acid is a component of the glycoprotein located in the carbohydrate chains attached to soluble proteins. This results in a high sialic acid content in the SE extract as compared to FS extracts. As mentioned earlier, the soluble and an insoluble fractions of the FS extract resulted from the incomplete breakdown of protein and may reduce the water-soluble bound sialic acid contents.

Both proteins and sialic acid are major nutraceutical ingredients in EBNs. In a study conducted using EBN constituents producing synergistic antioxidative effects, sialic acid was found to ameliorate the progression of atherosclerosis and other cardiovascular disease biomarkers [7]. Sialic acid in EBNs has also shown potential antiviral properties by inhibiting viral genes, strengthening the lungs, improving skin health, and showing anti-aging properties [4]. Taken together, the findings of crude protein, amino acids, and sialic acid analyzed in the half-cup and stripe-shaped EBNs show that EBNs are a good source of protein and glycoprotein, potentially exhibiting nutritional and medicinal properties.

3.5. Soluble Protein Concentration of EBN Hydrolysates

The high soluble protein found in the half-cup EBN as reported in Section 3.3 indicates a protein-rich sample that yielded a high amount of hydrolysate. The yield of the half-cup FSh (47.5%) was comparable to half-cup SEh (51.5%), as shown in Figure 4a. Enzymatic hydrolysis potentially breaks down complex protein and helps in releasing bioactive peptides from the inactive parent protein, which enables them to exhibit biological properties for therapeutic purposes [6]. Figure 4b shows an uptrend pattern of soluble protein concentration in EBN hydrolysates. Solubility is primarily dependent on the distribution of hydrophobic and hydrophilic amino acids on the surface of the protein and is affected by the protein–water interaction.

A similar concentration of FSh and SEh in the half-cup EBN showed a significant difference in soluble protein concentration ($p < 0.05$), except for at 10,000 ppm, where the soluble protein concentration of SEh (189.80 mg/mL) was comparable to the protein concentration of FSh (179.80 mg/mL). This may be due to the presence of a strong peptide bond in the protein solution of EBNs, whereby any further increase in the concentration of EBN hydrolysate (10,000 ppm) was not able to increase the solubility of the protein due to the saturated state.

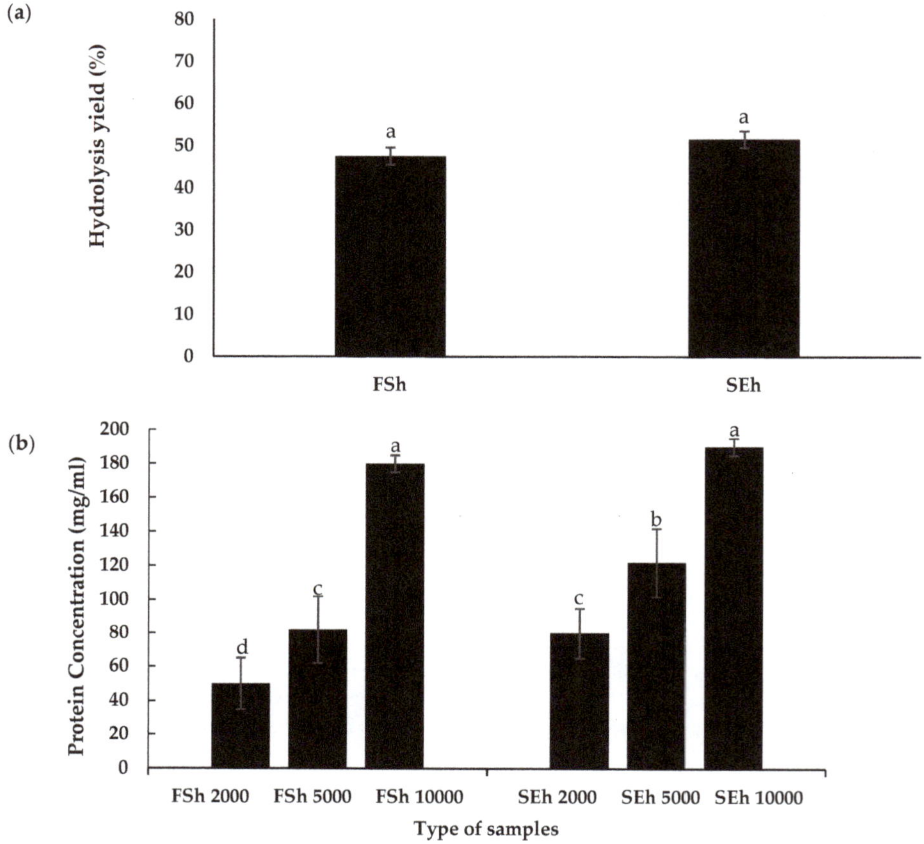

Figure 4. The (**a**) hydrolysis yield and (**b**) protein concentration of half-cup EBN hydrolysate. Data shown are means ± standard deviation of triplicates. FSh: full stew hydrolysate; SEh: stew hydrolysate. [a,b,c] Different letters indicate a significant difference between FS and SE hydrolysates at different concentrations ($p < 0.05$).

Enzymatic hydrolysis is often used to improve the solubility of the protein, and the protein degradation results in the formation of amino acids and peptides with smaller molecular masses easily absorbed by the digestive system. However, strong peptide–peptide interactions and the presence of glycoprotein that cannot be hydrolyzed result in less than 100% protein solubility [38]. Pepsin is a proteolytic enzyme secreted by the stomach, while pancreatin is produced by the pancreas, which contains other enzymes such as elastase, trypsin, and chymotrypsin proteases. Optimized extraction followed by simulated gastrointestinal digestion (pepsin and pancreatin hydrolysis) mimics the natural surroundings of the human digestive tract. This process could potentially increase the bioavailability of EBN bioactive peptides and subsequently absorb them through the gut to exert their functional effects.

The EBN hydrolysis process involves cleaving the peptide bonds within EBN proteins into bioactive glycopeptides using enzymes, thus improving the solubility of protein [4]. FS and SE extraction methods, in combination with enzymatic hydrolysis, increased the solubility of EBN extracts and hydrolysate in the respective samples. It is important for high protein food to be able to be well digested, metabolized, and absorbed by the human body, especially in athletes or patients with injuries or burns. Therefore, EBNs are a great alternative for substituting high protein sources into normal dietary intake.

3.6. Molecular Weight Distribution of Protein in Extracts and Hydrolysates of EBNs

Characterization of the protein in EBN extracts (FS and SE) and hydrolysates (FSh and SEh) were done by SDS-PAGE with Coomassie blue staining, as shown in Figure 5. The FS and SE extracts depicted a dominant band between 70 kDa and 180 kDa at the top of the gel and a smeared electrophoresis pattern down to 20 kDa, suggesting larger protein sizes had difficulty moving through the gel matrix. This is supported by [21], where the EBN protein band was clearly shown at the top of the SDS-PAGE gel matrix. These band patterns show that EBN extract contains large and complex molecular weight proteins. The protein molecular weight ranges from 100 to 135 kDa, indicating the presence of sialic acid, which has been reported as bound to protein with molecular weights of 106 and 128 kDa in EBN samples, known as 'sialo-glycoprotein' [18].

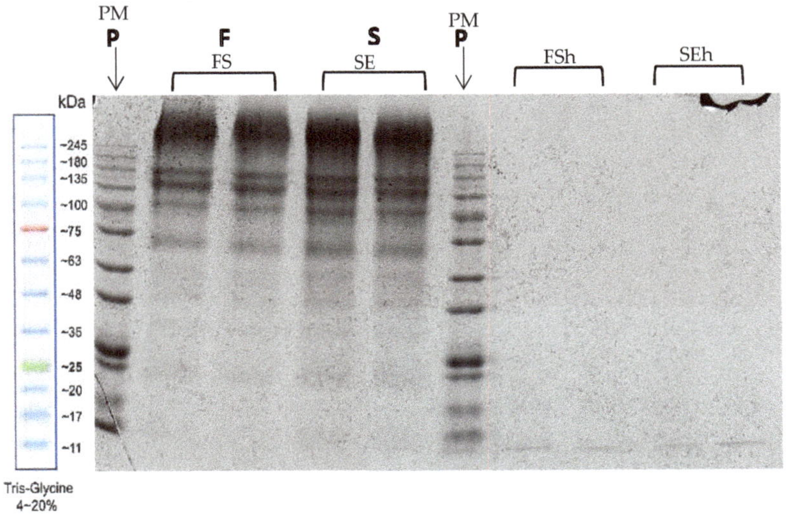

Figure 5. The molecular weight of protein for half-cup EBN extracts and hydrolysates. PM: Protein Marker; FS: Full stew; SE: Stew extract; FSh: Full stew hydrolysate; SEh: Stew hydrolysate.

The FSh and SEh band shows that almost all protein is present in smaller peptides with a molecular weight of less than 11 kDa. Enzymatic hydrolysis breaks down the extracted protein of EBNs by uncoiling the protein and peptide bond, thus producing small molecules with lower molecular weight, as observe using SDS-PAGE. This is also due to the 7.0 KDa cut-off molecular weight of the dialysis tubing used to separate the mixture. The protein size of EBN extract was 140.8, 64.8, and 21.1 kDa, whereas the molecular weight of EBN glycoprotein was 140.8 and 64.8 kDa [39]. Raw EBNs depict more and distinct bands as compared to processed EBNs, where the EBN processing may have reduced the original amount of intact protein, thus resulting in different protein profiles [4]. The different types of EBN depict different protein profiles in SDS-PAGE, with 37–52 kDa as the most abundant protein size found in EBNs from Malaysia [5,36].

3.7. Protein and Peptide Profile of EBN

The protein and peptide sequences identified in FS and SE extracts from half-cup EBNs using LC-MS/MS are shown in Table 3. There were seven parent proteins identified in EBNs, namely 78 kDa glucose-regulated protein, lysyl oxidase-3, mucin-5AC-like, acidic mammalian chitinase-like (AMCase-like), 45 kDa calcium-binding protein, nucleobindin-2, and ovoinhibitor-like, with a molecular weight ranging from 39.15 to 181.68 kDa.

Table 3. Protein and peptide sequences identified by LC-MS/MS in the full stew (FS) and stew extracts (SE) of half-cup EBNs.

Protein	MW (kDa)	Parent Protein, (Accession No.) [a]	Score [b]		Sample	Peptide Sequence [c]	Position [d]	mz^{-1}	Biological Function (Annotated in UniProt/SwissProt)
			Full Stew(FS)	Stew (SE)					
1	67.72	78 kDa glucose-regulated protein(A0A0A0B169)	204.84	39.01	FS/SE	[K]. SQIFSTASDNQPTVTIK. [V]	407–423	918.96	Stress response
2	53.97	Lysyl oxidase homolog 3 (A0A0A0B371)	538.63	538.63	FS/SE	[R]. QLPVTEGIVEVR. [Y]	97–108	670.38	Stabilization of collagen fibrils, elasticity of mature elastin
					FS/SE	[R]. IPGFKDSNVIETEQSHVEEVR. [L]	66–86	604.05	
					FS/SE	[R]. LRPVVSGAR. [R]	87–95	318.87	
					FS/SE	[K]. DSNVIETEQSHVEEVR. [L]	71–86	935.94	
					SE	**RQLPVTEGIVEVR. [Y]**	**96–108**	**499.29**	
3	181.68	Mucin-5AC (R7VT28)	668.96	326.51	FS/SE	[K]. GVLLTGWR. [S]	700–707	451.27	Gel-forming glycoprotein of gastric and respiratory tract epithelia
					FS/SE	[K]. TTSGVIEGTSAAFGNTWK. [T]	598–615	913.95	
					FS/SE	[K]. SPYEDFNIQIR. [R]	115–125	691.34	
					FS/SE	[R]. SQSVVGNVLEFANSWK. [V]	1064–1079	882.95	
					FS	**[R]. GSVLLDGK. [L]**	**152–159**	**394.72**	
4	42.04	Acidic mammalian chitinase-like (A0A0A0APZ4	198.81	213.39	FS/SE	[K]. LLVGFPTYGR. [N]	243–252	561.82	Chitin degradation, inflammatory response against pathogen
					FS/SE	[K]. FSTMVSTPQNR. [Q]	94–104	634.31	
					FS/SE	[K]. YPLITTLK. [N]	360–367	474.79	
5	39.15	45 kDa calcium-binding protein (A0A0A0AQY4)	119.81	147.78	FS/SE	[K]. NNEELKIDEETQEVLDNLKDR. [W]	153–173	636.82	Exocytosis
					FS/SE	[K]. LTLSEFISLPVGTVENQQAQ-DIDDDWVK. [D]	223–250	1054.19	
					FS/SE	[K]. TDEHFQEAVEENK. [M]	101–113	525.9	
					FS/SE	[K]. EMEEFEEDSEPR. [K]	56–67	763.8	
					FS/SE	[R]. AVDPDGDGHVSWDEYK. [I]	118–133	895.39	
					FS	**[K]. QMIAVADENQNHHLELEEILK. [Y]**	**291–311**	**619.31**	
					SE	**[K]. IKNNEELKIDEETQEVLDNLK. [D]**	**151–171**	**629.33**	
6	53.47	Nucleobindin-2 (A0A0A0AZD6)	38.59	73.77	FS/SE	[K]. EVWEEADGLDPNEFDPK. [T]	231–247	995.44	Calcium homeostasis
					FS/SE	[R]. LVTLEEFLR. [A]	312–320	560.32	
					FS/SE	[K]. AATSDLENYDK. [T]	161–171	613.78	
					FS	**[K]. LHDVNNDGFLDEQE-LEALFTK. [E]**	**252–272**	**816.39**	
					SE	**[K]. VENPDTGLYYDEYLR. [Q]**	**45–59**	**923.93**	
					SE	**[K]. QFEHLNHQNPDTFEPK. [D]**	**138–153**	**495.99**	
					SE	**[K]. LQTADIEEIK. [S]**	**75–84**	**580.31**	
7	43.22	Ovoinhibitor (A0A099ZXZ8)	46.77	41.4	FS/SE	[R]. QLMACTMIYDPVCGT-DGVTYASECTLCAHNLEHR. [T]	344–377	998.43	Anti-viral

MW: Molecular weight; [a] Accession number is a reference to *Charadrius vociferus* in SwissProt database; [b] Protein score contributed by each peptide matched in SwissProt database; [c] Peptide sequences were matched with *C. vociferous* protein databases in SwissProt; [d] Position of the peptide inside the parent protein; Peptide sequences with **BOLD** text only identified in respective extract.

About 29 peptide sequences were found matched to the seven identified parent proteins, which had between 8 and 34 amino acid sequences in length and a molecular weight between 318.87 and 1054.19 Da. In this study, 78 kDa glucose-regulated protein was identified in half-cup EBN extract by LC-MS/MS, which has not been reported elsewhere. Meanwhile, six similar parent proteins were reported by [21], with the best-scored proteins being AMCase-like, mucin-5AC-like, and ovoinhibitor-like proteins. These similar findings may be attributed to the same double-boiled method of water extraction used in the present study.

The most abundant proteins identified in both FS and SE extract were Lysyl oxidase-3 and Mucin-5AC-like protein, which were also frequently detected in EBN samples [18,37]. The authenticity of EBNs from Malaysia, Indonesia, and Vietnam was achieved and was comparable with genuine EBNs, while the outcome of adulterated EBNs (being mixed with faked EBNs) was the opposite [40]. This authenticity assessment concluded that Muc-5AC protein data could serve as an internal marker of EBN fingerprinting with the highest discriminative power, due to this protein abundance in EBNs. However, there were peptide sequences that were only present in either FS or SE half-cup EBNs, corresponding to their parent protein. This might contribute to the different extraction methods used in digestion, resulting in different bioactive peptides.

The 78 kDa glucose-regulated protein or heat shock protein was found only in this study. By referring to their chaperone role, 78 kDa glucose-regulated protein was found to be responsible for regulating cell survival and apoptosis by degradation of misfolded protein and regulation of apoptotic activity, which significantly increases the neuroprotective effects in 6-OHDA-treated neuroblastoma cell model SH-SY5Y [10]. Lysyl oxidase-3 is an enzyme that is essential for the stabilization of collagen fibrils and elasticity of elastin, supporting the evidence on skin complexion improvement in EBN-supplemented ovariectomized rats [9].

Mucin-5Ac-like protein, with a molecular weight of 181.68 kDa, is a gel-forming glycoprotein produced by epithelial tissues and is a key component in most gel-like secretion, like saliva. Muc5Ac is secreted in the gastric and respiratory tract and functions to protect the mucosa from infection and chemical damage [21]. The AMCase-like protein could be highly expressed in the salivary gland of swiftlets and secreted into the saliva for digestion of chitin-containing insects or fungi [10]. The expression of chitinase in the human body could be triggered in response to allergies, possibly accounting for the anti-inflammatory properties of EBNs [11].

The 45 kDa calcium-binding protein consists of 361 amino acids and is associated with cellular and extracellular functions, from calcium homeostasis to calcium signaling pathways [41]. This protein is present in the brain regions and is highly expressed in astrocytes, neurons, oligodendrocyte progenitor cells, myelinating oligodendrocyte, microglia/macrophage, and endothelial cells; it might have fundamental functions in a cell's calcium homeostasis [42].

Nucleobindin-2 is a calcium-binding protein with a suggested role in calcium level maintenance, eating regulation in the hypothalamus, and release of tumor necrosis factor from vascular endothelial cells. In another study on nueclobindinn-2, it was found able to control insulin sensitivity in the condition of caloric excess in the body; loss of this protein increased inflammation and insulin resistance upon high-fat feeding [43]. The protein ovoinhibitor is a serine protease inhibitor that can be found in egg whites. It plays a significant role in the antibacterial defense against *Bacillus subtilis* and *S. aureus* [44].

Proteomic analysis was influenced by the extraction method in projecting the biodiversity of the peptide profile. The application of different extraction methods for similar types of EBN resulted in a homogenous protein profile with respective matched peptide sequences. This input could be used as a future reference by researchers to explore the bioactive peptide and functional activities of EBNs. It is believed that there are still many unknown proteins, as limited databases are available for EBN proteins. Through the protein identification results, hidden potential bioactive peptides can be explored for nutritional and medicinal usage.

4. Conclusions

Numerous studies have proven the health benefits of EBN consumption, leading to an increase in the EBN farming business. However, the physicochemical and chemical characteristics of different shapes of EBNs have scarcely been reported. The findings from this study conclude that two different shapes (half-cup and stripe-shaped) of EBN comprise different physicochemical (physical, morphology, elemental composition, and microbial content) and chemical (crude protein content and amino acid profile) characteristics. In addition, optimum hydrolysis produces desirable effects on the extraction yield, soluble protein, sialic acid content, molecular weight, and protein profile. The extraction temperature (73 to 80 °C) applied in this study influences the yield, protein concentration, and solubility of EBNs. This contributes to the identification of potentially bioactive compounds, proteins, and peptides after different extraction methods. Thus, it is important to fully characterize EBNs before grading. Simultaneously, these findings can be a reference method for optimizing food preparation for EBN products by both industry and consumers. This research could be a reference for further studies on EBN biological activity.

Author Contributions: Conceptualization, N.N.M.N. and R.M.I.; Methodology, N.N.M.N. and R.M.I.; Software, N.N.M.N. and R.M.I.; Validation, N.N.M.N. and R.M.I.; Formal analysis, N.N.M.N. and R.M.I.; Investigation, N.N.M.N. and R.M.I.; Resources, R.M., M.Z.A.B. and N.A.A.R.; Data curation, N.N.M.N. and R.M.I.; Writing, N.N.M.N. and R.M.I.; Writing—review and editing, N.A.A.R.; Visualization, N.N.M.N. and R.M.I.; Supervision, R.M., M.Z.A.B. and N.A.A.R.; Project administration, R.M., M.Z.A.B. and N.A.A.R.; Funding acquisition, N.A.A.R. All authors have read and agreed to the published version of the manuscript.

Funding: This work was supported by the Research Management Centre, Universiti Putra Malaysia (UPM) under Putra Grant-Putra Young Initiative (IPM) [grant number: 9543900].

Data Availability Statement: The physicochemical analysis (physical, morphology, elemental composition, and microbial content), chemical analysis (crude protein and amino acid), sialic acid analysis and extraction effects on soluble protein and molecular weight data used to support the findings are provided within the article.

Acknowledgments: The authors are thankful to the Natural Medicine and Products Research Laboratory, Institute of Bioscience, UPM; Department of Chemistry, Faculty of Science, UPM; Technical Services & Technology Commercialization Centre, Malaysian Agricultural Research and Development Institute (MARDI) for providing infrastructure and technical support for the analysis. We are grateful for the financial support by the Research Management Centre of UPM.

Conflicts of Interest: The authors declare no conflict of interest. The funders had no role in the design of the study; in the collection, analyses, or interpretation of data; in the writing of the manuscript, or in the decision to publish the results.

References

1. Kamaruddin, R.; Engku Ismail, C.E.M.; Ahmad, S.A. Key Factors for the Sustainable Production of Swiftlet Birds' Nest Industry in Malaysia: A Case Study in Northern Peninsular Malaysia. *Supply Chain Manag.* **2019**, *8*, 724–733.
2. Halimi, N.M.; Kasim, Z.M.; Babji, A.S. Nutritional composition and solubility of edible bird nest (Aerodramus fuchiphagus). In Proceedings of the THE 2014 UKM FST POSTGRADUATE COLLOQUIUM: Proceedings of the Universiti Kebangsaan Malaysia, Faculty of Science and Technology 2014 Postgraduate Colloquium, Selangor, Malaysia, 9–11 April 2014; AIP Publishing: Melville, NY, USA, 2014; Volume 1614, pp. 476–481.
3. Ghassem, M.; Arihara, K.; Mohammadi, S.; Sani, N.A.; Babji, A.S. Identification of two novel antioxidant peptides from edible bird's nest (Aerodramus fuciphagus) protein hydrolysates. *Food Funct.* **2017**, *8*, 2046–2052. [CrossRef]
4. Daud, N.; Yusop, S.M.; Babji, A.S.; Lim, S.J.; Sarbini, S.R.; Yan, T.H. Edible Bird's Nest: Physicochemical Properties, Production, and Application of Bioactive Extracts and Glycopeptides. *Food Rev. Int.* **2021**, *37*, 177–196. [CrossRef]
5. Ma, F.; Liu, D. Sketch of the edible bird's nest and its important bioactivities. *Food Res. Int.* **2012**, *48*, 559–567. [CrossRef]
6. Nurfatin, M.H.; Etty Syarmila, I.K.; Nur'Aliah, D.; Zalifah, M.K.; Babji, A.S.; Ayob, M.K. Effect of Enzymatic Hydrolysis on Angiotensin Converting Enzyme (ACE) Inhibitory Activity in Swiftlet Saliva. *Int. Food Res.* **2016**, *23*, 141–146.
7. Hou, P.; Hu, S.; Wang, J.; Yang, Z.; Yin, J.; Zhou, G.; Guo, S. Exogenous supplement of N-acetylneuraminic acid improves macrophage reverse cholesterol transport in apolipoprotein E-deficient mice. *Lipids Health Dis.* **2019**, *18*, 24. [CrossRef]
8. Mahaq, O.; Rameli, M.A.P.; Edward, M.J.; Hanafi, N.M.; Aziz, S.A.; Abu Hassim, H.; Noor, M.H.M.; Ahmad, H. The effects of dietary edible bird nest supplementation on learning and memory functions of multigenerational mice. *Brain Behav.* **2020**, *10*. [CrossRef] [PubMed]

9. Matsukawa, N.; Matsumoto, M.; Bukawa, W.; Chiji, H.; Nakayama, K.; Hara, H.; Tsukahara, T. Improvement of Bone Strength and Dermal Thickness Due to Dietary Edible Bird's Nest Extract in Ovariectomized Rats. *Biosci. Biotechnol. Biochem.* **2011**, *75*, 590–592. [CrossRef] [PubMed]
10. Yew, M.Y.; Koh, R.Y.; Chye, S.M.; Othman, I.; Ng, K.Y. Edible bird's nest ameliorates oxidative stress-induced apoptosis in SH-SY5Y human neuroblastoma cells. *BMC Complement. Altern. Med.* **2014**, *14*, 391. [CrossRef] [PubMed]
11. Yida, Z.; Imam, M.U.; Ismail, M.; Hou, Z.; Abdullah, M.A.; Ideris, A.; Ismail, N. Edible Bird's Nest attenuates high fat diet-induced oxidative stress and inflammation via regulation of hepatic antioxidant and inflammatory genes. *BMC Complement. Altern. Med.* **2015**, *15*, 310. [CrossRef] [PubMed]
12. Chua, L.S.; Zukefli, S.N. A comprehensive review of edible bird nests and swiftlet farming. *J. Integr. Med.* **2016**, *14*, 415–428. [CrossRef]
13. Xie, Y.; Zeng, H.; Huang, Z.; Xu, H.; Fan, Q.; Zhang, Y.; Zheng, B. Effect of Maternal Administration of Edible Bird's Nest on the Learning and Memory Abilities of Suckling Offspring in Mice. *Neural Plast.* **2018**, *2018*, 7697261. [CrossRef] [PubMed]
14. Mohd Khan, A.; Etty Syarmila, I.K.; Nurfatin, M.H.; Farahniza, Z.; Engku Hanisah, E.U.; Norhasidah, S.; Masitah, E.H.; Masturah, A.K.; Nurul'Ain, M.; Maaruf, A.G.; et al. Antioxidative properties of ready-to-drink products incorporated with enzymatically hydrolysed edible bird nest. In Proceedings of the Edible Bird Nest Industry Conference, Marriot Hotel Putrajaya, Putrajaya, Malaysia, 25–26 November 2014.
15. Department of Standards Malaysia. Edible-birdnest (EBN)—Specification. In *Malaysian Standard MS 2334:2011*; Food Safety and Quality Division; Ministry of Health Malaysia: Kuala Lumpur, Malaysia, 2011.
16. Hun, L.T.; Wani, W.A.; Tan, E.T.T.; Adnan, N.A.; Ling, Y.L.; Abdul Aziz, R. Investigations into the physicochemical, biochemical, and antibacterial properties of edible bird's nest. *J. Chem. Pharm. Res.* **2015**, *7*, 228–247.
17. Shinners, T.C.; Tewari, J.P. Diversity in crystal production by some bird's nest fungi (Nidulariaceae) in culture. *Can. J. Chem.* **1997**, *75*, 850–856. [CrossRef]
18. Wong, S.F.; Lim, P.K.C.; Mak, J.W.; Ooi, S.S.; Chen, D.K.F. Molecular characterization of culturable bacteria in raw and commercial edible bird nests (EBNs). *Int. Food Res. J.* **2018**, *25*, 966–974.
19. Wan Syahidah, H.; Normah, M.; Albert, T. Protein content and amino acid composition of farmed edible bird's nests in Malaysia. *MJVR* **2014**, *5*, 342–346.
20. Najafian, L.; Babji, A. A review of fish-derived antioxidant and antimicrobial peptides: Their production, assessment, and applications. *Peptides* **2012**, *33*, 178–185. [CrossRef]
21. Imam, M.U.; Ismail, M.; Ooi, D.J.; Azmi, N.H.; Sarega, N.; Chan, K.W.; Bhanger, M.I. Are bioactive-rich fractions functionally richer? *Crit. Rev. Biotechnol.* **2015**, *36*, 585–593. [CrossRef]
22. Wong, Z.C.; Chan, G.K.; Wu, L.; Lam, H.H.; Yao, P.; Dong, T.T.; Tsim, K.W. A comprehensive proteomics study on edible bird's nest using new monoclonal antibody approach and application in quality control. *J. Food Compos. Anal.* **2018**, *66*, 145–151. [CrossRef]
23. U.S. Food and Drug Administration/Bacteriological Analytical Manual (FDA/BAM). Available online: https://www.fda.gov/food/laboratory-methods-food/bacteriological-analytical-manual-bam (accessed on 23–27 August 2021).
24. AOAC. Method 991.14. In *Official Methods of Analysis of AOAC International*, 17th ed; AOAC: Gaithersburg, MD, USA, 1995; pp. 22–23.
25. Horwitz, W. Methods 930.15, 942.05, 920.39, 954.01, 991.43, 50.1.14, 9.1.09. In *Official Methods of Analysis*, 17th ed.; Association of Official Analytical Chemists: Gaithersburg, MD, USA, 2000.
26. Mæhre, H.; Dalheim, L.; Edvinsen, G.; Elvevoll, E.; Jensen, I.-J. Protein Determination—Method Matters. *Foods* **2018**, *7*, 5. [CrossRef]
27. Murad, M.; Abdullah, A.; Mustapha, W.A.W. Antioxidant capacity and amino acid profiles of egg tofu. *Am. J. Appl. Sci.* **2013**, *10*, 1315–1324. [CrossRef]
28. Careena, S.; Sani, D.; Tan, S.N.; Lim, C.W.; Hassan, S.; Norhafizah, M.; Kirby, B.; Ideris, A.; Stanslas, J.; Bin Basri, H. Effect of Edible Bird's Nest Extract on Lipopolysaccharide-Induced Impairment of Learning and Memory in Wistar Rats. *Evid.-Based Complement. Altern. Med.* **2018**. [CrossRef] [PubMed]
29. Mohd Rosni, S.; Fisal, A.; Azwan, A.; Chye, F.Y.; Matanjun, P. Crude proteins, total soluble proteins, total phenolic contents and SDS-PAGE profile of fifteen varieties of seaweed from Semporna, Sabah, Malaysia. *Int. Food Res. J.* **2015**, *22*, 1483–1493.
30. Laemmli, U.K. Cleavage of Structural Proteins during the Assembly of the Head of Bacteriophage T4. *Nature* **1970**, *227*, 680–685. [CrossRef] [PubMed]
31. Yu, Y.-Q.; Gilar, M.; Lee, P.J.; Bouvier, E.S.P.; Gebler, J.C. Enzyme-Friendly, Mass Spectrometry-Compatible Surfactant for In-Solution Enzymatic Digestion of Proteins. *Anal. Chem.* **2003**, *75*, 6023–6028. [CrossRef] [PubMed]
32. WHO. Food Safety. Food Safe. 2020. Available online: https://www.who.int/newsroom/fact-sheets/detail/food-safety (accessed on 14 March 2021).
33. Kew, P.E.; Wong, S.F.; Lim, P.K.; Mak, J.W. Structural analysis of raw and commercial farm edible bird nests. *Trop. Biomed.* **2014**, *31*, 63–76. [PubMed]

34. Deveaux, A.; Pham, I.; West, S.G.; André, E.; Lantoine-Adam, F.; Bunouf, P.; Sadi, S.; Hermier, D.; Mathé, V.; Fouillet, H.; et al. L-Arginine Supplementation Alleviates Postprandial Endothelial Dysfunction When Baseline Fasting Plasma Arginine Concentration Is Low: A Randomized Controlled Trial in Healthy Overweight Adults with Cardiometabolic Risk Factors. *J. Nutr.* **2016**, *146*, 1330–1340. [CrossRef]
35. Bezerra, R.; Costa, F.G.P.; Givisiez, P.E.N.; Goulart, C.C.; Dos Santos, R.A.; De Lima, M.R. Glutamic acid supplementation on low protein diets for laying hens. *Acta Sci. Anim. Sci.* **2015**, *37*. [CrossRef]
36. Zukefli, S.N.; Chua, L.S.; Rahmat, Z. Protein Extraction and Identification by Gel Electrophoresis and Mass Spectrometry from Edible bird's Nest Samples. *Food Anal. Methods* **2016**, *10*, 387–398. [CrossRef]
37. Norhayati, M.K.; Azman, O.; Wan Nazaimoon, W.M. Preliminary study of the nutritional content of Malaysian edible bird's nest. Malays. *J. Nutr.* **2010**, *16*, 389–396.
38. Khushairay, E.S.I.; Ayub, M.K.; Babji, A.S. Effect of enzymatic hydrolysis of pancreatin and alcalase enzyme on some properties of edible bird's nest hydrolysate. In Proceedings of the THE 2014 UKM FST POSTGRADUATE COLLOQUIUM: Proceedings of the Universiti Kebangsaan Malaysia, Faculty of Science and Technology 2014 Postgraduate Colloquium, Selangor, Malaysia, 9–11 April 2014; AIP Publishing LLC: Melville, NY, USA, 2014; Volume 1614, pp. 427–432.
39. Utomo, B.; Rosyidi, D.; Radiati, L.; Puspaningsih, N.; Proborini, W. Protein Characterization of Extracted Water from Three Kinds of Edible Bird Nest Using SDS-PAGE CBB Staining and SDSPAGE Glycoprotein Staining and LC-MS/MS Analyses. *IOSR J. Agric. Veter-Sci.* **2014**, *7*, 33–38. [CrossRef]
40. Wong, C.-F.; Chan, G.K.-L.; Zhang, M.-L.; Yao, P.; Lin, H.-Q.; Dong, T.T.-X.; Li, G.; Lai, X.-P.; Tsim, K.W.-K. Characterization of edible bird's nest by peptide fingerprinting with principal component analysis. *Food Qual. Saf.* **2017**, *1*, 83–92. [CrossRef]
41. Gong, W.; Martin, T.A.; Sanders, A.J.; Jiang, A.; Sun, P.; Jiang, W.G. Location, function and role of stromal cell-derived factors and possible implications in cancer (Review). *Int. J. Mol. Med.* **2020**, *47*, 435–443. [CrossRef]
42. Kelemen, K.; Szilágyi, T. New Approach for Untangling the Role of Uncommon Calcium-Binding Proteins in the Central Nervous System. *Brain Sci.* **2021**, *11*, 634. [CrossRef] [PubMed]
43. Ravussin, A.; Youm, Y.-H.; Sander, J.; Ryu, S.; Nguyen, K.; Varela, L.; Shulman, G.I.; Sidorov, S.; Horvath, T.L.; Schultze, J.L.; et al. Loss of Nucleobindin-2 Causes Insulin Resistance in Obesity without Impacting Satiety or Adiposity. *Cell Rep.* **2018**, *24*, 1085–1092. [CrossRef] [PubMed]
44. Słowińska, M.; Liszewska, E.; Nynca, J.; Bukowska, J.; Hejmej, A.; Bilińska, B.; Szubstarski, J.; Kozłowski, K.; Jankowski, J.; Ciereszko, A.; et al. Isolation and Characterization of an Ovoinhibitor, a Multidomain Kazal-Like Inhibitor from Turkey (Meleagris gallopavo) Seminal Plasma1. *Biol. Reprod.* **2014**, *91*, 108. [CrossRef] [PubMed]

foods

MDPI

Article

Antioxidant Effect of Moroccan Pomegranate (*Punica granatum* L. Sefri Variety) Extracts Rich in Punicalagin against the Oxidative Stress Process

Lamiae Benchagra [1], Hicham Berrougui [1,2], Mohamed Obaidul Islam [2,3], Mhamed Ramchoun [1], Samira Boulbaroud [1], Abdelouahed Hajjaji [1], Tamas Fulop [2], Gianna Ferretti [3] and Abdelouahed Khalil [2,*]

1 Department of Biology, Polydisciplinary Faculty, University Sultan Moulay Slimane, Beni Mellal 23020, Morocco; benchagralamiae@gmail.com (L.B.); hichamberg@gmail.com (H.B.); ramchoun_10@yahoo.fr (M.R.); sboulbaroud@gmail.com (S.B.); abhajjaji@yahoo.fr (A.H.)

2 Department of Medicine, Geriatrics Service, Faculty of Medicine and Biological Sciences, University of Sherbrooke, Sherbrooke, QC J1H 4N4, Canada; m.o.islam@pm.univpm.it (M.O.I.); Tamas.Fulop@USherbrooke.ca (T.F.)

3 Department of Clinical Science and Odontostomatology (DISCO), Polytechnic University of Marche, I-60126 Ancona, Italy; g.ferretti@univpm.it

* Correspondence: a.khalil@usherbrooke.ca

Citation: Benchagra, L.; Berrougui, H.; Islam, M.O.; Ramchoun, M.; Boulbaroud, S.; Hajjaji, A.; Fulop, T.; Ferretti, G.; Khalil, A. Antioxidant Effect of Moroccan Pomegranate (*Punica granatum* L. Sefri Variety) Extracts Rich in Punicalagin against the Oxidative Stress Process. *Foods* **2021**, *10*, 2219. https://doi.org/10.3390/foods10092219

Academic Editor: Antonello Santini

Received: 28 June 2021
Accepted: 12 September 2021
Published: 18 September 2021

Abstract: Natural antioxidants products are widely distributed in food and medicinal plants. These natural antioxidants, especially polyphenols, exhibit a wide range of biological activities including anti-cancer, anti-inflammatory, and anti-atherosclerosis activities. Pomegranate (*Punica granatum* L.) is a rich source of polyphenolic components. The purpose of this study was to characterize the phenolic composition and flavonoids and anthocyanin content of different parts (peel and aril) of the *Sefri* variety of pomegranate. Our results showed that Peel extract was richer in these compounds than that of the Arils, especially in Punicalagin (A and B). DPPH free radical scavenging, reducing power (FRAP), β-carotene bleaching, and hydrogen peroxide scavenging assays revealed a greater dose-dependent activity of pomegranate peel phenolic extract (PPPE) compared to pomegranate aril phenolic extract (PAPE). PPPE was also more potent than PAPE concerning its ability to inhibit conjugated diene formation and to reduce α-tocopherol disappearance induced by CuSO4-mediated LDL peroxidation. Interestingly, both extracts (PPPE and PAPE) significantly inhibited lipid peroxidation and the formation of reactive oxygen species (ROS) in stressed J82 human bladder cancer cells. These results reflect the protective effects that this Moroccan variety of pomegranate can provide against the development of metabolic disorder, cancer, atherosclerosis, and cardiovascular disease. Given these properties, further studies should be undertaken to investigate possible applications of *Sefri* pomegranate extracts in the fields of food preservation and health supplements.

Keywords: *Punica granatum* L.; antioxidant activity; low density lipoprotein (LDL); J82 human bladder cell line; paraoxonase 1

1. Introduction

Pomegranate peel, seeds, juice, and arils are a rich source of several valuable bioactive compounds with considerable nutritional, antioxidant, and other beneficial properties [1–3]. Peel possesses a higher polyphenol content than seeds and juice [4]. These polyphenols include punicalagin, which exhibits high antioxidant activity. Pomegranate also contains other polyphenols, including anthocyanins (delphinidin, cyanidin, and pelargonidin 3-glucosides and 3,5-glucosides) as well as flavonols [5]. Pomegranate peel is known for its healing properties with respect to inflammatory diseases, diabetes, atherosclerosis, oxidative stress, cancer, and microbial infections [6–11]. Moreover, pomegranate fruit is used in the food industry such as dairy products, charcuterie and juice preparation, and conservation. Sweet pomegranates are consumed fresh while sour pomegranates with hard seeds are generally intended for processing [12]. Even if the consumption of arils leaves less waste, the fact remains that agri-food industry use of the pomegranate generates large amounts of peel-waste and by-products that are usually poorly exploited. Pomegranate peels represent 50% of total fruit weight and are a potential source of bioactive compounds, mainly phenolic compounds with a very broad spectrum of activity [13]. Indeed, several studies have reported the effects (antioxidant, anti-cancer, anti-inflammatory, lipid-lowering, anti-hypertensive) of phenolic extracts from the peel of the pomegranate, which makes them suitable as natural ingredients [13–17]. In addition, the natural product supplementation industry is increasingly interested in peel waste, manufacturing capsules containing concentrated phenolic compounds based on peels or whole fruit. Moreover, pomegranate peels are also used for their dyeing properties [18]. In fact, pomegranate peels may also be used as green antimicrobial agents to reduce inorganic nanoparticle consumption on wool yarns [19]. Pomegranate waste (peels and seeds) is also used in meat preparation and conservation to prevent bacterial development and the oxidation process [17,20,21].

Morocco's annual pomegranate production exceeds 58,000 tones from a total area of 5000 hectares [22,23] and half of this production is grown in central Morocco on the planes of the Middle Atlas (Beni Mellal-Khenifra area) and is mainly represented by the Sefri variety. There are many pomegranate varieties in Morocco. P. granatum also has considerable synonymy, in which the same genotype is known by different names in different regions.

In this study we investigated several beneficial health properties of various bioactive compounds from *Sefri* pomegranates, notably the anti-oxidative and physicochemical properties of pomegranate peel phenolic extract (PPPE) and pomegranate aril phenolic extract (PAPE) and evaluated their effects on atherosclerosis and bladder cancer cells.

LDL oxidation is considered to be a hallmark of early atherogenesis. Nutritional antioxidants such as phenolic compounds can markedly inhibit LDL oxidative damage by reducing free radicals generated during oxidative metabolism, preserving endogenous antioxidants in LDL (vitamin E and carotenoids), chelating transition metal ions, and modulating the oxidative state of the arterial cell wall. These properties act to inhibit cell-mediated oxidation of LDL and increase serum paraoxonase (PON1) activity [24].

Polyphenols and/or their derivatives are used to treat cancer. Cancer initiation may be modulated by an increase in ROS levels, which can damage DNA and stimulate pro-oncogenic signaling [25]. Oxidative stress regulates the progression of different types of cancer, including breast, liver, lung, colon, prostate, and bladder cancer.

In the Beni Mellal-Khenifra area, all parts of the pomegranate, especially peel, are used for health remedies to treat diarrhea, ulcers, nasal bleeding, and inflammation. Moreover, this fruit is highly consumed by patients with vascular disorders. This variety may have a more powerful health potential regard to its possible richness in phenolic compounds.

To the best of our knowledge, and even though various pomegranates are cultivated in different regions of Morocco, only a few studies have focused on the chemical composition and properties of the Sefri variety of pomegranate and even fewer on its biological properties, particularly in the prevention of diseases linked to oxidative stress such as cardiovascular disease and cancer. From this perspective, we conducted this study firstly to analyze the chemical composition of phenolic compounds from peels and arils of this variety. At the biological level, we took a particular interest in studying the relationship between antioxidant activity of the Sefri cultivar and its impact in preventing lipid peroxidation in human LDLs [26] and in modulating PON 1 activity and enzyme expression. On the other hand, since oxidative stress plays an important role in cancer development and progression [27], we evaluated the effect of sefri pomegranate polyphenols on reactive oxygen species (ROS) production in J82 human bladder cancer cells.

2. Materials and Methods

2.1. Chemicals

All chemicals were of analytical reagent grade and were purchased from Sigma-Aldrich Chemical Co. (Saint Louis, MO, USA), except for Trolox, $FeCl_2$-$4H_2O$, ethylene-di-amine-tetra-acetic acid (EDTA), butylated hydroxytoluene (BHT), ferrozine ascorbate, H_2O_2, potassium ferricyanide, phosphate buffer, ferric chloride, gallic acid, catechin, Folin-Ciocalteu reagent, sodium nitrite, aluminium chloride, glacial acetic acid, acetonitrile, and formic acid, which were obtained from Sigma-Aldrich Chemical Co. (Pool, UK). Ethanol, methanol, trichloroacetic acid (TCA), sodium carbonate, and sodium hydroxide were purchased from Fisher Scientific (Loughborough, UK). Synergi 4 µm Hydro-RP 80A columns (250 mm × 4.6 mm × 5 µm) and C18 5 µm columns (250 mm × 3.0 mm) were purchased from Phenomenex (Torrance, CA, USA).

2.2. Plant Materials

Pomegranate fruit was harvested from Sefri pomegranate trees (central Morocco; Latitude: 23°50′05″ E; Longitude: 6°48′98″ N). The authenticity of the variety was confirmed by Dr Abbas Younes, taxonomist, and a voucher specimen was conserved for further reference at our laboratory herbarium (Beni Mellal, Morocco). The fruit was washed and hand peeled. Arils were squeezed using a commercial blender to obtain pomegranate molasses. Air-dried pomegranate peels were ground to a fine powder, which was stored in a freezer.

2.3. Extraction of Pomegranate Phenolic Compounds

Peel powder and pomegranate aril molasses were macerated by sonication in 70% methanol/0.1% acetic acid for 48 h in the dark. The hydroalcoholic extracts were centrifuged for 10 min, and the solids were removed by vacuum filtration through a Whatman filter Grade GF 10. The supernatants were concentrated under vacuum and were then freeze-dried and stored at −80 °C until used.

2.4. Quantification of Total Phenolic Content

Total phenolic content (TPC) was determined using the modified Folin-Ciocalteu method [28]. Gallic acid was used as a standard equivalent. The reaction mixtures were prepared by mixing 100 µL of each sample with 500 µL of Folin-Ciocalteu reagent (1/10). The mixtures were incubated for 10 min, following which 400 µL of 20% w/v Na_2CO_3 was added. Following a 2-h incubation at room temperature in the dark, the absorbance was read at 760 nm using a UV-6300PC double beam spectrophotometer (VWR, Darmstadt, Germany). TPC was determined using the gallic acid calibration curve and was expressed as milligrams of gallic acid equivalents per gram of dry matter. All samples were analyzed in triplicate.

2.5. Quantification of Total Flavonoid Content

Total flavonoid content (TFC) was quantified using the method described by Woisky et al., with minor modifications [29]. Briefly, 50 µL of the sample was mixed with 1.5 mL of 95% methanol, 100 µL of 2% aluminium chloride, and 350 µL of distilled water. Following a 1-h incubation at room temperature, the absorbance was measured at 420 nm. A methanolic solution of quercetin was used as a reference, and TFC was expressed as milligrams of quercetin equivalents per gram of dry matter.

2.6. Determination of Total Anthocyanin Content

Total anthocyanin content (TAC) was determined by the pH differential method as described by Sellappan et al. [30] using two buffer systems: potassium chloride buffer

(25 mM, pH 1.0) and sodium acetate buffer (400 mM, pH 4.5). The absorbance of the buffers was read at 510 and 700 nm, respectively (1).

$$A = (A_{510nm} - A_{700nm})_{pH\,1.0} - (A_{510nm} - A_{700nm})_{pH\,4.5.} \quad (1)$$

TAC was calculated as cyanidin-3-glucoside equivalents (2):

$$TAC\left(\frac{mg}{L}\right) = \frac{A \times MW \times DF \times 100}{\varepsilon} \quad (2)$$

MW: molecular weight of cyanidin-3-glucoside (449.2 g/mol). DF: dilution factor. ε: Molar extinction coefficient (26,900 L. $cm^{-1} \cdot mol^{-1}$).

2.7. HPLC Analysis of Phenolic Compounds by UV Detection

An HPLC system composed of an autosampler (SIL-HTc), a degasser (DGU-14A), a column oven (CTO-10AS), and a diode array detector (SPD-M10A) (Shimadzu, Japan), as well as an Inertsil® WP300-C18 (250 × 4.6 mm, 5 μm) column and pre-column (Canadian Life Science, ON, Canada), were used. The chromatogram was monitored at 220–400 nm), with the spectra recorded continuously throughout the elution. The following two eluents were used: A, double distilled water containing 0.2% acetic acid (pH 3.0) and B, acetonitrile. The flow rate was 1 mL/min), and the gradient was optimized as follows: 5 min (0–5%) B; 5–10 min (5–13%) B; 10–13 min (13–18%) B; 13–20 min (18%) B; 20–23 min (18–25%) B; 23–35 min (25%) B; 35–40 min (25–30%) B; 40–41 min (40–85%) B; 41–50 min (85–90%); 50–54 min (90–100%); and B; 54–60 min (100–5%) B. The samples were filtered through 0.2-mm PTFE filters, and 10 μL of the sample were injected at a stable column temperature of 30 °C. The absorption wavelengths used to detect the polyphenols ranged from 200 to 400 nm. Peak areas were quantified using a calibration curve obtained using gallic acid, α and β punicalagin, and ellagic acid as external standards. For this purpose, calibration curves were prepared for each analytical standard. Linearity, the limits of detection and quantification, are cited in Table S1 (Supplementary Data).

2.8. Antioxidant Activity Measurement Using the DPPH Radical Scavenging Assay

The DPPH (2,2-diphenyl-1-picrylhydrazyl) assay uses the capacity of the DPPH radical to scavenge, as a measure of the antioxidant activity that prevents lipid peroxidation. The free-radical scavenging activities of PPPE and PAPE were evaluated using the DPPH assay based on the method of Zhang et al. [31], with slight modifications. A 0.06 mM solution of DPPH in methanol was prepared daily. This solution (2 mL) was mixed with 50 μL of PPPE or PAPE (0–0.4 mg/mL). The mixture was incubated in the dark at room temperature for 30 min. The decrease in absorbance was measured at 517 nm. The percentage of inhibition was calculated using the following equation:

$$\%\ Inhibition = [(A_0 - A_S)/A_0] \times 100 \quad (3)$$

where A_0 is the absorbance of the control reaction (containing all the reagents except the test compound) and A_S is the absorbance of the test compound. Ascorbic acid was used as a positive control.

The half-maximal extract concentration (IC_{50}) was calculated from the plotted graph of scavenging activity against the concentrations of each extract. All experiments were performed in triplicate.

2.9. Hydrogen Peroxide (H_2O_2) Scavenging Assay

The ability of PPPE and PAPE to scavenge H_2O_2 was determined using the method of Ruch et al. [32]. The concentration of the H_2O_2 solution (40 mM) prepared in 50 mM phosphate buffer (pH 7.4) was determined by measuring the absorption at 230 nm. The absorption of the assay mixture, which contained 500 μL of different concentrations of PPPE, PAPE, or a standard ascorbic acid solution (0–200 μg/mL) together with 1 mL of H_2O_2

was determined after 30 min against a blank solution containing phosphate buffer without H_2O_2. The percentage of H_2O_2 scavenged was calculated using the following formula:

$$H_2O_2 \text{ scavenged } (\%) = [(OD_{control} - OD_{test}/OD_{control})] \times 100 \quad (4)$$

2.10. Ferric Antioxidant Power (FRAP) Assay

Ferric antioxidant power was determined using the potassium ferricyanide-ferric chloride assay [33]. A 500 μL aliquot of PPPE or PAPE was mixed with 1 mL of phosphate buffer (0.2 M, pH 6.6) and 1 mL of 1% $K_3Fe(CN)$. The mixtures were shaken well and were incubated at 50 °C for 20 min, following which 1 mL of 10% TCA was added. They were then centrifuged at 3000 rpm for 10 min. The supernatants (1.5 mL) were mixed with 1.5 mL of distilled water and 0.1 mL of 0.1% $FeCl_3$. The absorbance was read at 700 nm. Ascorbic acid was used as a reference standard. The total antioxidant activity (TAA) determined by FRAP was expressed as mg of ascorbic acid equivalents per gram of dry matter (mg AAE/g dm).

2.11. β-Carotene-Linoleic Acid Bleaching (BCB) Assay

The antioxidant activities of PPPE and PAPE were evaluated using the β-carotene-linoleic acid (BCB) assay as per the method of Jayaprakasha et al. [34]. β-carotene (200 μg), 20 mg of purified linoleic acid, and 200 mg of Tween 40 were mixed in 0.5 mL of chloroform. After the chloroform was removed under vacuum, the emulsion was further diluted with 40 mL of distilled water. Aliquots (4 mL) of this solution were transferred into a series of tubes containing 0.2 mL of the extract. As soon as the emulsion was added to each tube, the zero-time (t = 0 min) absorbance was measured at 490 nm. The absorbance was then measured at 15-min intervals until the color of the β-carotene disappeared in the control tubes (t = 180 min). A mixture without β-carotene served as a blank. Butylated hydroxytoluene (BHT) was used as a control. The antioxidant activity (AA) of the extracts was determined in terms of β-carotene bleaching using the following formula:

$$AA = [1 - (A_0 - A_{180})/(A^\circ_0 - A^\circ_{180})] \times 100 \quad (5)$$

where A_0 and A°_0 are the absorbances measured at t = 0 min of the incubation of the test sample and control, respectively, and (A_{180}) and (A°_{180}) are the absorbances measured in the sample and control, respectively, after a 180-min incubation. The assay was carried out in triplicate.

2.12. Low Density Lipoprotein Isolation

LDL (low density lipoproteins) were isolated from fasting human heparinized plasma using the method of Sattler et al. [35]. Briefly, LDLs were isolated by ultracentrifugation (543,200 g) at 15 °C using a Beckman Optima TLX ultracentrifuge equipped with a TLA-100.4 rotor. The isolated LDL were dialyzed overnight at 4 °C in sodium phosphate buffer (10^{-2} M, pH 7). Protein concentrations were measured using commercial assay kits (Bio-Rad, Mississauga, ON, Canada) according to the manufacture's protocol and were expressed as LDL total protein concentration.

The present study was conducted according to the guidelines set out in the Declaration of Helsinki. The protocol was approved by the Ethics Committee of the Sherbrooke University Institute of Geriatrics (# 2009/19). Written informed consent was obtained from all subjects.

2.13. Copper-Mediated LDLs Oxidation

LDLs were oxidized using transition metal ions as oxidizing agents [36]. Briefly, 100 μg/mL of LDL were suspended in 10 mmol/L sodium phosphate buffer (pH 7) and were incubated in the presence or absence of 0.2 mg/mL of PPPE or PAPE at 37 °C and in the presence of 10 μmol/L cupric sulfates for 4 h. The oxidation reactions were stopped by cooling in an ice bath after adding 300 μmol/L of EDTA, and the resulting lipid peroxides were measured immediately.

2.14. Biochemical Markers of Lipid Peroxidation

Conjugated Diene Formation and α-Tocopherol Disappearance

Lipid peroxidation was evaluated by measuring conjugated diene formation and the disappearance of vitamin E (α-tocopherol). Oxidized LDL levels, alone or in the presence of PPPE or PAGE, were continuously monitored at 234 nm as previously described by Berrougui et al. 2006 [37], to measure conjugated diene formation.

The endogenous LDL content of vitamin E was assayed by measuring the α-tocopherol content following 2-h oxidation of LDL in the presence or absence of 0.2 mg/mL of PPPE or PAPE using reverse-phase HPLC and electrochemical (ESA Coulochem II 5010A electrochemical cell, company) and UV (at 292 nm) detection as previously described [37]. α-Tocopherol was assayed on a Sephasil peptide column (C18 5 μm ST 4.6/250) (Pharmacia Biotech, Piscataway, NJ, USA). Tocopherol acetate was used as an internal standard. The mobile phase was composed of methanol/ethanol/isopropanol (88/24/10 by volume) containing lithium perchlorate (20 μmol/L) at a flow rate of 1 mL/min.

2.15. Paraoxonase 1 (PON 1) Protein Expression and Activity Measurement

For intracellular staining, Fu5AH hepatic cells were fixed, permeabilized using staining Kit (abcam) and stained with anti-PON1 antibody (ab24261, 1μg/1 $\times$ 10^6 cells) for 30 min at RT. The secondary antibody used was DyLight® 488 goat anti-mouse IgG (ab96879) at 1/500 dilution for 30 min at RT. Flow cytometry data were collected on a CytoFLEX instrument (Beckman Coulter, Brea, CA, USA) and analyzed using FlowJo 10.2 software (Tree Star Inc., Ashland, OR, USA).

PON1 activity was measured in plasma samples treated or not for 2 h with 0 to 80 μg/mL of PPPE or PAPE using paraoxon as a substrate, as previously described [38]. Briefly, activity was measured by combining 50 μL of the sample with 1 mL of 100 mM Tris/HCl buffer (pH 8.0) containing 2 mM $CaCl_2$ and 5.5 mM paraoxon. The rate of 4-nitrophenol release was measured at 412 nm, and enzymatic activity was calculated using a molar extinction coefficient of 17,100 M^{-1} cm^{-1}. One unit of PON1 activity was defined as nM 4-nitrophenol formed per minute. Plasma was obtained from subjects who provided written informed consent before being enrolled.

2.16. Cell Culture

The J82 (HTB1™) human bladder cancer cell line (batch # 70002468) was purchased from the American Type Culture Collection (Manassas, VA, USA) via Cederlane® company (Burlington, ON, Canada). The cells were cultured in Dulbecco's modified Eagle's medium supplemented with 10% fetal bovine serum in a humidified 95% air/5% CO_2 atmosphere at 37 °C. Cells were trypsinized using 0.05% EDTA-0.02% trypsin. The Fu5AH (rat hepatom cell line) were kindly provided by Dr J. Genest's laboratory (University of McGill, Montreal, QC, Canada). Fu5AH cells were maintained in minimal essential medium containing 5% bovine serum and antibiotics [39,40].

2.17. Determination of Reactive Oxygen Species (ROS) in J82 Cells

J82 bladder cells were seeded into 24-well cell culture plates at a density of 2.5 $\times$ 10^4 cells/well. Following a 24-h incubation, the medium was replaced with a fresh medium containing 0.1 or 0.2 mg/mL of PPPE or PAPE. The cells were incubated for a further 24 h. They were then washed with cold PBS. DCFH-DA (10 μmol/L) was added to each well. The cells were incubated for 45 min at 37 °C and were washed with cold PBS to remove DCFH-DA that did not enter the cells. The cells were rinsed, 100 μM TBHP (tert-butyl hydroperoxide) was added, and the cells were incubated for a further 2 h. The fluorescence was immediately recorded using a Victor multilabel plate reader (PerkinElmer, Guelph, ON, Canada), and the fluorescence intensity was quantified using excitation and emission wavelengths of 492 and 517 nm, respectively [41].

2.18. Determination of Thiobarbituric Acid Reactive Substances (TBARS) in J82 Cells

The assay was performed as described previously, with slight modifications [42,43]. Briefly, J82 cells were pretreated or not with 0.1 or 0.2 mg/mL of PPPE or PAPE for 24 h and then with 100 μM TBHP for 3 h. The cell culture supernatants were collected and were cleared by centrifugation (13,000× *g* for 1 min at 4 °C). The supernatants were then mixed with 200 uL of 30% TCA and 200 uL of 200 mM Tris-HCl (pH 7.4) and were incubated for 10 min at room temperature. A solution of 2 M Na_2SO_4 and 55 mM TBA was then added, and the supernatants were incubated for 1 h at 95 °C. The supernatants were cooled on ice for 5 min and, after adding 70% TCA was centrifuged (13,000× *g* for 1 min at 4 °C). The absorbance of the supernatants was monitored at 532 nm. Total cell protein was determined using a BCA assay. The calculation of the TBARS concentration was based on the malondialdehyde standard curve calculation. Each experiment was repeated at least in triplicate.

2.19. Data and Statistical Analysis

The results of the experiments were expressed as means ± standard error of the mean (SEM). Mean values were compared using an unpaired t-test or ANOVA (Dunnett's multiple comparisons test). Statistical analyses were performed using GraphPad Prism program 8 (GraphPad Software®, Inc., La Jolla, CA, USA). Results were considered to be significant at $p < 0.05$.

3. Results and Discussion

3.1. Total Phenolic, Flavonoid, and Anthocyanin Content

In the present study, we focused mainly on investigating bioactive compounds of pomegranate and their effects in preventing some disorders and pathologies such as bladder cancer, their cardioprotective functions, and their ability to modulate oxidative-related diseases [44–46]. We studied their phytochemical-related biological properties, with an emphasis on anthocyanins, phenolic acids, flavonoids, and hydrolysable tannins.

Our results showed that the levels of phenolic compounds were different in different parts of pomegranates (Table 1). PPPE had a higher TPC than PAPE (283.86 ± 17.89 vs. 166.9 ± 18.10 mg GAE/g dw, respectively, $p < 0.05$). We also found that PPPE was significantly richer in flavonoids than PAPE (185.37 ± 3.05 vs. 57.43 ± 0.41 mg QE/g dw, respectively, $p < 0.001$). However, the total anthocyanin content of PPPE was non significantly richer than that of PAPE (102.97 ± 9.19 vs. 81.26 ± 18.39 mg cy-3-glu/100 g dw, respectively).

Table 1. Total phenolic, flavonoid, and anthocyanin content of PPPE and PAPE. Results are expressed as the mean ± sem of at least three independent assays of each sample. * $p < 0.05$ and *** $p < 0.001$ indicate significant differences compared to the control. ns: not significant.

Plant Extract	Polyphenols (mg GAE/g dw)	Flavonoids (mg QE/g dw)	Total Anthocyanin (mg cy-3-glu/ 100 g dw)	α-Punicalagin (mg/g dw)	β-Punicalagin (mg/g dw)	Gallic Acid (mg/g dw)	Ellagic Acid (mg/g dw)
PPPE	283.86 ± 17.89 *	185.37 ± 3.05 ***	102.97 ± 9.19, ns	148.95 ± 2.43 ***	302.38 ± 7.26 ***	5.87 ± 0.08	18.85 ± 0.41
PAPE	166.90 ± 18.10	57.43 ± 0.41	81.26 ± 18.39	40.40 ± 2.67	3.03 ± 0.44	3.88 ± 0.04	14.43 ± 0.21

Our results also showed that pomegranates contain high levels of phenolic compounds and anthocyanidin, especially peel and aril extracts from the Sefri variety, compared to Italian, Iranian, Turkish, Indian, and Tunisian varieties (Table S2) [47]. Russo et al. reported that six old Italian varieties, as well as the international cultivar "Wonderful," all Gaeta varieties, are quantitatively the richest in terms of TPC in the peel. However, the Moroccan Sefri variety contains even more TPC (202.22 mg GAE/g vs. 283.86 mg GAE/g Sefri) [15]. Derakhchan et al. also investigated the TPC of peel and reported that the peel of the Natanz variety has a lower TPC than the peel of the Sefri variety (276.36 vs. 283.86 mg GAE/g Sefri) [48]. All these results, including ours, corroborate the fact that the peel extract is richer in TPC than the aril extract. The total anthocyanin content of our samples was also higher than those reported in other studies conducted with other varieties of pomegranate [49,50]. However, the difference in TPC may be influenced by variations in phytochemical composition, the cultivar studied, the extraction methods, and the experimental and environmental conditions.

3.2. Polyphenol HPLC Analysis

It is well documented that pomegranates, especially peels, are rich in ellagitannins, which can make up 66% of the total polyphenols [51]. We quantified punicalagin isomers (α and β), ellagic acid, and gallic acid by HPLC-UV and compared the results with pure standards. The chromatographic patterns of the phenolic fractions showed that α and β-punicalagin levels were higher than those of other phenolic compounds in both PPPE and PAPE (Figure 1). The retention times of gallic acid, α-punicalagin, β-punicalagin, and ellagic acid were 5.32, 14.42, 16.84, and 23.45 min, respectively. The results given in Table 2 are expressed as mg/g dry weight extract and show that the β-punicalagin content was twice as high as that of α-punicalagin (309.88 ± 13.81 mg/g dw vs. 148.95 ± 2.43 mg/g dw, respectively). These results are consistent with those obtained by Sabraoui et al. [3], who showed in a comparative study of three Moroccan pomegranate varieties that the range of punicalagin concentrations varied from 120.9 to 210.6 mg/g dw. Studies on other pomegranate cultivars also showed that the range of punicalagin concentrations varies depending on the cultivar.

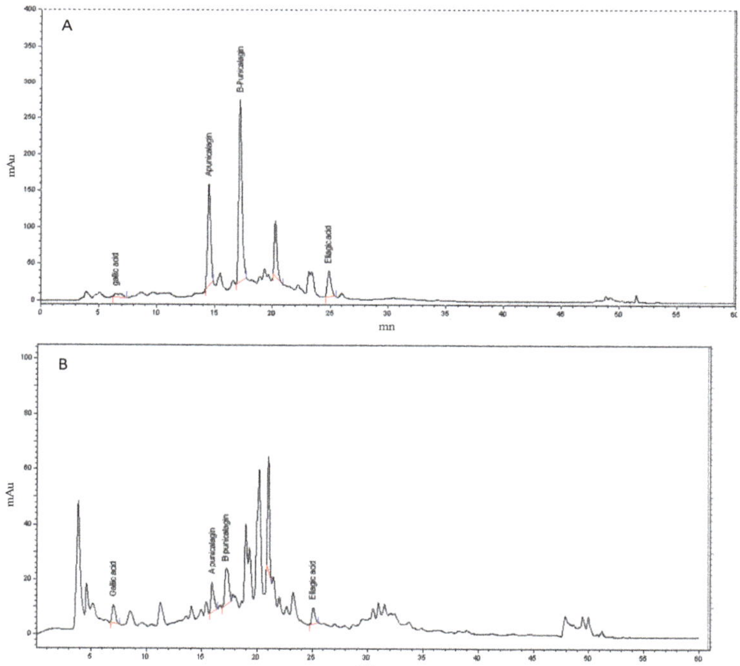

Figure 1. High-performance liquid chromatography-photodiode array (HPLC-PDA) chromatogram of bioactive molecules in (**A**): PPPE and (**B**): PAPE. The vertical red/blue lines correspond to the integration limits of each peak. Horizontal red lines correspond to peak detection threshold.

Table 2. Antioxidant activities of PPPE and PAPE. Results are expressed as the mean ± sem of at least three independent assays of each sample. * $p < 0.05$ and *** $p < 0.001$ indicate significant differences compared to the control.

Plant Extract	Antioxidant Activities			
	DPPH (IC_{50} Values, µg/mL)	H_2O_2 (IC_{50} Values, µg/mL)	FRAP (mg AAE/g dw)	BCB (%)
PPPE	12.49 ± 0.60 *	19.96 ± 0.02 ***	374.83 ± 16.85 ***	86.83 ± 1.22 ***
PAPE	21.58 ± 4.44	37.06 ± 0.05	189.83 ± 5.29	55.64 ± 1.14

3.3. Antioxidant Activities

DPPH radical scavenging activity was assayed using increasing concentrations of PPPE and PAPE, with ascorbate as an internal standard (0 to 0.4 mg/mL). Our results (Table 2) showed that both PPPE and PAPE exhibit significant free radical scavenging activity in a dose-dependent manner. However, PPPE exhibited 1.73-fold higher activity than PAPE, as shown by the IC_{50} values (PPPE: IC_{50} = 12.49 ± 0.60 µg/mL, PAPE: IC_{50} = 21.58 ± 4.44 µg/mL, $p < 0.05$). Ascorbic acid (IC_{50} = 6.61 µg/mL) was used as a positive control and exhibited 1.89-fold higher activity than PPPE and 3.27-fold higher activity than PAPE. The IC_{50} values obtained for DPPH in our study were much lower than those reported by Ali et al., who obtained an IC_{50} of 14.75 µg/mL for a peel extract and 128.27 µg/mL for an aril extract [52], indicating that the antioxidant effect was much greater in our study. Sabraoui et al. [3] recently reported that three Moroccan varieties had lower antioxidant activities (EC_{50} ranging from 42.71 to 65.55 µg/mL) than the Moroccan Sefri variety. However, another study conducted by Guo et al. [53], showed that a pomegranate peel extract of a Chinese variety had a high scavenging activity for hydrogen peroxide, with an IC_{50} of 0.032 µg/mL, which is much lower than our results. These differences could be explained by the growing conditions or the analytical methods used. The DPPH scavenging activity of pomegranate fruit is associated with their total phenolic, anthocyanin, and flavonoid contents, as shown in Table 1, which is why the hydro-alcoholic peel extract exhibited a higher radical scavenging activity than the aril extract. Interestingly, if TPC, especially punicalagin, gallic acid, and ellagic acid, is taken into account there is a positive correlation between TPC and the antioxidant power of PPPE, which is higher than that of PAPE [9,54]. This is in agreement with our results showing that PPPE is significantly richer in phenolic compounds than PAPE.

Our results also showed that PPPE has a better capacity to scavenge H_2O_2 free radicals (IC_{50} = 19.96 ± 0.02 µg/mL) than PAPE (IC_{50} = 37.06 ± 0.05 µg/mL ($p < 0.001$). In terms of antioxidant activity (AA%) at the same concentration, ascorbate exhibited the highest AA% followed by PPPE and then PAPE. At 200 ug/mL, the AA% was 98.21%, 87.8%, and 64.12%, respectively, for ascorbic acid, PPPE, and PAPE. H_2O_2 scavenging activity was likely affected by the concentration of phenolic compounds. Since phenolic compounds are powerful chain-breaking antioxidants, they may accelerate the decomposition of H_2O_2 to H_2O and oxygen [55]. H_2O_2 is highly reactive and contributes to the formation of transition metal ion-dependent OH radical-mediated oxidative DNA, protein, and lipid damage [56].

The specificity and sensitivity of the DPPH and H_2O_2 assays did not confirm the antioxidant activities of the pomegranate extracts. Given this, FRAP and BCB assays were conducted to provide a reliable assessment of the antioxidant properties of pomegranate compounds.

Our results from the FRAP assay showed that PPPE has a higher capacity to reduce the ferric tripyridyl-triazine complex (Fe(III)-TPTZ) to a ferrous complex (Fe(II)-TPTZ) due to the electron-donating abilities of its rich phenolic compounds as expressed by TAA (374.83 ± 16.85 mg AAE/g dw) than PAPE (189.83 ± 5.29 mg AAE/g dw, $p < 0.001$) (Table 2). A similar trend was observed with the results obtained by Li et al. [57] and Sabraoui et al. [3], who found that the reducing power of a peel extract was higher than that of an aril extract. Zeljka et al. also reported that peel extracts exhibit strong antioxidant activity (100.25–176.60 µmol Trolox Eq/100 g) in reducing the Fe(III)-TPTZ complex [58]. The reducing power of pomegranate fruit parts is probably due to the action of the hydroxyl groups of phenolic compounds, which may act as electron donors. Antioxidant compounds that act as reducing agents exert their effect by donating a hydrogen atom to the ferric complex, thus breaking the radical chain reaction [57].

We evaluated the antioxidant potential of PPPE and PAPE for lipid peroxidation using the β-carotene/linoleic acid bleaching assay, which is based on a decrease in the color of β-carotene following its reaction with radicals generated when linoleic acid is oxidized. Table 2 shows that the phenolic extracts tested cause a decrease in linoleic acid oxidation. The addition of PPPE or PAPE prevented the generation of free radicals by the coupled oxidation of linoleic acid and β-carotene. At 2 mg/mL PPPE exhibited significantly higher antioxidant activity (86.83 ± 1.22%) than PAPE (55.64 ± 1.14%, $p < 0.001$). This is consistent with a study carried out by Singh et al. [59], who also reported that peel extract exhibited higher antioxidant activity (83%) than aril extract (22,6%). On the other hand, Derakhshan et al. [48] reported that a peel extract exhibited 58% antioxidant activity compared to 54% for an aril extract, while Orak et al. [60] reported that there was no significant difference in antioxidant activity between pomegranate juice (47.87%) and a peel extract (46.24%). In the β-carotene-linoleic acid system, the oxidation of linoleic acid generates radical species due to hydrogen abstraction, which occurs in the methylene groups of linoleic acid. The free radicals oxidize β-carotene by hydroperoxides. The presence of antioxidants in the extract neutralizes the linoleic free radicals as well as any other free radicals formed within the system. The oxidation of β-carotene thus depends on the antioxidant activity of the extracts [61]. Our results show clearly that PPPE and PAGE reduce the oxidation of β-carotene.

3.4. LDL Oxidation and Paraoxonase 1 (PON1) Activity

To gain more insight into the antioxidant mechanism of pomegranate phenolic-rich extracts, we analyzed the ability of PPPE and PAPE to inhibit copper-induced LDL oxidation, prevent the disappearance of vitamin E (α-tocopherol), and promote PON1 activity. The copper-induced LDL oxidation (ox-LDL) results showed that both PPPE and PAPE significantly inhibit conjugated diene formation ($p < 0.0001$), and that PPPE was 8% more efficient than PAPE (0.154 ± 0.0017 and 0.167 ± 0.002, $p = 0.011$, respectively). This effect was confirmed by measuring the α-tocopherol content of ox-LDL in the presence or absence of PPPE or PAPE. As shown in Figure 2, the oxidation of LDL alone increased the α-tocopherol disappearance rate after 4 h of oxidation (3.29 ± 0.04 and 2.21 ± 0.06, $p < 0.001$, for non-ox-LDL and Ox-LDL, respectively). However, the phenolic pomegranate extracts significantly prevented α-tocopherol degradation. The results in Figure 2 show that PPPE exhibits a more potent effect than PAPE (3.28 ± 0.12, $p < 0.01$, and 2.81 ± 0.17, $p < 0.05$, respectively, compared to the Ox-LDL values). These results are in agreement with those obtained by Aviram et al. [62], who showed that, after 12 months of pomegranate juice consumption, oxidized serum LDL and LDL susceptibility to copper ion-induced oxidation were significantly reduced by 90% and 59%, respectively. In vitro, pomegranate peel and aril extracts exert their antioxidative activities by scavenging free radicals and inhibiting copper ion-induced LDL oxidation [63,64]. Polyphenols are significant antioxidants in pomegranate extracts. However, the differences in their antioxidative capacity can be attributed to different types of polyphenols and the polyphenol content of the various extracts. PPPE possessed a higher antioxidant capacity than PAPE. The strong antioxidant potency of PPPE may be due to its higher potential phenolic and flavonoid content as indicated by our HPLC analysis, which showed that PPPE has a higher punicalagin, gallic acid, and ellagic acid content than PAPE. These results suggest that peel polyphenols are major contributors to the antioxidative capacity of pomegranates [12,31,65].

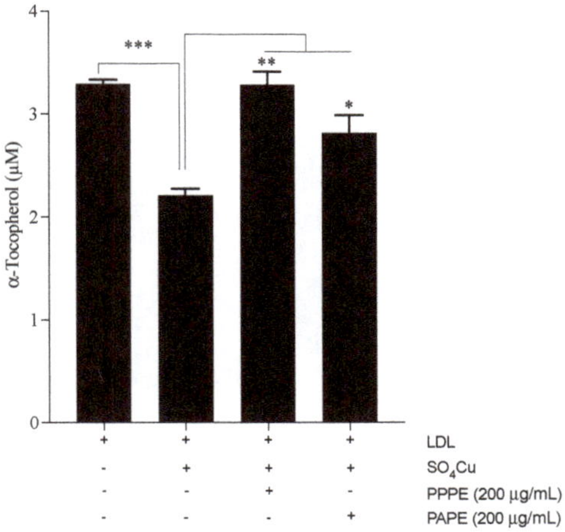

Figure 2. Effect of PPPE and PAPE on endogenous α-tocopherol disappearance during 4 h of $CuSO_4$-induced low-density lipoprotein (LDL) oxidation. Results are expressed as the means ± sem of at least three independent assays. *** $p < 0.001$, ** $p < 0.01$ and * $p < 0.05$ indicate significant differences compared to the control.

PON1 activity was evaluated in the presence of PPPE or PAPE. PON1, an HDL-associated esterase, hydrolyzes oxidized lipids, which are inactivated under oxidative stress. The incubation of human plasma with 100 μg/mL of PPPE or PAPE for 4 h at 37 °C showed that PPPE and PAPE both significantly increase plasma PON1 activity and protein expression, with PPPE exhibiting the strongest effect (Figures 3 and 4, respectively). In the present study, we showed that PPPE and PAPE slightly decrease serum oxidative stress. This may be related to an increase in serum PON1 activity. PPPE was more potent ($p < 0.01$) (Figure 3) than PAPE and significantly increased serum PON1 activity ($p < 0.05$) as well as PON1 protein expression ($p < 0.001$, $p < 0.0001$; respectively) (Figure 4). These results showed that polyphenolic compounds in the pomegranate extracts, especially in peels (punicalagin, gallic acid, and ellagic acids), have a potent effect on serum PON1 activity and protein expression and are major contributors to its beneficial effects. This protection is probably the result of the ability of PON1 to hydrolyze specific oxidized lipids in oxidized lipoproteins [24]. These beneficial effects of pomegranate consumption on serum PON1 stability and activity may contribute to a delay in the development of atherosclerosis. The administration of pomegranate extracts to Apo-E deficient mice increased serum PON1 activity, with whole fruit juice being more efficient than aril juice [63]. In obese mice, daily pomegranate juice supplementation reduces oxidative stress and increases serum PON1 expression and activity [12,66]. Aviram et al. [63] showed that pomegranate juice and aril consumption resulted in a significant 43% and 22% increase in serum PON1 arylesterase activity, respectively, whereas pomegranate peel had no significant effect. They showed that pomegranate juice can preserve and enhance PON1 activity during lipoprotein oxidation. Moreover, a recent clinical study by Estrada-Luna et al. and a study using New Zealand rabbits by Dorantes-Morales et al. both showed that supplementation with an aril preparation significantly enhances PON1 activity [67,68]. In the same vein, Bentazos-Cabrera et al. reported that the consumption of fresh pomegranate juice increases PON1 activity in mice fed a high-fat diet [69]. These results indicate that there is an inverse association between serum PON1 activity and lipid peroxidation [70]. Our results suggest that pomegranate may be a source of dietary phenolic compounds that can prevent atherosclerosis and cardiovascular disease development by inhibiting lipoprotein oxidation, reducing peroxide content, and increasing PON1 activity.

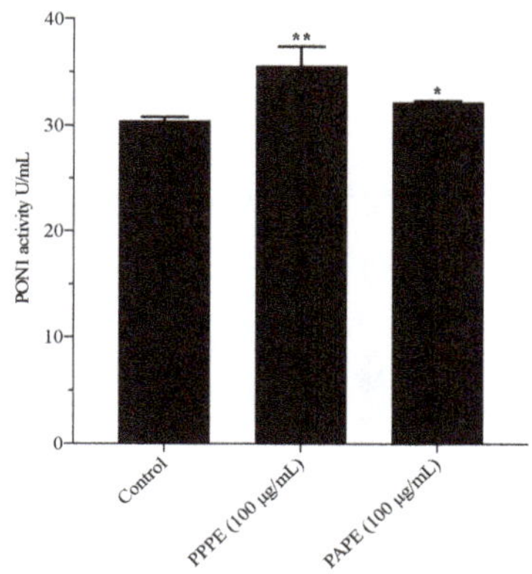

Figure 3. Pomegranate polyphenols improve PON1 activity. PON1 activity was measured in PPPE- or PAPE -enriched (80 μg/mL) plasma for 2 h. Results are expressed as the means ± sem of three independent assays. * $p < 0.05$, ** $p < 0.01$ and indicate significant differences compared to the control.

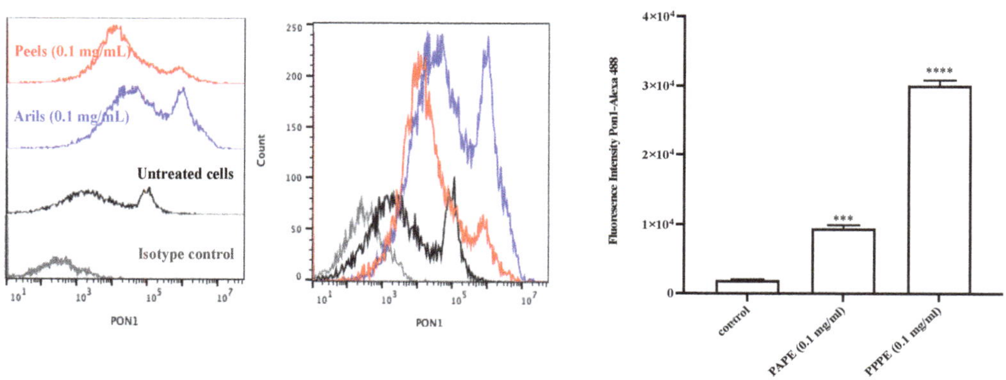

Figure 4. The extract of pomegranate peel and aril induces PON1 expression in Fu5AH cells. Fu5AH cells were cultured for 4 h in the presence (100 ug/mL) or absence of the extract of pomegranate's peel or aril. The cells were washed and labelled with anti-PON1 mAbs. Expression of PON1 was determined by multi-color flow cytometry analysis in cells exposed or not to peels or arils extracts. Mean fluorescence intensities (MFI) values of FACS profiles are shown. Data are representative of three independent experiments. The asterisks indicate statistically significant differences determined by one-way ANOVA tests. *** $p < 0.001$ and **** $p < 0.0001$.

3.5. Effect of Phenolic-Rich Pomegranate Extracts on Antioxidant Activities in J82 Cells

Cancer cells grow better in oxidative stress conditions as this increases their survival potential via various pathways that induce redox signaling activation, that in turn may lead to the suppression of tumor guardian and suppressor genes [71], the activation of survival factors such as AP-1 and NFκB, or the activation of point mutation [72]. Excess cellular ROS production may result in many harmful effects, including oxidative modifications to lipids, proteins, and DNA, that can cause various diseases.

In the present study, we examined the effects of PPPE and PAPE on ROS production and lipid peroxidation in J82 human bladder cancer cells. Our results showed that both PPPE and PAPE induce a significant decrease in the ROS content of J82 cells in a dose-dependent manner compared to the control (Figure 5A). PPPE (100–200 μg/mL) decreased intracellular ROS levels compared to the H_2O_2 control by 32.04% and 37.95%, respectively ($p < 0.001$). On the other hand, 100 μg/mL of PAPE caused no significant decrease in intracellular ROS levels whereases 200 μg/mL of PAPE reduced intracellular ROS formation by 27.83% ($p < 0.001$). However, in terms of an in vitro antioxidant effect, PPPE reduced intracellular ROS formation 10% more than PAPE. These results are in agreement with those reported by Rosenblat et al., who showed that pomegranate polyphenols, especially punicalagin and gallic acid, markedly reduce ROS formation in J774 cells [73].

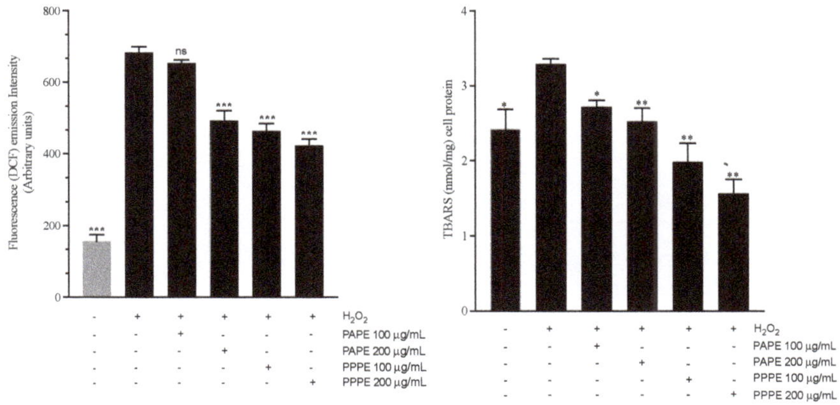

Figure 5. (**A**): Intracellular radical scavenging activity of PPPE and PAPE. J82 cells were treated with 100 or 200 µg/mL of PPPE or PAPE. Cells were labelled with 10 µmol/L DCFH-DA. The DCF fluorescence intensities were measured. The results are expressed as the means ± sem of more than three independent assays. *** $p < 0.001$ indicates a significant difference compared to the control. (**B**): Effects of PPPE and PAPE on TBARS levels in J82 cells. TBARS levels were assessed using a spectrophotometer. All values are expressed as the means ± sem of three independent assays. ** $p < 0.01$ and * $p < 0.05$ indicates a significant difference compared to the control (untreated cells).

Lipid peroxidation, which is another approach for evaluating oxidative damage, was quantified by measuring TBARS (Figure 5B). The two methanolic extracts used in the present study both caused a decrease in lipid peroxidation compared to the control. PPPE (100 and 200 µg/mL) caused a significant reduction in MDA levels of 39.81% and 52.58% ($p < 0.01$), respectively, compared to the control. PAPE (100 and 200 µg/mL) caused a much lower reduction in TBARS of 17.32% ($p < 0.05$) and 23.4% ($p < 0.01$), respectively, than PPPE. Park et al. [74] showed that a PPPE decreased ROS levels in THP-1 monocytic cells exposed to particulate matter 10-induced cytotoxicity. Zaid et al. [75] reported that a polyphenol-rich pomegranate fruit extract (POMx) decreased lipid peroxidation in human immortalized HaCaT keratinocytes following UVB-induced oxidative stress and photoaging. POMx significantly reduced peroxide accumulation, suggesting that POMx can scavenge ROS and inhibit lipid peroxidation due to its antioxidant activity. However, a study by Elango et al. [76] showed that 20 µg/mL of gallic acid (GA) isolated from pomegranate peel extracts increased ROS levels and induced the apoptosis of A549 cells through an intrinsic pathway.

4. Conclusions

Fruits are a rich source of vitamins, minerals, and biologically active compounds. However, they are often consumed without the peels although some fruit peels are rich in polyphenolic compounds, flavonoids, ascorbic acid, and other biologically active compounds that have a positive effect on health. Our results show that pomegranate peel and aril extracts increase serum PON1 activity, which may delay the development and progression of atherosclerosis. We also show that pomegranate extracts attenuate ROS production and lipid peroxidation in J82 human bladder cancer cells. These results indicate that pomegranate extracts from peel and arils protect against oxidative stress and exhibit anticarcinogenic activity against J82 cells. Finally, pomegranates may be a promising source of cancer-preventing agents due to their high phenolic content. However, pomegranate peel extract exhibited a better potential effect than aril extract.

Supplementary Materials: The following are available online at https://www.mdpi.com/2304-8158/10/9/2219/s1, Table S1: Regression equations, linearity range, r, LOD and LOQ of all analytes, Table S2: Comparative evaluation of polyphenolic. flavonoid and anthocyanin content of peels and aril extracts of Sefri and various pomegranate (*Punica granatum*) cultivars.

Author Contributions: Conceptualization, H.B. and L.B.; methodology, H.B., A.K., L.B., M.R., M.O.I.; software, H.B., M.R. and L.B.; validation, H.B., M.R. and A.K.; formal analysis, H.B., M.R. and L.B.; investigation, H.B. and L.B.; resources, H.B., S.B.; writing—original draft preparation, L.B., H.B., A.K. and S.B.; visualization.; H.B., G.F., A.H. and T.F.; A.K. and H.B. supervision, project administration and funding acquisition, H.B. All authors have read and agreed to the published version of the manuscript.

Funding: This research was funded by Agence Nationale des Plantes Médicinales et Aromatiques), (Centre National de Recherche Scientifique et Technique) and (Université Sultan Moulay Slimane), Morocco. 2020–2023 (H.B.) and by Canadian Institutes of Health Research (grant Number # PJT-162366) (A.K.).

Institutional Review Board Statement: The study was conducted according to the guidelines of the Declaration of Helsinki and the protocol was approved by the Ethics Committee of the Sherbrooke University Institute of Geriatrics (# 2009/19). Informed consent was obtained from all subjects involved in the study.

Informed Consent Statement: Informed consent was obtained from all subjects involved in the study.

Data Availability Statement: Data are contained within the article.

Conflicts of Interest: The authors declare no conflict of interest.

References

1. Di Stefano, V.; Pitonzo, R.; Novara, M.E.; Bongiorno, D.; Indelicato, S.; Gentile, C.; Avellone, G.; Bognanni, R.; Scandurra, S.; Melilli, M.G. Antioxidant activity and phenolic composition in pomegranate (*Punica granatum* L.) genotypes from south Italy by UHPLC-Orbitrap-MS approach. *J. Sci. Food Agric.* **2019**, *99*, 1038–1045. [CrossRef] [PubMed]
2. Russo, M.; Fanali, C.; Tripodo, G.; Dugo, P.; Muleo, R.; Dugo, L.; De Gara, L.; Mondello, L. Analysis of phenolic compounds in different parts of pomegranate (*Punica granatum*) fruit by HPLC-PDA-ESI/MS and evaluation of their antioxidant activity: Application to different Italian varieties. *Anal. Bioanal. Chem.* **2018**, *410*, 3507–3520. [CrossRef]
3. Sabraoui, T.; Khider, T.; Nasser, B.; Eddoha, R.; Moujahid, A.; Benbachir, M.; Essamadi, A. Determination of Punicalagins Content, Metal Chelating, and Antioxidant Properties of Edible Pomegranate (*Punica granatum* L.) Peels and Seeds Grown in Morocco. *Int. J. Food Sci.* **2020**, *2020*, 8885889. [CrossRef] [PubMed]
4. Faria, A.; Calhau, C. The Bioactivity of Pomegranate: Impact on Health and Disease. *Crit. Rev. Food Sci. Nutr.* **2011**, *51*, 626–634. [CrossRef] [PubMed]
5. Wachtel-Galor, S.; Benzie, I.F.F. (Eds.) Herbal Medicine: An Introduction to Its History, Usage, Regulation, Current Trends, and Research Needs. In *Herbal Medicine: Biomolecular and Clinical Aspects*, 2nd ed.; CRC Press: Boca Raton, FL, USA, 2011.
6. Ismail, T.; Sestili, P.; Akhtar, S. Pomegranate peel and fruit extracts: A review of potential anti-inflammatory and anti-infective effects. *J. Ethnopharmacol.* **2012**, *143*, 397–405. [CrossRef] [PubMed]
7. Turrini, E.; Ferruzzi, L.; Fimognari, C. Potential Effects of Pomegranate Polyphenols in Cancer Prevention and Therapy. *Oxidative Med. Cell. Longev.* **2015**, *2015*, 938475. [CrossRef]
8. Song, B.; Li, J.; Li, J. Pomegranate peel extract polyphenols induced apoptosis in human hepatoma cells by mitochondrial pathway. *Food Chem. Toxicol.* **2016**, *93*, 158–166. [CrossRef]
9. Kalaycıoğlu, Z.; Erim, F.B. Total phenolic contents, antioxidant activities, and bioactive ingredients of juices from pomegranate cultivars worldwide. *Food Chem.* **2017**, *221*, 496–507. [CrossRef]
10. Kojadinovic, M.I.; Arsic, A.C.; Debeljak-Martacic, J.D.; Ristić, A.K.; Kardum, N.D.; Popovic, T.B.; Glibetic, M.D. Consumption of pomegranate juice decreases blood lipid peroxidation and levels of arachidonic acid in women with metabolic syndrome. *J. Sci. Food Agric.* **2017**, *97*, 1798–1804. [CrossRef]
11. Razani, Z.; Dastani, M.; Kazerani, H.R. Cardioprotective Effects of Pomegranate (*Punica granatum*) Juice in Patients with Ischemic Heart Disease. *Phytother. Res.* **2017**, *31*, 1731–1738. [CrossRef]
12. Kandylis, P.; Kokkinomagoulos, E. Food Applications and Potential Health Benefits of Pomegranate and its Derivatives. *Foods* **2020**, *9*, 122. [CrossRef]

13. Alexandre, E.; Silva, S.; Santos, S.; Silvestre, A.J.; Duarte, M.F.; Saraiva, J.A.; Pintado, M. Antimicrobial activity of pomegranate peel extracts performed by high pressure and enzymatic assisted extraction. *Food Res. Int.* **2019**, *115*, 167–176. [CrossRef]
14. Sood, A.; Gupta, M. Extraction process optimization for bioactive compounds in pomegranate peel. *Food Biosci.* **2015**, *12*, 100–106. [CrossRef]
15. Mastrogiovanni, F.; Bernini, R.; Basiricò, L.; Bernabucci, U.; Campo, M.; Romani, A.; Santi, L.; Lacetera, N. Antioxidant and anti-inflammatory effects of pomegranate peel extracts on bovine mammary epithelial cells BME-UV1. *Nat. Prod. Res.* **2020**, *34*, 1465–1469. [CrossRef] [PubMed]
16. Chaves, F.M.; Pavan, I.C.B.; Da Silva, L.G.S.; De Freitas, L.B.; Rostagno, M.A.; Antunes, A.E.C.; Bezerra, R.M.N.; Simabuco, F.M. Pomegranate Juice and Peel Extracts are Able to Inhibit Proliferation, Migration and Colony Formation of Prostate Cancer Cell Lines and Modulate the Akt/mTOR/S6K Signaling Pathway. *Plant Foods Hum. Nutr.* **2020**, *75*, 54–62. [CrossRef]
17. Gullón, P.; Astray, G.; Gullón, B.; Tomasevic, I.; Lorenzo, J.M. Pomegranate Peel as Suitable Source of High-Added Value Bioactives: Tailored Functionalized Meat Products. *Molecules* **2020**, *25*, 2859. [CrossRef] [PubMed]
18. Mansour, R.; Dhouib, S.; Sakli, F. UV Protection and Dyeing Properties of Wool Fabrics Dyed with Aqueous Extracts of Madder Roots, Chamomiles, Pomegranate Peels, and Apple Tree Branches Barks. *J. Nat. Fibers* **2020**, *2020*, 1–11. [CrossRef]
19. Sadeghi-Kiakhani, M.; Tehrani-Bagha, A.R.; Gharanjig, K.; Hashemi, E. Use of pomegranate peels and walnut green husks as the green antimicrobial agents to reduce the consumption of inorganic nanoparticles on wool yarns. *J. Clean. Prod.* **2019**, *231*, 1463–1473. [CrossRef]
20. Turgut, S.S.; Soyer, A.; Işıkçı, F. Effect of pomegranate peel extract on lipid and protein oxidation in beef meatballs during refrigerated storage. *Meat Sci.* **2016**, *116*, 126–132. [CrossRef]
21. Paul, A.; Radhakrishnan, M. Pomegranate seed oil in food industry: Extraction, characterization, and applications. *Trends Food Sci. Technol.* **2020**, *105*, 273–283. [CrossRef]
22. Martínez, J.J.; Hernández, F.; Abdelmajid, H.; Legua, P.; Martínez, R.; El Amine, A.; Melgarejo, P. Physico-chemical characterization of six pomegranate cultivars from Morocco: Processing and fresh market aptitudes. *Sci. Hortic.* **2012**, *140*, 100–106. [CrossRef]
23. Hmid, I.; Elothmani, D.; Hanine, H.; Oukabli, A.; Mehinagic, E. Comparative study of phenolic compounds and their antioxidant attributes of eighteen pomegranate (*Punica granatum* L.) cultivars grown in Morocco. *Arab. J. Chem.* **2017**, *10*, S2675–S2684. [CrossRef]
24. Aviram, M.; Kaplan, M.; Rosenblat, M.; Fuhrman, B. Dietary Antioxidants and Paraoxonases against LDL Oxidation and Atherosclerosis Development. In *Handbook of Experimental Pharmacology*; Springer: Berlin/Heidelberg, Germany, 2005. [CrossRef]
25. Saha, S.K.; Lee, S.B.; Won, J.; Choi, H.Y.; Kim, K.; Yang, G.-M.; Dayem, A.A.; Cho, S.-G. Correlation between Oxidative Stress, Nutrition, and Cancer Initiation. *Int. J. Mol. Sci.* **2017**, *18*, 1544. [CrossRef]
26. Lee, C.-K.; Liao, C.-W.; Meng, S.-W.; Wu, W.-K.; Chiang, J.-Y.; Wu, M.-S. Lipids and Lipoproteins in Health and Disease: Focus on Targeting Atherosclerosis. *Biomedicines* **2021**, *9*, 985. [CrossRef]
27. Reuter, S.; Gupta, S.C.; Chaturvedi, M.M.; Aggarwal, B.B. Oxidative stress, inflammation, and cancer: How are they linked? *Free Radic. Biol. Med.* **2010**, *49*, 1603–1616. [CrossRef] [PubMed]
28. Singleton, V.L.; Orthofer, R.; Lamuela-Raventós, R.M. [14] Analysis of total phenols and other oxidation substrates and antioxidants by means of folin-ciocalteu reagent. In *Methods in Enzymology*; Academic Press: Cambridge, MA, USA, 1999; Volume 299, pp. 152–178.
29. Woisky, R.G.; Salatino, A. Analysis of propolis: Some parameters and procedures for chemical quality control. *J. Apic. Res.* **1998**, *37*, 99–105. [CrossRef]
30. Sellappan, S.; Akoh, C.C.; Krewer, G. Phenolic Compounds and Antioxidant Capacity of Georgia-Grown Blueberries and Blackberries. *J. Agric. Food Chem.* **2002**, *50*, 2432–2438. [CrossRef] [PubMed]
31. Zhang, W.-W.; Duan, X.-J.; Huang, H.-L.; Zhang, Y.; Wang, B.-G. Evaluation of 28 marine algae from the Qingdao coast for antioxidative capacity and determination of antioxidant efficiency and total phenolic content of fractions and subfractions derived from *Symphyocladia latiuscula* (Rhodomelaceae). *Environ. Boil. Fishes* **2006**, *19*, 97–108. [CrossRef]
32. Bhatti, M.Z.; Ali, A.; Ahmad, A.; Saeed, A.; Malik, S.A. Antioxidant and phytochemical analysis of *Ranunculus arvensis* L. extracts. *BMC Res. Notes* **2015**, *8*, 279. [CrossRef] [PubMed]
33. El Jemli, M.; Kamal, R.; Marmouzi, I.; Zerrouki, A.; Cherrah, Y.; Alaoui, K. Radical-Scavenging Activity and Ferric Reducing Ability of *Juniperus thurifera* (L.), *J. oxycedrus* (L.), *J. phoenicea* (L.) and *Tetraclinis articulata* (L.). *Adv. Pharmacol. Sci.* **2016**, *2016*, 6392656. [CrossRef] [PubMed]
34. Jayaprakasha, G.K.; Rao, L.J. Phenolic Constituents from the Lichen Parmotrema stuppeum (Nyl.) Hale and Their Antioxidant Activity. *Z. Nat. C* **2000**, *55*, 1018–1022. [CrossRef] [PubMed]
35. Sattler, W.; Mohr, D.; Stocker, R. [50] Rapid isolation of lipoproteins and assessment of their peroxidation by high-performance liquid chromatography postcolumn chemiluminescence. In *Methods in Enzymology*; Academic Press: Cambridge, MA, USA, 1994; Volume 233, pp. 469–489.
36. Berrougui, H.; Grenier, G.; Loued, S.; Drouin, G.; Khalil, A. A new insight into resveratrol as an atheroprotective compound: Inhibition of lipid peroxidation and enhancement of cholesterol efflux. *Atherosclerosis* **2009**, *207*, 420–427. [CrossRef] [PubMed]
37. Berrougui, H.; Cloutier, M.; Isabelle, M.; Khalil, A. Phenolic-extract from argan oil (*Argania spinosa* L.) inhibits human low-density lipoprotein (LDL) oxidation and enhances cholesterol efflux from human THP-1 macrophages. *Atherosclerosis* **2006**, *184*, 389–396. [CrossRef] [PubMed]

38. Jaouad, L.; de Guise, C.; Berrougui, H.; Cloutier, M.; Isabelle, M.; Fulop, T.; Payette, H.; Khalil, A. Age-related decrease in high-density lipoproteins antioxidant activity is due to an alteration in the PON1's free sulfhydryl groups. *Atherosclerosis* **2006**, *185*, 191–200. [CrossRef] [PubMed]
39. Asztalos, B.F.; de la Llera-Moya, M.; Dallal, G.E.; Horvath, K.V.; Schaefer, E.J.; Rothblat, G.H. Differential effects of HDL subpopulations on cellular ABCA1- and SR-BI-mediated cholesterol efflux. *J. Lipid Res.* **2005**, *46*, 2246–2253. [CrossRef] [PubMed]
40. Posadas-Sánchez, R.; Posadas-Romero, C.; Mendoza-Pérez, E.; Caracas-Portilla, N.A.; Cardoso-Saldaña, G.; Medina-Urrutia, A.; Jorge-Galarza, E.; Juárez-Rojas, J.G. Cholesterol Efflux and Metabolic Abnormalities Associated With Low High-Density-Lipoprotein-Cholesterol and High Triglycerides in Statin-Treated Coronary Men With Low-Density Lipoprotein-Cholesterol <70 mg/dl. *Am. J. Cardiol.* **2021**, *109*, 636–641.
41. Yang, Q.-Q.; Cheng, L.-Z.; Zhang, T.; Yaron, S.; Jiang, H.-X.; Sui, Z.-Q.; Corke, H. Phenolic profiles, antioxidant, and antiproliferative activities of turmeric (*Curcuma longa*). *Ind. Crops Prod.* **2020**, *152*, 112561. [CrossRef]
42. Botsoglou, N.A.; Fletouris, D.J.; Papageorgiou, G.E.; Vassilopoulos, V.N.; Mantis, A.J.; Trakatellis, A.G. Rapid, Sensitive, and Specific Thiobarbituric Acid Method for Measuring Lipid Peroxidation in Animal Tissue, Food, and Feedstuff Samples. *J. Agric. Food Chem.* **1994**, *42*, 1931–1937. [CrossRef]
43. Venkatesan, T.; Choi, Y.-W.; Kim, Y.-K. Effect of an extraction solvent on the antioxidant quality of Pinus densiflora needle extract. *J. Pharm. Anal.* **2019**, *9*, 193–200. [CrossRef]
44. Bowtell, J.; Kelly, V. Fruit-Derived Polyphenol Supplementation for Athlete Recovery and Performance. *Sports Med.* **2019**, *49*, 3–23. [CrossRef] [PubMed]
45. Kashi, D.S.; Shabir, A.; Da Boit, M.; Bailey, S.J.; Higgins, M.F. The Efficacy of Administering Fruit-Derived Polyphenols to Improve Health Biomarkers, Exercise Performance and Related Physiological Responses. *Nutrients* **2019**, *11*, 2389. [CrossRef]
46. Akhtar, S.; Ismail, T.; Layla, A. Pomegranate Bioactive Molecules and Health Benefits. In *Reference Series in Phytochemistry*; Springer: Berlin/Heidelberg, Germany, 2018. [CrossRef]
47. Singh, B.; Singh, J.P.; Kaur, A.; Singh, N. Phenolic compounds as beneficial phytochemicals in pomegranate (*Punica granatum* L.) peel: A review. *Food Chem.* **2018**, *261*, 75–86. [CrossRef] [PubMed]
48. Derakhshan, Z.; Ferrante, M.; Tadi, M.; Ansari, F.; Heydari, A.; Hosseini, M.S.; Conti, G.O.; Sadrabad, E.K. Antioxidant activity and total phenolic content of ethanolic extract of pomegranate peels, juice and seeds. *Food Chem. Toxicol.* **2018**, *114*, 108–111. [CrossRef]
49. Tehranifar, A.; Zarei, M.; Nemati, Z.; Esfandiyari, B.; Vazifeshenas, M.R. Investigation of physico-chemical properties and antioxidant activity of twenty Iranian pomegranate (*Punica granatum* L.) cultivars. *Sci. Hortic.* **2010**, *126*, 180–185. [CrossRef]
50. Caliskan, O.; Bayazit, S. Phytochemical and antioxidant attributes of autochthonous Turkish pomegranates. *Sci. Hortic.* **2012**, *147*, 81–88. [CrossRef]
51. Sun, Y.-Q.; Tao, X.; Men, X.-M.; Xu, Z.-W.; Wang, T. In Vitro and In Vivo antioxidant activities of three major polyphenolic compounds in pomegranate peel: Ellagic acid, punicalin, and punicalagin. *J. Integr. Agric.* **2017**, *16*, 1808–1818. [CrossRef]
52. Ali, S.; El-Baz, F.; El-Emary, G.A.E.; Khan, E.; Amin Mohamed, A. HPLC-analysis of polyphenolic compounds and free radical scavenging activity of pomegranate fruit (*Punica granatum* L.). *Int. J. Pharm. Clin. Res.* **2014**, *6*, 348–355.
53. Guo, S.; Deng, Q.; Xiao, J.; Xie, B.; Sun, Z. Evaluation of Antioxidant Activity and Preventing DNA Damage Effect of Pomegranate Extracts by Chemiluminescence Method. *J. Agric. Food Chem.* **2007**, *55*, 3134–3140. [CrossRef]
54. Zaouay, F.; Mena, P.; Garcia-Viguera, C.; Mars, M. Antioxidant activity and physico-chemical properties of Tunisian grown pomegranate (*Punica granatum* L.) cultivars. *Ind. Crops Prod.* **2012**, *40*, 81–89. [CrossRef]
55. Bendary, E.; Francis, R.; Ali, H.; Sarwat, M.; El Hady, S. Antioxidant and structure–activity relationships (SARs) of some phenolic and anilines compounds. *Ann. Agric. Sci.* **2013**, *58*, 173–181. [CrossRef]
56. Gülçin, I.; Huyut, Z.; Elmastaş, M.; Aboul-Enein, H. Radical scavenging and antioxidant activity of tannic acid. *Arab. J. Chem.* **2010**, *3*, 43–53. [CrossRef]
57. Li, Y.; Guo, C.; Yang, J.; Wei, J.; Xu, J.; Cheng, S. Evaluation of antioxidant properties of pomegranate pulp extract. *Food Chem.* **2006**, *96*, 254–260. [CrossRef]
58. Peršurić, Ž.; Saftić Martinović, L.; Malenica, M.; Gobin, I.; Pedisić, S.; Dragović-Uzelac, V.; Kraljević Pavelić, S. Assessment of the Biological Activity and Phenolic Composition of Ethanol Extracts of Pomegranate (*Punica granatum* L.) Peels. *Molecules* **2020**, *25*, 5916. [CrossRef] [PubMed]
59. Singh, R.; Murthy, K.; Jayaprakasha, G. Studies on the Antioxidant Activity of Pomegranate (*Punica granatum*) Peel and Seed Extracts Using in Vitro Models. *J. Agric. food Chem.* **2002**, *50*, 81–86. [CrossRef] [PubMed]
60. Orak, H.H.; Yagar, H.; Isbilir, S.S. Comparison of antioxidant activities of juice, peel, and seed of pomegranate (*Punica granatum* L.) and inter-relationships with total phenolic, Tannin, anthocyanin, and flavonoid contents. *Food Sci. Biotechnol.* **2012**, *21*, 373–387. [CrossRef]
61. Mariod, A.; Ibrahim, R.M.; Ismail, M.; Ismail, N. Antioxidant activity and phenolic content of phenolic rich fractions obtained from black cumin (*Nigella sativa*) seedcake. *Food Chem.* **2009**, *116*, 306–312. [CrossRef]
62. Aviram, M.; Rosenblat, M.; Gaitini, D.; Nitecki, S.; Hoffman, A.; Dornfeld, L.; Volkova, N.; Presser, D.; Attias, J.; Liker, H.; et al. Pomegranate juice consumption for 3 years by patients with carotid artery stenosis reduces common carotid intima-media thickness, blood pressure and LDL oxidation. *Clin. Nutr.* **2004**, *23*, 423–433. [CrossRef] [PubMed]

63. Aviram, M.; Volkova, N.; Coleman, R.; Dreher, M.; Reddy, M.K.; Ferreira, D.; Rosenblat, M. Pomegranate phenolics from the peels, arils, and flowers are antiatherogenic: Studies in vivo in atherosclerotic apolipoprotein e-deficient (E^0) mice and In Vitro in cultured macrophages and lipoproteins. *J. Agric. Food Chem.* **2008**, *56*, 1148–1157. [CrossRef]
64. Ambigaipalan, P.; de Camargo, A.C.; Shahidi, F. Phenolic Compounds of Pomegranate Byproducts (Outer Skin, Mesocarp, Divider Membrane) and Their Antioxidant Activities. *J. Agric. Food Chem.* **2016**, *64*, 6584–6604. [CrossRef]
65. Ricci, D.; Giamperi, L.; Bucchini, A.; Fraternale, D. Antioxidant activity of *Punica granatum* fruits. *Fitoterapia* **2006**, *77*, 310–312. [CrossRef]
66. BinMowyna, M.; Binobead, M.; Badr, N.; AlSedairy, S.; Elredh, I.; Alqahtani, W. Effect of Saudi and Egyptian pomegranate polyphenols in regulating the activity of PON1, PON2 and lipid profile for preventing coronary heart disease. *J. Regen. Biol. Med.* **2019**. [CrossRef]
67. Estrada-Luna, D.; Carreón-Torres, E.; Bautista-Pérez, R.; Betanzos-Cabrera, G.; Dorantes-Morales, A.; Luna-Luna, M.; Vargas-Barrón, J.; Mejía, A.M.; Fragoso, J.M.; Carvajal-Aguilera, K.; et al. Microencapsulated Pomegranate Reverts High-Density Lipoprotein (HDL)-Induced Endothelial Dysfunction and Reduces Postprandial Triglyceridemia in Women with Acute Coronary Syndrome. *Nutrients* **2019**, *11*, 1710. [CrossRef]
68. Dorantes-Morales, A.; Estrada-Luna, D.; Bautista-Pérez, R.; Betanzos-Cabrera, G.; Luna-Luna, M.; Flores-Castillo, C.; Vargas-Alarcón, G.; Fragoso, J.M.; Pérez-Méndez, Ó.; Carreón-Torres, E. Microencapsulated Pomegranate Modifies the Composition and Function of High-Density Lipoproteins (HDL) in New Zealand Rabbits. *Molecules* **2020**, *25*, 3297. [CrossRef]
69. Estrada-Luna, D.; Martínez-Hinojosa, E.; Cancino-Diaz, J.C.; Belefant-Miller, H.; López-Rodríguez, G.; Betanzos-Cabrera, G. Daily supplementation with fresh pomegranate juice increases paraoxonase 1 expression and activity in mice fed a high-fat diet. *Eur. J. Nutr.* **2017**, *57*, 383–389. [CrossRef]
70. Aviram, M.; Dornfeld, L.; Rosenblat, M.; Volkova, N.; Kaplan, M.; Coleman, R.; Hayek, T.; Presser, D.; Fuhrman, B. Pomegranate juice consumption reduces oxidative stress, atherogenic modifications to LDL, and platelet aggregation: Studies in humans and in atherosclerotic apolipoprotein E–deficient mice. *Am. J. Clin. Nutr.* **2000**, *71*, 1062–1076. [CrossRef] [PubMed]
71. Sun, Y.; Oberley, L.W. Redox regulation of transcriptional activators. *Free Radic. Biol. Med.* **1996**, *21*, 335–348. [CrossRef]
72. Ripple, M.O.; Henry, W.F.; Schwarze, S.R.; Wilding, G.; Weindruch, R. Effect of antioxidants on androgen-induced AP-1 and NF-kappaB DNA-binding activity in prostate carcinoma cells. *J. Natl. Cancer Inst.* **1999**, *91*, 1227–1232. [CrossRef] [PubMed]
73. Rosenblat, M.; Volkova, N.; Aviram, M. Pomegranate phytosterol (β-sitosterol) and polyphenolic antioxidant (punicalagin) addition to statin, significantly protected against macrophage foam cells formation. *Atherosclerosis* **2013**, *226*, 110–117. [CrossRef]
74. Park, S.; Seok, J.K.; Kwak, J.Y.; Suh, H.-J.; Kim, Y.M.; Boo, Y.C. Anti-Inflammatory Effects of Pomegranate Peel Extract in THP-1 Cells Exposed to Particulate Matter PM10. *Evid.-Based Complement. Altern. Med.* **2016**, *2016*, 6836080. [CrossRef] [PubMed]
75. Abu Zaid, M.; Afaq, F.; Syed, D.N.; Dreher, M.; Mukhtar, H. Inhibition of UVB-mediated Oxidative Stress and Markers of Photoaging in Immortalized HaCaT Keratinocytes by Pomegranate Polyphenol Extract POMx. *Photochem. Photobiol.* **2007**, *83*, 882–888. [CrossRef]
76. Elango, S.; Balwas, R.; Padma, V.V. Gallic Acid Isolated from Pomegranate Peel Extract Induces Reactive Oxygen Species Mediated Apoptosis in A549 Cell Line. *J. Cancer Ther.* **2011**, *2*, 638–645. [CrossRef]

foods

MDPI

Article

Study of the Lipolysis Effect of Nanoliposome-Encapsulated *Ganoderma lucidum* Protein Hydrolysates on Adipocyte Cells Using Proteomics Approach

Sucheewin Krobthong [1], Yodying Yingchutrakul [2], Wonnop Visessanguan [3], Thanisorn Mahatnirunkul [4], Pawitrabhorn Samutrtai [5], Chartchai Chaichana [6], Phakorn Papan [7] and Kiattawee Choowongkomon [1,8,9,*]

1 Interdisciplinary Graduate Program in Genetic Engineering, Kasetsart University, Bangkok 10900, Thailand; sucheewin82@gmail.com
2 National Omics Center, NSTDA, Pathum Thani 12120, Thailand; yodying.yin@nstda.or.th
3 National Center for Genetic Engineering and Biotechnology, NSTDA, Pathum Thani 12120, Thailand; wonnop@biotec.or.th
4 National Nanotechnology Center, NSTDA, Pathum Thani 12120, Thailand; thanisorn.mah@nanotec.or.th
5 Department of Pharmaceutical Sciences, Faculty of Pharmacy, Chiang Mai University, Chiang Mai 50200, Thailand; pawitrabhorn.s@cmu.ac.th
6 Siriraj Center of Research Excellence for Diabetes and Obesity (SiCORE-DO), Faculty of Medicine Siriraj Hospital, Mahidol University, Bangkok 10700, Thailand; chartchai.chi@mahidol.ac.th
7 Department of Chemistry, Faculty of Science, Chiang Mai University, Chiang Mai 50200, Thailand; phakorn_papan@cmu.ac.th
8 Department of Biochemistry, Faculty of Science, Kasetsart University, Bangkok 10900, Thailand
9 Center for Advanced Studies in Nanotechnology for Chemical, Food and Agricultural Industries, KU Institute for Advanced Studies, Kasetsart University, Bangkok 10900, Thailand
* Correspondence: fsciktc@ku.ac.th; Tel.: +66-2-562-5444 (ext. 2046)

Citation: Krobthong, S.; Yingchutrakul, Y.; Visessanguan, W.; Mahatnirunkul, T.; Samutrtai, P.; Chaichana, C.; Papan, P.; Choowongkomon, K. Study of the Lipolysis Effect of Nanoliposome-Encapsulated *Ganoderma lucidum* Protein Hydrolysates on Adipocyte Cells Using Proteomics Approach. *Foods* **2021**, *10*, 2157. https://doi.org/10.3390/foods10092157

Academic Editor: Antonello Santini

Received: 22 July 2021
Accepted: 10 September 2021
Published: 12 September 2021

Abstract: Excessive lipid accumulation is a serious condition. Therefore, we aimed at developing safe strategies using natural hypolipidemic products. Lingzhi is an edible fungus and potential lipid suppression stimulant. To use Lingzhi as a functional hyperlipidemic ingredient, response surface methodology (RSM) was conducted to optimize the time (X_1) and enzyme usage (X_2) for the hydrolysate preparation with the highest degree of hydrolysis (DH) and % yield. We encapsulated the hydrolysates using nanoscale liposomes and used proteomics to study how these nano-liposomal hydrolysates could affect lipid accumulation in adipocyte cells. RSM analysis revealed X_1 at 8.63 h and X_2 at 0.93% provided the highest values of DH and % yields were 33.99% and 5.70%. The hydrolysates were loaded into liposome particles that were monodispersed. The loaded nano-liposomal particles did not significantly affect cell survival rates. The triglyceride (TG) breakdown in adipocytes showed a higher TG increase compared to the control. Lipid staining level upon the liposome treatment was lower than that of the control. Proteomics revealed 3425 proteins affected by the liposome treatment, the main proteins being TSSK5, SMU1, GRM7, and KLC4, associated with various biological functions besides lipolysis. The nano-liposomal Linzghi hydrolysate might serve as novel functional ingredients in the treatment and prevention of obesity

Keywords: RSM; Lingzhi; hypolipidemic activity; peptides; 3T3-L1

1. Introduction

Modern functional food products are available on the market, ranging from isolated nutrients, dietary supplements, and specific products to processed or engineered foods. Peptides from foodstuff are candidates for functional food ingredients due to their beneficial health aspects such as immune-boosting, anti-oxidative stress, hypolipidemic and tumor suppressing activity [1,2]. One of the above-mentioned beneficial aspects is the hypolipidemic activity on adipocytes, affecting lipid storage, directly associated with obesity, a contemporary health problem. Obesity is caused by excessive triacylglycerol (TAG) accumulation in the adipocytes. Increasing TAG breakdown or hypolipidemic activity might contribute to reducing body fat and triglyceride (TG) levels. Several natural-sourced peptides could be combined with foodstuffs, and their effective delivery could display beneficial aspects [3].

Most edible mushrooms such as *Volvariella volvacea*, *Lentinula edodes*, and *Ganoderma lucidum* are beneficial for health. They have been generally consumed as basic food, as they provide plenty of dietary nutrients including fibers, minerals, and vitamins. They are also excellent sources of proteins [4]. Beyond their role as foodstuffs, edible mushrooms feature in certain types of holistic or alternative medicine. *G. lucidum*, locally known as Lingzhi, is defined as a medicinal mushroom for the prevention of various diseases, as well as for recuperation and health improvement. Ganoderma species are generally found all over the world. Lingzhi exhibits the prevailing features of being an excellent nutrient source of proteins, lipids, and carbohydrates [5]. Lingzhi has been consumed widely in East-Asia as a traditional remedy for centuries [6]. Many of its pharmacological effects have been widely reported such as immune modulation enhancement, soothing the nerves, inflammatory response reduction, cancer growth suppression, cell-aging deceleration, oxidative stress reduction, and anti-aging and lipid accumulation suppressive effects [7–9].

Foodstuff-derived protein hydrolysates contain a high level of functional peptides. The dominant features of the protein hydrolysates are their lower molecular weight and relative lack of high-order structure, as well as the increased number of functionally ionizable and exposed hydrophobic groups compared to those of intact proteins. These features denote that their surface interactions, water-solubility, host-receptors, and biological activities might be different from those of proteins. This includes the transduction triggering capability of various signaling pathways, leading to the activation or deactivation of regulators and biological activities above their generic nutritional value [10]. However, the major obstacle in introducing peptides into functional food ingredients is their functional stability during commercial processing and under human physiological conditions [11]. Therefore, functional peptides might partially or completely lose their activity before reaching the target cells or organs [12]. Hence, choosing a delivery system that is highly compatible with human physiological conditions would alleviate this problem.

Liposome encapsulation is a well-known compatibility delivery approach for foodstuff hydrolysates. The advantage of encapsulation within small particles is the stability and bioactivity enhancement of the protein hydrolysates [13]. This approach is suitable for protein and peptide delivery as their molecules possess various polar and non-polar regions similar to their liposome properties [14]. Several studies revealed the potential of liposomal encapsulation of peptides. For example, a pharmacological study of ghrelin, the appetite-stimulating peptide hormone, indicated increasing ghrelin stability and circulation period in the blood [15]. Both the pharmaceutical and cosmetic industries generally use liposome-based carriers to store and deliver functional proteins and peptides for specific purposes [16,17]. Although the use of liposomal encapsulation can be observed in a small number of products in the food industry market, liposomal encapsulation would be a promising approach as its safety and efficiency are proven by the pharmaceutical and food industries.

In this study, we established liposome carriers for protein hydrolysates to enhance the biological activities and stability of the latter. In addition, we also investigated the lipolysis-stimulating activity of the encapsulated Lingzhi protein hydrolysates on 3T3-L1 adipocyte cells. A possible signaling pathway for the encapsulated hydrolysates on the stimulation of lipid breakdown was also investigated using quantitative proteomic analysis. Finally, the possible beneficial mechanisms of the nano-liposomal hydrolysates are clarified and their value as a functional food additive supported.

2. Materials and Methods

2.1. Ganoderma Lucidum Hydrolysate Preparation

Dried Lingzhi (200 g) was powdered using an ultra-centrifugal mill (Retsch Co., Haan, Germany) equipped with a sieve (diameter = 1 mm^3) at 8000 rounds per minute (rpm). The powdered mushroom was heated using a modified Pressurized Hot Water Extraction method [18]. Briefly, the Lingzhi was mixed with deionized water at a ratio of 1:2 (*w*/*v*) and incubated at 121 °C, 15 psi for 20 min. The extracted Lingzhi was left to cool down and hydrolyzed with pepsin as the first independent factor (X_1) at 0.25%, 0.5%, and 1% in 0.1% of HCl for digestion times of 3, 6, and 9 h as the second independent factor (X_2) at a constant temperature of 37 °C. Next, the crude was filtrated with a 0.22-µm nylon membrane and fractionated through Vivaspin-20 (GE Healthcare Co., Amersham, UK), with a molecular weight cut-off of 3 kDa. Peptides of <3 kDa were subjected to Solid-Phase Extraction (SPE) (Waters Co., Milford, MA, USA). An amount of 3 mg of small peptides was loaded on an equilibrated SPE column (Sep-Pak C18) and eluted using acetonitrile: water (1:1, *v*/*v*). The supernatant was dried using a freeze-drying machine.

2.2. Lingzhi Protein Hydrolysate Optimization by Response Surface Methodology (RSM)

The two independent variable factors used in this study were the digestion time (X_1) and the enzyme concentration (X_2). The experimental outputs were the degree of hydrolysis (DH) (Y_1) and the product yield (Y_2). The DH determination was performed according to the method of Nielsen et al. [19] and the product yield was calculated as a percentage of the proteins found in the hydrolysates divided by the raw protein content. While calculating the optimal condition of an independent factor, the values of the other independent factors were fixed. An experimental design was set with 11 conditions, including 9 experimental conditions and 2 central points. The correlation of the independent factors and experimental outputs was used to generate RSM by the following equation:

$$y = \beta_0 + \varepsilon + \sum_{i=1}^{k} \beta_i x_i + \sum_{i=1}^{k}\sum_{j=1}^{k} \beta_{ij} x_i x_j + \sum_{i=1}^{k} \beta_{ii} x_{ij}^2 \quad (1)$$

where y is the experimental output; β_0 is constant intercept value; β_i, β_{ii}, and β_{ij} are the linear, quadratic, and interaction coefficients, respectively; and x_i and x_j are the independent variable factors. Three-dimensional response surface plots were drawn to illustrate the correlation between the levels of the process variable factors and the outcome results.

2.3. Nano-Liposome Carrier Preparation and Characterization

Soybean lecithin (Sigma Aldrich Co., St. Louis, MO, USA) and cholesterol (Sigma Aldrich Co., St. Louis, MO, USA) (8:1, *w*/*w*) were dissolved in 10 mL of diethyl ether in a 50-mL round bottom flask for 5 min. Once the lipids were thoroughly mixed in diethyl ether, the solvent was removed to yield a lecithin-cholesterol film layer by rotary evaporation (Buchi Co., Flawil, Switzerland) at 100 rpm under reduced pressure. The hydration of the lecithin-cholesterol film layer was accomplished by adding 10 mL of Lingzhi extract and agitating on an orbital shaker at 220 rpm for 6 h at 28 °C to obtain a vesicular white suspension. The vesicular suspension was forced through a membrane filter with a defined pore size of 200 nm by an extruder (GE Healthcare Co., Amersham, UK). After day 7, the loading efficiency of the loaded nanoliposome was determined by a protein-based spectrophotometric analysis. We mixed 100 µL samples of loaded liposomes with 1% Triton X-100 (Sigma Aldrich Co.) and sonicated for 10 min (10 s-interval) to disassemble the liposomes and release the extract. Afterward, the protein content of the clearance solution was assessed by Lowry protein assay using Bovine Serum albumin (Sigma Aldrich Co.) as a reference. The loading efficiency was calculated using the following equation:

Extracted loading Efficiency (*w*/*w*) (%) = (protein extracted of which encapsulated in

$$\text{liposomes (mg)} \div \text{protein content of extracted Lingzhi (mg)}) \times 100 \quad (2)$$

The hydrodynamic diameter of the liposomal formulations in deionized water was measured by dynamic light scattering (DLS) using ZetaSizer Nano-ZS (Malvern Instruments, Worcestershire WR, UK), in which the zeta potential was also examined (n = 3).

2.4. Effect of Loaded Nanoliposomes on 3T3-L1 Adipocyte Cells

Cell cytotoxicity of the loaded liposome and unloaded liposome control was evaluated through an MTT assay. Human fibroblasts (American Type Culture Collection., Manassas, VA, USA)) and 3T3-L1 mouse differentiated adipocyte cells (induced by an adipogenic cocktail containing 2.5 mM dexamethasone, 0.5 mM 3-isobutyl-1-methylxanthine, and 10 g/mL insulin for 8 days) were tested for cytotoxicity at various concentrations (104.68, 52.34, 26.17, 13.09, 6.54, 3.27, 1.64, 0.82, 0.41, and 0.20 μg/mL) of loaded liposomes and unloaded liposome as control for 24 h. Next, we measured the optical absorbance at 570 nm using a microplate reader and transformed the results into cell survival rate percentage [20].

The lipolytic effect of the loaded nanoliposome was used to quantify glycerol, a byproduct of lipolysis (EnzyChrom™ Glycerol Assay Kit, BioAssay Systems, Hayward, CA., USA) in cell culture supernatant after 24 h of treatment with the loaded nanoliposome. To determine the intracellular TG content, the differentiated 3T3-L1 cells were treated with the loaded nanoliposomes, as described previously, for 24 h. The cells were washed twice with PBS and fixed with 4% paraformaldehyde for 1 hour at room temperature. Next, the cells were washed once with PBS and isopropanol 60% (v/v), then they were allowed to dry. Next, the cells were stained with 0.5% (v/v) Oil Red O (ORO) (Sigma Aldrich Co.) in an isopropanol solution of 60% for 1 hour. After staining, the unstained dye was removed by rinsing with distilled water. The stained lipid droplets were observed under a stereomicroscope. The stained oil droplets indicating lipid accumulation were solubilized by absolute isopropanol for 15 min and their absorbance was measured at 510 nm using a microplate reader (Multiskan Go, Thermo Scientific, Waltham, MA, USA).

2.5. Proteomic Analysis and Data Processing

To investigate the adipocyte protein expression profiles after the exposure to the loaded liposomes, the cells were lysed by a lysis buffer solution (10 mM HEPES-NaOH pH 8.0 and 0.5% Triton X-100) supplemented with a protease inhibitor cocktail (Thermo Scientific Co.). The supernatant was collected by centrifugation, followed by ice-cold acetone precipitation (1:5 v/v). After precipitation, the protein pellet was reconstituted in 0.2% RapidGest SF (Waters Co.) in 10 mM of Ammonium bicarbonate (Sigma Aldrich Co.). The total protein (50 μg) was subjected to gel-free based digestion. Next, sulfhydryl bond reduction was performed using 5 mM DTT (Sigma Aldrich Co.) in 10 mM ammonium bicarbonate at 72 °C for 1 h and sulfhydryl alkylation using IAA (Sigma Aldrich Co.) at room temperature for 30 min in the dark. The solution was cleaned up using a Desalting Zebra-spin column (Thermo Scientific Co.). The flow-through solution was enzymatically digested by Trypsin (Promega Co., Madison, WI, USA) at a ratio of 1:50 (enzyme: protein) and incubated at 37 °C for 3 h. The digested solution was dried and reconstituted in 0.1% formic acid before being subjected to tandem-mass spectroscopy using a nanoLC-system coupled with high resolution 6600 TripleTOF™ (AB-Sciex, Concord, ON, Canada). The LC conditions were as follows: mobile phase A and B were used, with mobile phase A being composed of 0.1% formic acid in water and mobile phase B comprising 95% acetonitrile with 0.1% formic acid. The LC-method parameters comprised a 135-min long process for a single injection. The analytical column was maintained at 55 °C. Using the data-dependent acquisition mode of mass spectroscopy, the MS scans over a mass range of 400–1600 m/z, selecting the top 20 most abundant peptide ions with charge state in the range of 2–5 (positive mode) for fragmentation. The dynamic exclusion duration was set at 15 s. The raw MS-spectra resulting (.wiff) file was extracted and annotated with protein sequences using the Paragon™ Algorithm by ProteinPilot™ Software [21]. The *Mus musculus* protein database, retrieved from UniProtKB (16,477 sequences) and used in Paragon™, was assembled in FASTA format and downloaded in May 2021. We set a detected protein threshold of (Unused ProtScore (Conf)) ≥ 0.05 with 1% false discovery rate (FDR) with ≥10 peptides/protein. The protein and peptide comparisons exhibiting >20% coefficient of variation (C.V.) between the replicates were rejected. Both library and SWATH-MS data were imported into SWATH™ processing microapp in PeakView® software. The normalization of the relative protein abundances was performed using the R package, NormalyzerDE [22], in which Quantile-normalization was applied to expression data analysis, after adding 1 to all expression values to avoid errors upon log transformation.

2.6. Statistical Analysis

All experiments were carried out in at least three independent replicates (n = 3), and all data were expressed as the means ± standard deviation. The statistical significance was determined by Duncan's multiple range test (p-values < 0.05). For the RSM analysis, the generated 3D surface was determined from the fitted polynomial equation, and significant coefficients (p < 0.01) were used in the model. The variance table was generated from both independent variables and experimental outputs using the Design Expert statistical software (version 11.0; State-Ease Inc., Minneapolis, MN, USA). For the pairwise comparisons during the proteomic analysis, we performed One-Way analysis of variance (One-Way ANOVA) at the protein-level analysis with two multiple testing correction methods including the Bonferroni and the Benjamini–Hochberg FDR corrections using the ProteinPilot™ Software.

3. Results

3.1. Lingzhi-Derived Protein Hydrolysate Optimization

The biological activity of the hydrolysates depends on the processing conditions. The activities of various foodstuff hydrolysates were reportedly directly dependent on the degree of hydrolysis, protease activity, and amino acid arrangement [23]. The optimum conditions for the Lingzhi hydrolysate regarding DH and product yield for functional food product manufacturing have not yet been established. Therefore, the present study was aimed at Lingzhi hydrolyzing proteins using RSM to study the effect of the processing conditions including time, enzyme usage on DH, and product yield of the resulting hydrolysates. We applied quadratic analysis statistics to fit an RSM model for independent variable factors. The experimental design using two independent variable factors with two center points (experiment no. 10 and 11) in RSM generation resulted in the observed DH and yield as displayed in Table 1. The RSM generation-related statistical value is shown in Appendix A.

Table 1. The experimental design and experimental outputs of the independent factors for the degree of hydrolysate and yield produced from Lingzhi proteins.

Experiment No.	Independent Factors		Experimental Outputs	
	x_1; Time (Hour)	x_2; Enzyme (%)	y_1; DH (%)	y_2; Yield (%)
1	3	0.25	28.11 ± 1.03	4.16 ± 0.13
2	3	0.50	29.83 ± 1.30	4.57 ± 0.17
3	3	1.00	29.36 ± 1.28	4.22 ± 0.15
4	6	0.50	33.21 ± 1.03	5.25 ± 0.12
5	6	1.00	32.91 ± 1.37	5.32 ± 0.14
6	6	2.00	32.03 ± 0.76	5.21 ± 0.22
7	9	0.25	33.96 ± 1.14	5.58 ± 0.16
8	9	0.50	33.17 ± 1.29	5.67 ± 0.09
9	9	1.00	34.18 ± 1.12	5.70 ± 0.20
10	6	0.50	33.16 ± 0.58	5.24 ± 0.09
11	6	0.50	32.92 ± 0.32	5.21 ± 0.13

As outputs from the overall experimental design, the DH and product yield ranged from 28.11% ± 1.03% to 34.18% ± 1.12% and 4.16% ± 0.13% to 5.70% ± 0.20%, respectively.

The difference in the DH and yield could be due to the difference in the digestion time and enzyme concentration. The equation for multiple regression analysis during the RMS was performed to resolve the coefficients of the independent factors of the linear (x_1, x_2), quadratic (x_1^2, x_2^2), and two-factor relation ($x_{1 \times 2}$) to fit the RSM. According to the multiple regression analysis, the explanatory model equation of the DH (y_1) and percentage of product yield (y_2) is given as follows in Table 2.

Table 2. The experimental design and experimental outputs of the independent factors for the degree of hydrolysate and yield produced from Lingzhi proteins.

Responding	Quadratic Model	R^2	*p*-Value
y_1	$y_1 = 33.14 + 2.08x_1 - 0.497x_2 - 0.283x_1x_2 - 1.53x_1^2 - 0.635x_2^2$	0.96	0.0019
y_2	$y_2 = 5.293 + 0.726x_1 - 0.027x_2 - 0.066x_1x_2 - 0.264x_1^2 - 0.073x_2^2$	0.97	0.0010

The total coefficient value (R^2) was used to imply the model suitability. The R^2 of the DH and the product % yield were 0.958 and 0.968, respectively. This result indicated that the variation in the experimental data was lower than 5% (within 95% level of confidence). The 3-dimensional response model surfaces (3D-RMS) for each variable are illustrated in Figure 1.

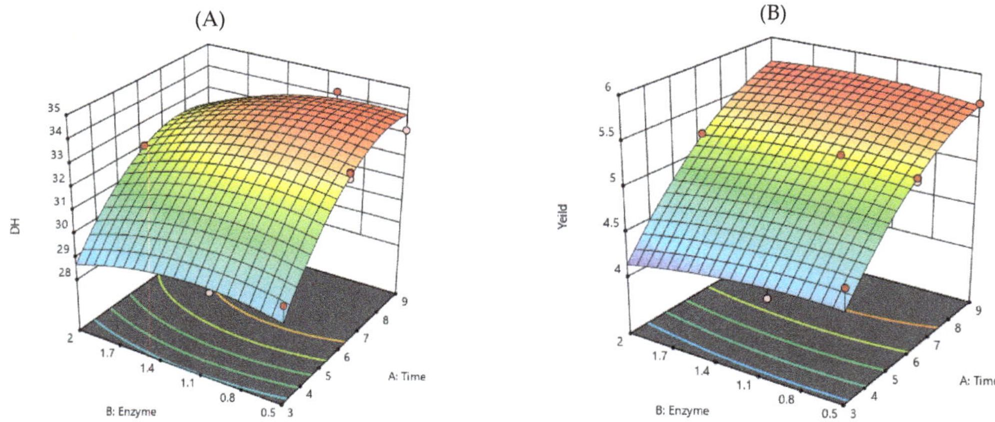

Figure 1. 3D-RMS plots showing the interactive effects of different factors on DH and yield. (**A**) DH of Lingzhi protein hydrolysate on digestion time versus enzyme usage, and (**B**) yield of Lingzhi protein hydrolysate on digestion time.

The experimental outputs of the processing related to both independent factors, DH (Figure 1A) and % yield (Figure 1B) indicating that the hydrolysate processing depended on the digestion time and enzyme usage. The 3D-RMS for the DH of hydrolysate as a function of digestion time, at fixed enzyme usage, revealed that DH was dependent on the digestion time. Also, DH increased with enzyme usage at the fixed digestion time, suggesting that DH was also dependent on the enzyme usage. Yield also had correlative results, dependent on the digestion time and enzyme usage. In order to obtain the highest DH and product yield, the RSM model was optimized by setting the highest value of response variable factors. As a result, X_1 was 8.63 h and X_2 was 0.93%, and the highest values of y_1 and y_2 were 33.99% and 5.70%, respectively. These characteristics of DH and yield curves were associated with feedback inhibition during the hydrolysis, where products may act as an inhibitor to protease [24]. The curves strongly suggested that the processing at different conditions and factors were involved. The independent factors, both time and enzyme concentration, had the optimum range for hydrolysate production to gain the maximum DH and yield. To endorse the reliability and validity of the model for processing, the assays were performed under those optimal conditions. The actual experimental values for DH and product yield were 32.71 ± 0.17% and 5.44 ± 0.14%, respectively; the experimental values fitted with the values that were predicted by the model within a 95% confidence interval. These results confirmed that the model was suitable for Lingzhi protein hydrolysate processing for use as functional ingredients regarding cost- and time-efficiency.

3.2. Encapsulation Efficiency and Loaded Liposome Size, Polydispersity Index, and Zeta Potential

The encapsulation efficiency of the liposomal formulation was estimated. The liposomes would passively entrap the protein hydrolysate in their hydrophilic region. However, many factors influence the entrapping efficiency such as lipid molar ratios, molecular size, charge, and molecule stability. To evaluate the entrapping efficiency, we used a non-ionic detergent, Triton X-100, as a neutral detergent to disrupt the liposome shell structure, thereby allowing the leakage of the encapsulated Lingzhi protein hydrolysate [25]. Based on the encapsulation condition, 61.24 ± 3.18% of the encapsulation efficiency was achieved. The encapsulation efficiency showed that the liposomal preparation for protein hydrolysate moderates the encapsulated level. The protein hydrolysate has a mixture of peptides with a variety of molecular weights, sizes, charges, and structures. Middle-sized peptides might interact with the lipid layer and form an oligomerization structure like a beta-barrel. This could disrupt the entrapped protein hydrolysate inside the core structure of the liposome [26]. Another reason was the fluctuation in electrostatic interaction between the charges of various peptides and the liposome surface, which might negatively affect the encapsulation efficiency.

The diameter of the nanoliposome in the closest realistic physiological condition was determined. Dynamic light scattering (DLS) analysis showed loaded liposome diameters in the PBS solution were at 149.84 ± 0.58 nm (Figure S1). Low polydispersity index (PdI) of =0.048 ± 0.014 supported that particles were monodispersed. In addition, the low PdI value also reflected that the particle exhibits a narrow size distribution, providing a very high surface area that would be ideal for the correct order. This evidence suggested the homogeneity of the loaded liposome. The overall charge of loaded liposomes was neutral. Zeta (ζ)-potential of the loaded liposome was −3.75 ± 0.25 mV (Figure S2). This could suggest that the overall structure of the liposome exhibited neutral charge particle, due to the value of ζ-potential ranging from −10 to +10 mV, is considered neutral [27]. The hydrodynamic size of the loaded liposome was roughly 140 nm, indicating that the liposome was characterized in the nanoscale. As the efficiency of cellular uptake relates to the particle size, a small particle size of around 100–160 nm would have great potential for cellular uptake into the blood steam via clathrin-dependent mechanisms [28]. Beneficial properties of the negative value of ζ-potential were particle stability under physiological conditions and the prevention of cellular fusion and aggression of phagocytosis, responding less than the positive value of ζ-potential [29]. Therefore, the hydrodynamics of loaded liposome size and negative ζ-potential are the two key criteria that have been considered for various applications.

3.3. Effect of Loaded Nanoliposome on 3T3-L1 Adipocyte Cells

The safety of using the loaded liposomes is a crucial factor for establishing commercialized products. Therefore, we investigated cell cytotoxicity to evaluate the safety of loaded liposomes using human fibroblasts as normal cell controls and the differentiated 3T3-L1 adipocyte cell line as a lipid storage cell model. Cell viability was measured through an MTT assay and illustrated in Figure 2.

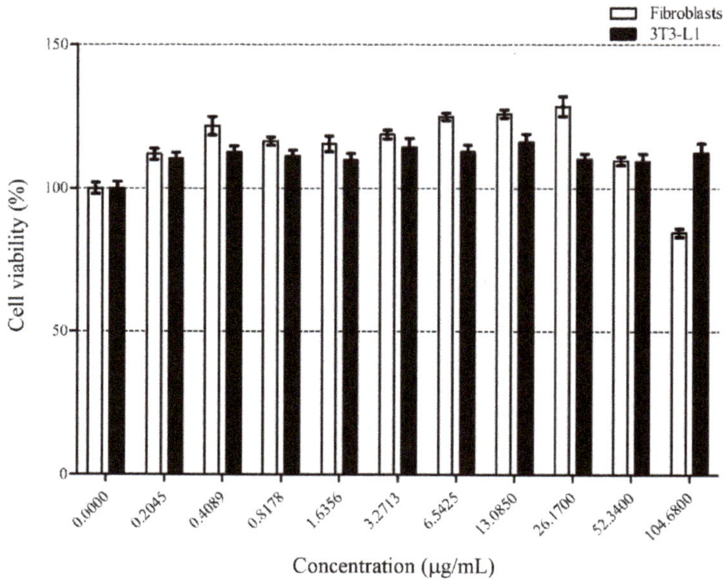

Figure 2. Fibroblast and differentiated adipocyte cells were treated with increasing concentration of loaded liposomes for 24 h. % cell viability was measured by MTT assay. Symbols, (■) and (□), represent the differentiated 3T3-L1 cells and fibroblast cells, respectively. *y*-axis represents the percentage of cell viability and *x*-axis represents concentrations of the loaded liposome. Data are shown as the mean ± SD from triplicate results.

As a result, the loaded liposomes did not significantly affect the viability of either cell lines at concentrations up to 52.34 μg/mL. However, a further increment (104.68 μg/mL) resulted in slight cytotoxic effects on the fibroblast cells. Therefore, we considered the cytotoxicity-related no-observed-adverse-effect level of the loaded liposomes was 52.34 μg/mL for further experiments. Oral delivery of liposomal protein and peptide is the easy and convenient route. The liposome particles made by cholesterol and lecithin were moderately stable (~80% stability measured by particle leakage) in gastric environment (pH 2) at 37 °C at 1 h and stable (~95% stability measured by particle leakage) in pancreatin [30]. These results indicate that our liposome formulations may be suitable as oral delivery particles due to their stable behavior through the oral route. As the potential application of the loaded liposome would be in functional food ingredients, this concentration was used in the determination of lipolysis activity and proteomics.

The lipolysis process is a metabolic process that breaks down TGs to free fatty acid (FA) and glycerol. It controls the energy homeostasis by regulating the breakdown of TGs [31]. Therefore, the effect of 52.34 μg/mL loaded liposome on the TG breakdown in adipocyte cells was investigated through the measurement of glycerol released into the medium culture. In the present study, the loaded liposome significantly increased glycerol release and reduced lipid accumulation. The loaded nanoliposome affected the adipocytes by inducing the TG breakdown, as we observed the release of glycerol at 1.63 ± 0.25-fold greater than that in the control ($p < 0.01$). The intracellular lipid exposed by the loaded nanoliposome was visualized by ORO staining where the lower staining intensity represented the lower lipid accumulation (Figure 3).

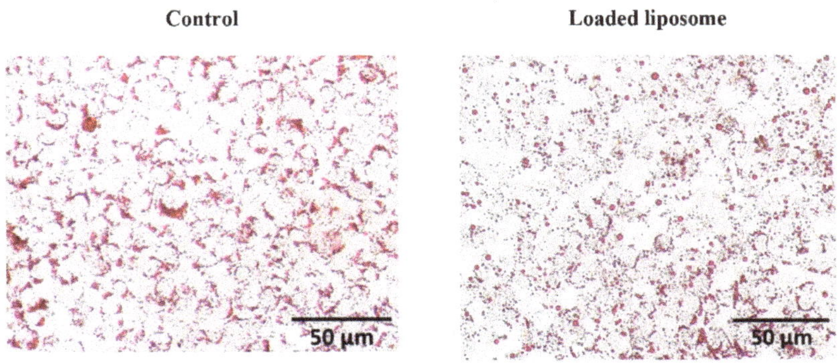

Figure 3. Lipolysis effects of the loaded liposome on the differentiated adipocyte cells. The ORO lipid staining of 3T3-L1 adipocytes was observed using a stereomicroscope at 5× magnification. The cells with no treatment were used as a control (Control).

The ORO staining demonstrated lower intracellular lipid accumulation in cells exposed to loaded liposomes compared to the control. The loaded liposome increased glycerol release corresponding to 50% release at 13.085 µg/mL. ORO staining revealed the most pronounced TG clearance at a peak concentration (52.34 µg/mL), with lower staining severity representing lower lipid aggregation (Figure 3). This evidence implied that the loaded nanoliposomes were able to reduce the lipid accumulation as determined by the reduced ORO staining level and the free glycerol level increase. Therefore, we next applied a label-free proteomics approach to study the molecular mechanisms of lipid breakdown activity that could potentially lead to the reduced lipid accumulation in the adipocytes for a better understanding of the loaded liposome-induced lipolytic pathways.

3.4. *Quantitative Proteomic Analysis*

We used a proteomics approach to investigate the signaling pathways that could be potentially affected by the loaded liposomes in the adipocyte cells. The LC-MS/MS analysis revealed a total number of 3425 proteins among the loaded liposome and the control groups. The interpretation of the quantitative proteomics and bioinformatics data showed that 439 proteins were affected by the loaded liposomes as shown in Figure 4. Although we used differentiated adipocytes from mice, this was a widely accepted cell-based model [32]. The raw data from the LC-MS/MS analysis showed a small difference in the total ion count between each LC-MS injection. Therefore, data normalization of the raw dataset was strongly required prior to further analysis. After the log transformation and VSN normalization, pooled intragroup median absolute deviation (PMAD) of the identified proteins among replicates was lower than 0.22 (Control and loaded liposomes $n = 3$ and 3, respectively; Figure S3). In general, a PMAD value of ≤ 0.3 was accepted as the superior precision dataset [33]. According to the normalized proteomic analysis, the volcano plot of the differential protein expression identifying the most significant protein expression changes is depicted in Figure 4. Each spot represents the protein expression ratio (loaded liposome: control) according to their $\log_{10}$ p-values. The differentially expressed proteins associated with these spots are listed in the proteomics table (Appendix B).

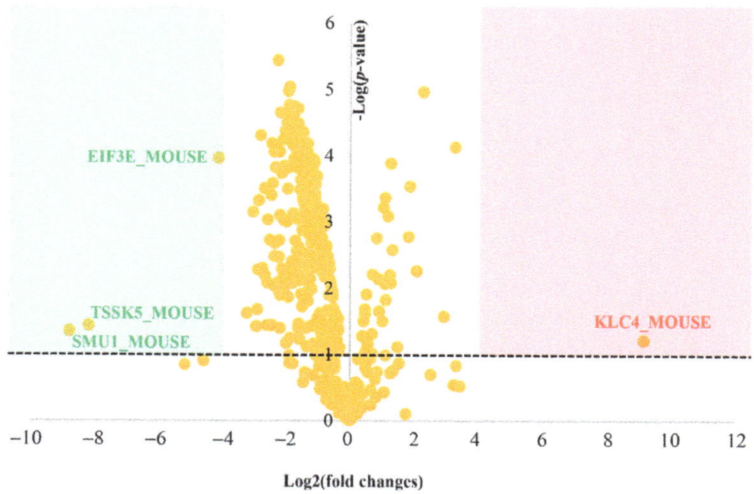

Figure 4. Quantitative proteomic analysis visualized by a volcano plot. The plot shows a negative natural log of the *p*-values plotted against the base2 logs of the change in each protein compared between the loaded liposome and control groups. Statistically significant results ($p < 0.05$) are plotted above the dashed line in the green and red regions. Proteins significantly up- and down-regulated upon the loaded liposome treatment are shown as red and green dots, respectively.

We identified four significantly different proteins, compared between the loaded liposome and control groups. The global protein expression changes were mostly down-regulated (79.37%; for 350 of 441 proteins). Specifically, three significantly different proteins ($p < 0.05$ and $-4 > \log_2$ (fold change) > 4) were down-regulated (green region, Figure 4) whereas one was up-regulated (red region, Figure 3). Considering the biological functions of the significantly different proteins, the down-regulated ones were Testis-specific serine/threonine-protein kinase 5 (TSSK5_MOUSE), WD40 repeat-containing protein SMU1 (SMU1_MOUSE), and metabotropic glutamate receptor 7 (GRM7_MOUSE), whereas the up-regulated one was Kinesin light chain 4 (KLC4_MOUSE). The detailed description and function of these proteins are presented in Table 3.

Table 3. The description and functions of the top 4 significant proteins uniquely identified in the liposome-encapsulated hydrolysate treatment group. This information was obtained from the UniProtKB database.

Accession	Protein Name	Biological Process
TSSK5_MOUSE	Testis-specific serine/threonine-protein kinase 5	Cell differentiation, intracellular signal transduction, multicellular organism development, protein phosphorylation, and spermatogenesis
SMU1_MOUSE	WD40 repeat-containing protein SMU1	mRNA splicing, via spliceosome, regulation of alternative mRNA splicing, via spliceosome, and RNA splicing
GRM7_MOUSE	Metabotropic glutamate receptor 7	adenylate cyclase-inhibiting G protein-coupled glutamate receptor signaling pathway, chemical synaptic transmission, and regulation of neuron death
KLC4_MOUSE	Kinesin light chain 4	-

The biological functions of these proteins were variable, including cell differentiation, intracellular signal transduction, organism development, protein phosphorylation, spermatogenesis, mRNA splicing, cAMP-related G protein inhibition, chemical synapsis-related activities, and the regulation of neuronal death. Notably, the liposome-encapsulated protein hydrolysates affected the 3T3-L1 cells in various biological functions beyond lipolysis.

Although these significant proteins were not directly associated with lipolysis, differentially expressed proteins in lipid biosynthesis and lipolysis could also be identified. Our investigation detected that fatty acid synthase (FAS; FAS_MOUSE), the major actor of lipogenesis, was suppressed more than 5-fold ($\log_2$ fold change as 2.35) in the loaded liposome group (supplementary data 2). The lipogenesis works via FAS to synthesize the long-chain FA from acetyl-CoA, malonyl-CoA, and NADPH. Hence, FAS downregulation could imply that cellular lipogenesis might be reduced due to the decrease in its abundance and activity. FAS-down regulation, an increased rate of lipolysis, and TG release could lead to a net TG loss on the cellular level. Moreover, another protein that elongates the long-chain fatty acids, protein 5 (ELOV5_MOUSE), was also down-regulated. Elov5, known as PUFA elongase, is a major PPARα-regulated enzyme functioning in monounsaturated and polyunsaturated fatty acid synthesis [34].

4. Conclusions

The concordance between the proteomics results and the cellular lipidemic activity could imply that the Lingzhi protein hydrolysate-loaded nano-liposomes induced cellular lipolysis without affecting cell viability. The proteomic study also indicated that loaded liposomes exhibited lipid accumulation with the suppression of FAS and ELOV5. Finally, other proteins including TSSK5, SMU1, GRM7_MOUSE, and KLC4, were identified in the loaded liposome treatment group, associated with various biological mechanisms beyond lipid metabolism. Therefore, the nano-liposomal hydrolysates might serve as novel functional ingredients in the treatment and prevention of obesity.

Supplementary Materials: The following are available online at https://www.mdpi.com/2304-8158/10/9/2157/s1, Figure S1: DLS analysis showed size distribution, Figure S2: DLS analysis showed zeta-potential distribution, Figure S3: Intragroup variation analysis.

Author Contributions: Conceptualization, S.K. and Y.Y.; methodology, T.M., C.C. and P.S.; software, S.K. and Y.Y.; validation, K.C., Y.Y. and K.C.; formal analysis, S.K. and Y.Y.; investigation, P.P.; resources, W.V. and K.C.; data curation, C.C. and K.C.; writing—original draft preparation, S.K.; writing—review and editing, S.K. and Y.Y.; visualization, T.M.; supervision, P.S. and C.C.; project administration, K.C.; funding acquisition, K.C. All authors have read and agreed to the published version of the manuscript.

Funding: This study was supported by the National Research Council of Thailand (NRCT) under the Royal Golden Jubilee Ph.D. Program (Grant No. NRCT5-RGJ63002-042) and Kasetsart University Research and Development Institute, KURDI (FF (KU)25.64).

Institutional Review Board Statement: Not applicable.

Informed Consent Statement: Not applicable.

Data Availability Statement: Not applicable.

Acknowledgments: This research project was supported by National Research Council of Thailand (NRCT): NRCT5-RGJ63002-042. Additional support for core research facilities was provided by faculty of pharmacy, Chiang-Mai university.

Conflicts of Interest: The authors declare no conflict of interest. The funders had no role in the design of the study; in the collection, analyses, or interpretation of data; in the writing of the manuscript, or in the decision to publish the results.

Appendix A

Table A1. The observed effects due to the hydrolysis variables including digestion time (A) and enzyme usage (B) on the response values for degree of hydrolysis (DH) and %yield.

1. DH						
Model	39.47	5	7.89	22.9	0.0019	significant
A-Time	4.68	1	4.68	13.56	0.0143	
B-Enzyme	0.769	1	0.769	2.23	0.1955	
AB	0.083	1	0.083	0.2407	0.6445	
A^2	5.7	1	5.7	16.54	0.0097	
B^2	1.06	1	1.06	3.06	0.1405	
Residual	1.72	5	0.3447			
Lack of Fit	1.68	3	0.5585	23.24	0.0415	significant
Pure Error	0.0481	2	0.024			
Cor Total	41.19	10				
Std. Dev.	0.5871		**R^2**			0.9582
Mean	32.08		**Adjusted R^2**			0.9163
C.V. %	1.83		**Predicted R^2**			0.4581
			Adeq Precision			11.99244
2. Yield						
Model	2.87	5	0.5747	30.13	0.001	significant
A-Time	0.5672	1	0.5672	29.74	0.0028	
B-Enzyme	0.0023	1	0.0023	0.1184	0.7448	
AB	0.0046	1	0.0046	0.2399	0.645	
A^2	0.1691	1	0.1691	8.87	0.0309	
B^2	0.0139	1	0.0139	0.7302	0.4318	
Residual	0.0954	5	0.0191			
Lack of Fit	0.0945	3	0.0315	72.69	0.0136	significant
Pure Error	0.0009	2	0.0004			
Cor Total	2.97	10				
Std. Dev.	0.1381		**R^2**			0.9679
Mean	5.1		**Adjusted R^2**			0.9358
C.V. %	2.71		**Predicted R^2**			−0.5277
			Adeq Precision			14.0722

Appendix B

Table A2. Differentially expressed proteins list.

Protein Name	*p*-Value	Log2 Intensity						
		Control01	Control02	Control03	Liposome01	Liposome02	Liposome03	log(*p*-Value)
HNRH1_MOUSE	3.65×10^{-6}	16.58	16.46	16.41	14.17	14.23	14.26	5.44
VIME_MOUSE	9.17×10^{-6}	19.58	19.48	19.52	17.64	17.55	17.70	5.04
TBA8_MOUSE	1.06×10^{-5}	19.94	19.82	19.70	17.87	17.87	17.90	4.98
HIG1A_MOUSE	1.72×10^{-5}	13.17	13.12	12.98	11.23	11.27	11.04	4.76
UBA1_MOUSE	1.80×10^{-5}	17.44	17.45	17.35	15.74	15.70	15.54	4.75
IF4A1_MOUSE	1.96×10^{-5}	16.22	16.32	16.30	14.65	14.57	14.71	4.71
KPYM_MOUSE	2.20×10^{-5}	20.80	21.00	20.79	18.90	19.10	18.83	4.66
PSMD2_MOUSE	2.24×10^{-5}	15.41	15.44	15.06	12.99	13.21	13.17	4.65
TBA4A_MOUSE	2.42×10^{-5}	20.35	20.20	20.09	18.36	18.42	18.53	4.62
ATPA_MOUSE	3.24×10^{-5}	18.45	18.15	18.20	16.13	16.46	16.29	4.49
IPO5_MOUSE	3.30×10^{-5}	15.47	15.28	15.53	13.84	13.79	13.80	4.48
RAN_MOUSE	3.41×10^{-5}	18.54	18.40	18.76	16.53	16.64	16.78	4.47
RANT_MOUSE	3.41×10^{-5}	18.54	18.40	18.76	16.53	16.64	16.78	4.47
RGL2_MOUSE	3.44×10^{-5}	15.83	15.62	15.75	13.64	13.99	13.91	4.46
MYH9_MOUSE	3.81×10^{-5}	17.63	17.64	17.86	15.73	15.94	16.04	4.42
ADT2_MOUSE	4.25×10^{-5}	18.79	19.10	18.81	16.78	17.10	16.74	4.37
ADT1_MOUSE	4.37×10^{-5}	18.12	18.30	18.16	16.21	16.60	16.37	4.36
RHOA_MOUSE	4.43×10^{-5}	15.32	15.23	15.38	13.93	13.79	13.77	4.35
GLRX3_MOUSE	4.53×10^{-5}	16.10	16.22	15.99	14.55	14.36	14.56	4.34
FLNA_MOUSE	4.90×10^{-5}	17.85	18.20	18.00	16.40	16.38	16.38	4.31
SYSC_MOUSE	4.95×10^{-5}	12.81	13.18	12.30	9.98	10.00	9.94	4.31

Table A2. *Cont.*

Protein Name	p-Value	Log2 Intensity						
		Control01	Control02	Control03	Liposome01	Liposome02	Liposome03	log(p-Value)
SDHA_MOUSE	5.34×10^{-5}	15.09	14.78	15.07	13.39	13.18	13.18	4.27
PRS6B_MOUSE	5.61×10^{-5}	15.29	15.71	15.39	13.70	13.69	13.52	4.25
CAN2_MOUSE	5.62×10^{-5}	12.78	13.08	12.52	10.78	10.59	10.80	4.25
ATAD1_MOUSE	5.75×10^{-5}	16.37	16.22	15.98	14.27	14.20	13.83	4.24
RS2_MOUSE	6.36×10^{-5}	16.51	16.43	16.18	14.73	14.83	14.62	4.20
NONO_MOUSE	6.55×10^{-5}	15.03	15.20	15.10	13.09	12.32	12.58	4.18
PGM1_MOUSE	6.58×10^{-5}	15.08	15.23	15.20	13.77	13.89	13.85	4.18
EF2_MOUSE	6.83×10^{-5}	19.20	19.04	19.08	17.60	17.80	17.57	4.17
FINC_MOUSE	6.87×10^{-5}	16.44	16.08	16.53	14.48	14.55	14.63	4.16
K22O_MOUSE	7.50×10^{-5}	15.65	15.76	14.98	18.79	19.12	18.30	4.13
TCPB_MOUSE	8.00×10^{-5}	15.64	15.75	15.50	14.27	14.01	14.12	4.10
SYVC_MOUSE	8.03×10^{-5}	13.57	14.21	13.62	11.53	11.61	11.32	4.10
PRDX2_MOUSE	8.32×10^{-5}	16.59	16.25	16.50	14.71	14.65	14.95	4.08
SERPH_MOUSE	8.67×10^{-5}	17.83	18.03	17.99	16.55	16.70	16.54	4.06
APT_MOUSE	8.76×10^{-5}	13.68	13.87	14.07	11.55	12.03	11.42	4.06
MYH10_MOUSE	8.77×10^{-5}	14.63	14.66	14.52	11.93	12.05	12.69	4.06
HS90A_MOUSE	9.32×10^{-5}	19.67	19.82	19.74	18.19	18.48	18.32	4.03
EIF3E_MOUSE	0.000109	12.02	13.07	13.27	8.21	9.18	8.62	3.96
COF1_MOUSE	0.000111	20.53	20.48	20.50	19.26	19.40	19.20	3.95
TBB5_MOUSE	0.000112	18.99	18.79	18.92	17.67	17.69	17.69	3.95
TBA1C_MOUSE	0.000112	21.61	21.08	21.35	19.63	19.46	19.68	3.95
RL6_MOUSE	0.000116	14.96	14.92	14.97	13.88	13.76	13.76	3.94
TBA1B_MOUSE	0.000118	21.62	21.10	21.37	19.64	19.50	19.72	3.93
TBB4B_MOUSE	0.000118	18.99	18.77	18.91	17.62	17.68	17.68	3.93
TBA1A_MOUSE	0.000122	21.32	20.84	21.12	19.51	19.37	19.53	3.91
TERA_MOUSE	0.000125	19.24	19.22	19.26	18.07	18.18	18.12	3.90
PP1B_MOUSE	0.000129	16.74	16.73	16.63	15.62	15.53	15.55	3.89
K1C15_MOUSE	0.000132	15.09	14.81	14.93	16.24	16.17	16.23	3.88
K1C17_MOUSE	0.000132	15.09	14.81	14.93	16.24	16.17	16.23	3.88
ATPB_MOUSE	0.000137	19.50	19.44	19.37	18.08	18.18	18.33	3.86
HS90B_MOUSE	0.000139	20.21	20.04	20.30	18.75	18.97	18.77	3.86
IF4G1_MOUSE	0.000144	13.62	13.50	13.27	11.78	11.10	11.33	3.84
CPSF5_MOUSE	0.000148	14.38	14.59	14.62	13.23	13.35	13.31	3.83
MYADM_MOUSE	0.000151	13.23	13.16	13.01	11.29	10.72	10.41	3.82
S35E3_MOUSE	0.000152	12.63	13.06	12.52	11.02	11.11	11.07	3.82
PUR6_MOUSE	0.000153	18.17	17.71	17.96	15.97	16.31	15.75	3.82
ANXA2_MOUSE	0.000156	20.59	20.84	20.61	19.12	19.44	19.25	3.81
RS3_MOUSE	0.000157	18.82	18.55	18.71	16.61	16.87	16.09	3.81
PDIA1_MOUSE	0.000158	20.30	20.30	20.23	18.96	19.21	19.09	3.80
HNRPF_MOUSE	0.00016	16.04	15.88	16.21	14.59	14.28	14.65	3.79
FLNB_MOUSE	0.000165	16.61	16.89	16.76	15.53	15.25	15.21	3.78
LDHA_MOUSE	0.000169	19.05	19.17	19.18	17.69	18.00	17.67	3.77
RL11_MOUSE	0.000175	17.54	17.32	17.35	15.90	16.21	15.90	3.76
FLNC_MOUSE	0.000184	16.60	16.87	16.57	15.41	15.23	15.42	3.74
MVP_MOUSE	0.000184	15.70	15.74	15.43	13.92	13.14	13.40	3.73
HSP7C_MOUSE	0.000191	19.22	19.02	18.99	17.98	17.88	17.86	3.72
ALDOA_MOUSE	0.000206	18.40	18.48	18.51	17.48	17.40	17.39	3.69
PP1G_MOUSE	0.000222	16.26	16.32	16.27	15.30	15.05	15.16	3.65
PP1A_MOUSE	0.000222	16.26	16.32	16.27	15.30	15.05	15.16	3.65
PSMD5_MOUSE	0.000233	15.59	15.92	15.73	14.00	14.46	14.11	3.63
PROF1_MOUSE	0.000243	20.22	19.93	19.78	18.46	18.31	18.66	3.61
XPO2_MOUSE	0.000262	12.68	12.49	12.25	10.12	10.52	9.52	3.58
LMNA_MOUSE	0.00027	16.59	16.69	16.44	15.47	15.54	15.47	3.57
TDRD1_MOUSE	0.000295	13.59	13.62	13.36	15.73	15.45	14.98	3.53
TIF1B_MOUSE	0.000299	16.56	16.31	16.41	15.27	15.25	14.94	3.52
RPN2_MOUSE	0.000314	15.97	16.00	16.01	13.80	12.59	13.57	3.50
HNRPU_MOUSE	0.000319	17.21	17.24	17.36	14.85	15.72	15.38	3.50
RL18A_MOUSE	0.000329	14.71	14.44	14.93	13.18	12.89	13.32	3.48
ERO1A_MOUSE	0.000386	15.67	15.38	15.62	14.37	14.48	14.54	3.41
TCPZ_MOUSE	0.000399	16.61	16.71	16.55	14.43	14.60	13.47	3.40
SCMC1_MOUSE	0.00041	14.28	14.38	14.72	13.11	13.18	12.81	3.39
TM183_MOUSE	0.00044	15.32	15.38	15.66	16.48	16.56	16.60	3.36

Table A2. *Cont.*

Protein Name	p-Value	Log2 Intensity						
		Control01	Control02	Control03	Liposome01	Liposome02	Liposome03	log(p-Value)
HSP74_MOUSE	0.000448	15.18	15.18	15.29	13.44	13.68	14.09	3.35
SMD1_MOUSE	0.000461	16.87	16.91	17.00	15.65	15.81	16.02	3.34
CH10_MOUSE	0.000468	16.69	16.69	16.73	15.41	15.80	15.57	3.33
UGDH_MOUSE	0.000477	14.07	14.57	13.84	11.66	10.55	11.72	3.32
CAP1_MOUSE	0.000498	16.30	16.31	16.47	14.58	15.23	14.87	3.30
ENPL_MOUSE	0.000579	18.40	18.37	18.66	17.38	17.53	17.48	3.24
K1C10_MOUSE	0.000616	19.19	18.83	19.03	19.97	20.10	20.12	3.21
2AAB_MOUSE	0.000622	16.25	16.35	16.31	15.06	15.08	15.45	3.21
IF5A1_MOUSE	0.000629	19.07	18.46	18.51	17.27	17.30	17.22	3.20
GDIR1_MOUSE	0.000645	18.34	18.29	18.31	17.37	17.53	17.51	3.19
AATM_MOUSE	0.000648	17.71	17.39	17.46	16.14	16.46	16.46	3.19
RTN4_MOUSE	0.000686	18.65	18.56	18.47	17.45	17.67	17.70	3.16
IMA3_MOUSE	0.000702	14.57	14.31	14.34	13.11	12.85	13.40	3.15
VINC_MOUSE	0.000715	15.15	15.11	15.34	14.37	14.18	14.12	3.15
IPO9_MOUSE	0.000716	13.49	13.79	13.48	9.63	10.70	11.33	3.15
PTBP1_MOUSE	0.000751	16.80	16.77	16.35	15.15	15.03	14.37	3.12
TCPE_MOUSE	0.000793	15.29	15.48	15.26	12.43	13.52	13.49	3.10
ROA2_MOUSE	0.000823	17.53	17.74	17.68	16.65	16.82	16.47	3.08
DYH17_MOUSE	0.000832	15.79	15.55	16.03	16.81	17.02	17.08	3.08
CLH1_MOUSE	0.000871	16.53	16.87	16.48	15.33	15.18	14.61	3.06
PDIA6_MOUSE	0.0009	17.40	17.61	17.45	16.51	16.69	16.65	3.05
TCPH_MOUSE	0.000902	15.71	15.81	15.83	14.01	14.55	14.69	3.04
TSN_MOUSE	0.000904	13.78	14.16	13.83	12.71	12.19	11.89	3.04
BIP_MOUSE	0.000924	19.73	19.42	19.53	18.43	18.63	18.67	3.03
ARP2_MOUSE	0.000936	13.07	13.22	12.74	11.18	9.70	10.44	3.03
AP2B1_MOUSE	0.000944	13.83	13.51	14.01	12.35	11.43	11.32	3.02
DDX3X_MOUSE	0.001	14.39	14.57	14.65	13.03	11.91	12.73	3.00
ALBU_MOUSE	0.001032	17.91	17.81	17.73	16.88	17.06	17.03	2.99
RSSA_MOUSE	0.001081	18.25	18.16	18.23	17.10	17.49	17.15	2.97
2AAA_MOUSE	0.001131	16.75	16.82	16.86	15.95	16.02	16.14	2.95
CATB_MOUSE	0.001131	15.43	15.32	15.81	14.43	14.49	14.20	2.95
NNRE_MOUSE	0.001157	11.06	10.70	11.31	9.55	9.59	9.94	2.94
TBB2B_MOUSE	0.001168	18.07	17.91	18.20	17.23	17.25	17.22	2.93
TBB2A_MOUSE	0.001168	18.07	17.91	18.20	17.23	17.25	17.22	2.93
ACTN4_MOUSE	0.001328	17.08	16.94	17.30	16.31	16.09	16.14	2.88
ERO1B_MOUSE	0.001349	14.30	13.93	14.27	12.28	12.63	13.09	2.87
6PGL_MOUSE	0.001531	15.37	15.81	15.86	14.54	14.71	14.65	2.81
RAP1B_MOUSE	0.001542	15.48	15.07	15.33	14.30	13.64	14.04	2.81
RAP1A_MOUSE	0.001542	15.48	15.07	15.33	14.30	13.64	14.04	2.81
ANXA6_MOUSE	0.001692	16.80	16.23	16.23	15.02	15.16	15.38	2.77
S10AB_MOUSE	0.001703	13.22	13.31	14.28	15.26	15.56	15.45	2.77
PPIB_MOUSE	0.001762	15.44	15.65	15.39	14.45	14.78	14.62	2.75
H14_MOUSE	0.00179	17.02	16.91	17.20	17.87	18.00	17.75	2.75
API5_MOUSE	0.001801	14.71	14.07	14.21	12.94	13.28	13.09	2.74
MDGA1_MOUSE	0.001891	19.07	18.32	17.69	16.43	15.32	15.80	2.72
SRPRA_MOUSE	0.001922	13.29	12.49	12.48	10.53	11.16	9.90	2.72
FKB1A_MOUSE	0.001939	13.86	13.87	14.02	13.22	12.72	12.76	2.71
CDC42_MOUSE	0.001946	17.24	17.13	17.09	14.77	15.74	15.80	2.71
FAS_MOUSE	0.002116	13.77	13.67	14.08	10.58	12.16	11.74	2.67
PLAK_MOUSE	0.002129	14.54	14.51	14.25	13.56	13.25	12.85	2.67
MDHM_MOUSE	0.002326	20.02	20.21	20.00	19.18	19.44	19.34	2.63
ANXA1_MOUSE	0.002341	16.96	16.78	17.14	15.77	16.14	16.13	2.63
PLCB1_MOUSE	0.002343	17.38	17.61	17.54	16.79	16.68	16.88	2.63
KAP0_MOUSE	0.002469	13.88	14.01	13.80	12.76	11.66	12.46	2.61
RAGP1_MOUSE	0.00262	14.51	14.81	14.25	13.50	12.75	12.53	2.58
BIRC2_MOUSE	0.002678	16.61	15.81	16.19	17.24	17.66	17.66	2.57
MOES_MOUSE	0.002759	15.45	15.26	15.56	13.43	14.44	13.37	2.56
KAD2_MOUSE	0.003069	13.79	14.07	13.67	12.41	12.90	11.92	2.51
R51A1_MOUSE	0.0031	12.95	13.00	13.56	11.02	12.05	11.58	2.51
RS7_MOUSE	0.003174	17.66	16.56	16.77	15.44	15.54	15.31	2.50
RLA0_MOUSE	0.0033	18.50	18.44	17.93	16.74	17.30	17.22	2.48
K2C1_MOUSE	0.003341	18.33	18.24	18.33	17.56	17.78	17.46	2.48
SF3B1_MOUSE	0.003346	12.56	13.20	12.99	10.58	9.66	11.43	2.48
SON_MOUSE	0.003363	15.63	16.38	16.05	14.59	15.01	14.28	2.47
USO1_MOUSE	0.003429	12.17	12.17	12.45	9.20	10.95	9.98	2.46
IMB1_MOUSE	0.003496	17.29	17.51	16.97	16.38	16.41	16.06	2.46

Table A2. *Cont.*

Protein Name	*p*-Value	Log2 Intensity						
		Control01	Control02	Control03	Liposome01	Liposome02	Liposome03	log(*p*-Value)
S10A6_MOUSE	0.003578	20.89	21.05	20.66	19.87	19.80	19.06	2.45
PPIA_MOUSE	0.003862	20.57	20.18	20.41	19.71	19.67	19.62	2.41
RL14_MOUSE	0.003872	16.09	15.94	15.68	14.82	14.68	15.23	2.41
RS8_MOUSE	0.003989	15.56	15.41	15.45	13.37	14.24	14.45	2.40
PRDX1_MOUSE	0.003992	18.09	18.49	18.10	17.08	17.53	16.93	2.40
MBB1A_MOUSE	0.00401	13.93	13.88	13.57	12.52	12.26	11.16	2.40
EF1A1_MOUSE	0.004196	18.52	18.47	18.60	17.73	18.02	17.90	2.38
SEPT2_MOUSE	0.004277	15.38	14.98	15.32	14.00	13.59	14.44	2.37
ARF1_MOUSE	0.004524	16.52	16.29	16.32	13.92	15.25	14.93	2.34
ADK_MOUSE	0.004525	14.30	14.42	14.35	12.92	13.04	13.73	2.34
ATPK_MOUSE	0.004911	16.16	15.57	14.97	13.98	14.08	14.28	2.31
VDAC3_MOUSE	0.004948	15.00	15.20	14.34	13.66	13.76	13.72	2.31
RACK1_MOUSE	0.005006	15.85	16.09	16.52	14.62	13.34	14.72	2.30
VATL_MOUSE	0.005044	14.01	14.17	13.35	12.02	11.46	9.83	2.30
PRDX4_MOUSE	0.005054	17.40	17.89	17.57	16.65	16.98	16.66	2.30
NUCL_MOUSE	0.005137	19.18	18.99	18.89	18.37	18.33	18.48	2.29
SRP68_MOUSE	0.00519	13.00	13.38	13.26	11.72	10.03	11.51	2.28
4F2_MOUSE	0.005313	14.17	13.88	13.85	12.39	13.22	12.15	2.27
RIC8B_MOUSE	0.005318	14.21	14.14	14.62	10.00	12.37	11.96	2.27
K1C14_MOUSE	0.005548	14.01	14.36	12.72	15.38	16.00	15.93	2.26
UBA6_MOUSE	0.005564	11.49	11.66	10.14	9.14	9.39	9.00	2.25
RL7A_MOUSE	0.005709	15.36	15.63	14.99	14.33	14.52	14.49	2.24
SAP_MOUSE	0.006308	18.39	18.62	18.60	19.01	19.41	19.32	2.20
MYL6_MOUSE	0.006405	15.38	15.40	16.33	16.91	16.93	17.00	2.19
GTR1_MOUSE	0.006448	14.20	13.70	14.27	10.09	12.49	11.59	2.19
CSN4_MOUSE	0.006496	11.88	11.84	12.08	9.60	10.34	11.04	2.19
UGGG1_MOUSE	0.006589	12.08	13.06	13.32	10.28	11.08	11.42	2.18
TCPD_MOUSE	0.006742	14.59	14.85	14.80	14.23	13.67	13.94	2.17
IMA7_MOUSE	0.007097	12.27	12.58	12.13	9.33	11.25	10.13	2.15
K1C13_MOUSE	0.007257	17.30	16.66	16.85	17.75	17.99	17.70	2.14
VDAC2_MOUSE	0.007331	15.12	15.16	15.24	14.16	13.27	14.32	2.13
GRP75_MOUSE	0.007397	18.49	18.38	18.09	17.63	17.72	17.69	2.13
HEAT3_MOUSE	0.007572	12.46	11.35	12.59	10.59	10.68	10.72	2.12
SMRC1_MOUSE	0.007688	11.23	11.33	11.38	9.91	10.38	8.91	2.11
CDK1_MOUSE	0.007805	9.17	10.81	10.28	7.99	8.33	8.49	2.11
AL7A1_MOUSE	0.008283	14.00	13.86	13.60	10.32	12.37	12.01	2.08
PGAM1_MOUSE	0.008302	20.10	19.97	20.04	19.21	19.51	19.57	2.08
ACON_MOUSE	0.008378	14.44	14.12	13.91	11.95	10.17	12.61	2.08
UB2D2_MOUSE	0.008403	13.84	14.98	14.95	12.17	12.82	13.37	2.08
UB2D3_MOUSE	0.008403	13.84	14.98	14.95	12.17	12.82	13.37	2.08
K1C42_MOUSE	0.008429	15.76	16.13	15.29	16.58	17.00	17.44	2.07
K22E_MOUSE	0.008676	17.43	17.32	17.02	17.85	18.75	18.42	2.06
DPY30_MOUSE	0.009099	14.38	14.41	14.24	13.73	13.58	13.94	2.04
CAND1_MOUSE	0.009439	13.46	14.71	15.00	12.52	12.57	11.02	2.03
PYRG1_MOUSE	0.00988	13.91	14.16	13.52	11.86	10.76	12.78	2.01
AT5G1_MOUSE	0.010402	14.30	14.61	14.59	13.56	13.68	12.63	1.98
AT5G3_MOUSE	0.010402	14.30	14.61	14.59	13.56	13.68	12.63	1.98
AT5G2_MOUSE	0.010402	14.30	14.61	14.59	13.56	13.68	12.63	1.98
TLN1_MOUSE	0.010823	12.88	13.67	13.19	11.53	12.54	11.78	1.97
HNRPL_MOUSE	0.011041	14.09	14.70	14.00	12.95	13.54	12.50	1.96
CKAP4_MOUSE	0.011183	15.39	15.64	15.70	14.98	14.96	15.14	1.95
RPN1_MOUSE	0.011709	13.95	13.88	14.23	12.28	12.69	10.48	1.93
PFKAL_MOUSE	0.01242	14.66	15.33	14.61	13.56	12.54	13.92	1.91
H13_MOUSE	0.01257	17.64	17.48	17.53	18.10	18.20	17.90	1.90
CSN7A_MOUSE	0.012775	13.73	13.81	13.95	13.46	13.15	12.94	1.89
LKHA4_MOUSE	0.012846	12.57	12.89	11.55	9.83	11.25	9.36	1.89
RLA2_MOUSE	0.013592	19.15	19.23	19.07	18.41	18.77	18.66	1.87
EIF3F_MOUSE	0.01375	13.49	13.62	13.54	11.75	12.62	10.47	1.86
C1QBP_MOUSE	0.01467	15.43	15.09	15.42	14.74	13.61	13.23	1.83
TAGL2_MOUSE	0.0152	14.60	14.88	15.73	16.05	16.15	16.36	1.82
DESP_MOUSE	0.015762	13.61	12.78	12.85	11.29	12.42	11.54	1.80
OTUB1_MOUSE	0.019088	14.60	14.23	13.91	12.65	11.96	13.63	1.72
CALX_MOUSE	0.020499	16.20	15.75	16.00	15.51	15.44	15.45	1.69
H12_MOUSE	0.021046	16.68	16.42	16.57	16.88	17.14	17.05	1.68
ATPG_MOUSE	0.02109	14.11	13.07	13.84	10.70	12.56	9.07	1.68
LAMP2_MOUSE	0.021862	16.01	15.97	15.44	14.86	15.13	15.31	1.66

Table A2. *Cont.*

Protein Name	p-Value	Log2 Intensity						
		Control01	Control02	Control03	Liposome01	Liposome02	Liposome03	log(p-Value)
PSA5_MOUSE	0.022314	15.01	15.16	15.13	14.48	14.81	14.20	1.65
TRY2_MOUSE	0.022394	23.38	23.70	23.06	24.10	24.78	23.98	1.65
TCPA_MOUSE	0.023869	14.28	14.09	14.44	8.73	12.79	11.64	1.62
RL12_MOUSE	0.024021	15.04	14.26	14.99	14.18	13.84	13.44	1.62
BASP1_MOUSE	0.024344	14.75	14.62	14.68	14.39	14.03	14.27	1.61
K1C19_MOUSE	0.024816	20.16	20.03	20.09	20.34	20.63	20.70	1.61
EF1D_MOUSE	0.026795	15.14	15.40	15.53	13.81	14.69	14.81	1.57
K2C4_MOUSE	0.026904	16.58	16.50	16.10	17.07	20.58	20.30	1.57
PRDX5_MOUSE	0.028629	15.67	14.57	15.15	14.44	14.17	13.68	1.54
THIO_MOUSE	0.029175	17.68	17.83	17.73	17.41	17.43	17.31	1.53
YBOX2_MOUSE	0.029181	16.07	15.87	16.09	15.42	15.41	15.77	1.53
YBOX3_MOUSE	0.029181	16.07	15.87	16.09	15.42	15.41	15.77	1.53
TOM20_MOUSE	0.030869	13.51	12.89	13.21	12.85	12.20	11.86	1.51
CH60_MOUSE	0.031589	18.43	18.23	18.18	17.81	17.99	17.84	1.50
CD63_MOUSE	0.03168	13.65	13.30	13.68	13.86	14.66	15.15	1.50
K2C8_MOUSE	0.03168	17.18	16.87	17.34	16.74	16.58	16.62	1.50
TPIS_MOUSE	0.031746	18.13	18.18	18.42	18.56	18.80	18.66	1.50
SODC_MOUSE	0.033924	16.50	16.81	17.01	16.47	15.51	15.88	1.47
SYAC_MOUSE	0.034719	14.61	14.49	14.41	14.20	12.04	12.45	1.46
SMU1_MOUSE	0.036374	12.11	11.53	11.24	0.00	10.41	0.00	1.44
C1TM_MOUSE	0.036422	12.82	13.00	13.37	11.36	8.63	11.91	1.44
ARF4_MOUSE	0.036656	16.11	15.77	15.77	11.12	14.59	14.17	1.44
ANXA5_MOUSE	0.036715	19.32	19.28	19.18	18.85	18.98	18.45	1.44
PARK7_MOUSE	0.037258	13.64	14.02	13.39	12.77	13.09	11.55	1.43
P4HA1_MOUSE	0.037299	14.18	14.14	14.57	13.57	14.02	13.65	1.43
THIKA_MOUSE	0.037977	13.61	13.17	13.46	12.23	8.06	11.17	1.42
MTAP_MOUSE	0.038542	15.04	14.85	14.92	14.72	14.39	14.51	1.41
THIC_MOUSE	0.038635	10.87	11.76	11.13	10.35	10.69	10.57	1.41
LAP2B_MOUSE	0.039043	14.57	14.75	14.52	14.24	14.14	14.35	1.41
NU155_MOUSE	0.039897	12.47	11.90	10.75	10.88	10.26	9.88	1.40
TKT_MOUSE	0.039985	13.72	14.07	13.84	12.88	10.86	13.00	1.40
CO1A1_MOUSE	0.040121	13.56	13.79	12.40	11.31	12.66	11.63	1.40
CAPZB_MOUSE	0.040224	14.17	14.21	13.87	10.09	13.29	12.32	1.40
VDAC1_MOUSE	0.042968	17.38	16.30	17.24	16.26	16.19	15.97	1.37
NP1L1_MOUSE	0.043459	16.89	17.08	17.00	16.56	16.77	16.57	1.36
TSSK5_MOUSE	0.044033	16.69	16.56	16.64	0.00	11.28	12.35	1.36
PEBP1_MOUSE	0.044407	16.86	16.87	17.32	16.39	16.63	16.60	1.35
H4_MOUSE	0.044792	18.83	19.42	18.62	17.86	18.65	17.62	1.35
NPM_MOUSE	0.049164	16.09	15.74	16.14	16.16	16.87	17.01	1.31
EF1B_MOUSE	0.04968	16.12	16.36	16.28	15.74	15.48	16.08	1.30
PSD13_MOUSE	0.051249	11.93	11.08	11.28	10.38	11.13	10.10	1.29
PAIRB_MOUSE	0.05299	15.41	15.12	15.11	15.45	15.65	15.68	1.28
PLEC_MOUSE	0.05689	13.93	14.45	14.22	13.50	12.90	10.57	1.24
H2B1K_MOUSE	0.058515	12.26	12.50	13.57	10.16	12.41	11.01	1.23
H2B1C_MOUSE	0.058515	12.26	12.50	13.57	10.16	12.41	11.01	1.23
H2B1H_MOUSE	0.058515	12.26	12.50	13.57	10.16	12.41	11.01	1.23
H2B1F_MOUSE	0.058515	12.26	12.50	13.57	10.16	12.41	11.01	1.23
H2B3B_MOUSE	0.058515	12.26	12.50	13.57	10.16	12.41	11.01	1.23
H2B3A_MOUSE	0.058515	12.26	12.50	13.57	10.16	12.41	11.01	1.23
H2B1P_MOUSE	0.058515	12.26	12.50	13.57	10.16	12.41	11.01	1.23
H2B2B_MOUSE	0.058515	12.26	12.50	13.57	10.16	12.41	11.01	1.23
H2B2E_MOUSE	0.058515	12.26	12.50	13.57	10.16	12.41	11.01	1.23
H2B1B_MOUSE	0.058515	12.26	12.50	13.57	10.16	12.41	11.01	1.23
H2B1M_MOUSE	0.058515	12.26	12.50	13.57	10.16	12.41	11.01	1.23
HNRPK_MOUSE	0.060011	18.52	18.54	18.47	17.59	18.44	17.70	1.22
KLC4_MOUSE	0.060749	12.56	0.00	14.57	17.84	17.96	18.79	1.22
1433Z_MOUSE	0.061261	18.97	19.31	19.48	18.83	18.90	18.83	1.21
GSHR_MOUSE	0.067255	11.26	11.97	11.15	11.44	8.39	8.57	1.17
RS23_MOUSE	0.067883	14.28	13.20	14.33	13.23	12.76	13.34	1.17
IF4B_MOUSE	0.069049	14.02	14.03	13.57	13.24	13.33	11.59	1.16
CALM3_MOUSE	0.069229	18.81	18.84	18.56	18.90	19.25	19.72	1.16
CALM2_MOUSE	0.069229	18.81	18.84	18.56	18.90	19.25	19.72	1.16
CALM1_MOUSE	0.069229	18.81	18.84	18.56	18.90	19.25	19.72	1.16
ATP5J_MOUSE	0.072899	13.97	14.29	14.26	13.62	13.68	14.03	1.14
COPG1_MOUSE	0.074631	14.63	14.90	14.06	12.58	14.17	13.74	1.13
CPNE1_MOUSE	0.07629	9.39	10.50	11.33	9.91	8.84	8.39	1.12

Table A2. *Cont.*

Protein Name	p-Value	Log2 Intensity						
		Control01	Control02	Control03	Liposome01	Liposome02	Liposome03	log(p-Value)
RAB1B_MOUSE	0.077103	15.28	14.29	14.56	13.56	14.43	13.86	1.11
RAB1A_MOUSE	0.077103	15.28	14.29	14.56	13.56	14.43	13.86	1.11
PCNA_MOUSE	0.077431	14.30	12.74	13.33	12.28	11.76	13.01	1.11
RBBP7_MOUSE	0.077454	13.35	15.62	15.59	15.96	16.83	16.23	1.11
RBBP4_MOUSE	0.077454	13.35	15.62	15.59	15.96	16.83	16.23	1.11
H2AV_MOUSE	0.07878	18.16	18.27	18.00	17.72	17.79	17.97	1.10
H2AZ_MOUSE	0.07878	18.16	18.27	18.00	17.72	17.79	17.97	1.10
AN32A_MOUSE	0.09057	19.92	19.71	19.91	20.11	20.81	20.08	1.04
K2C5_MOUSE	0.090801	18.63	18.91	18.73	17.67	18.00	18.75	1.04
TEBP_MOUSE	0.092832	14.31	14.27	14.67	14.25	13.87	14.03	1.03
AN32B_MOUSE	0.094555	19.89	19.70	19.87	20.08	20.79	20.05	1.02
G3P_MOUSE	0.097292	19.35	19.40	19.66	17.89	19.16	19.09	1.01
LRC59_MOUSE	0.097584	14.27	15.08	14.15	14.70	15.56	15.31	1.01
XPO1_MOUSE	0.099533	13.56	13.24	13.89	13.07	13.38	12.77	1.00
UBE2N_MOUSE	0.100923	15.22	14.93	14.57	14.29	13.85	14.79	1.00
RS3A_MOUSE	0.103532	16.06	15.61	15.98	11.72	15.25	14.96	0.98
CNBP_MOUSE	0.103689	12.35	14.37	14.12	15.04	14.54	14.59	0.98
PHB2_MOUSE	0.104427	14.67	14.13	13.97	13.95	13.60	12.01	0.98
SC61B_MOUSE	0.106335	13.45	13.48	12.63	13.93	14.12	13.38	0.97
P5CS_MOUSE	0.108547	9.54	10.84	10.65	9.90	9.66	8.91	0.96
LBH_MOUSE	0.116272	14.11	13.63	12.91	13.96	14.14	14.49	0.93
ML12B_MOUSE	0.118359	17.10	16.86	16.90	17.04	17.58	17.28	0.93
ENOA_MOUSE	0.119211	19.91	19.66	19.75	19.97	20.15	19.97	0.92
RTRAF_MOUSE	0.120351	9.09	10.08	9.42	9.90	10.19	10.06	0.92
PLBL2_MOUSE	0.125355	10.46	9.89	10.73	9.65	0.00	7.75	0.90
K2C6B_MOUSE	0.134139	18.83	19.04	18.86	17.95	18.42	18.93	0.87
K2C6A_MOUSE	0.134139	18.83	19.04	18.86	17.95	18.42	18.93	0.87
K2C75_MOUSE	0.134139	18.83	19.04	18.86	17.95	18.42	18.93	0.87
RINI_MOUSE	0.135641	10.54	13.30	12.66	8.46	11.63	10.75	0.87
TOM70_MOUSE	0.136482	8.39	10.68	11.45	12.25	12.00	10.84	0.86
NDKA_MOUSE	0.137404	15.14	15.82	16.17	16.20	16.09	16.32	0.86
K1C12_MOUSE	0.138179	14.33	12.40	13.09	11.04	9.90	13.51	0.86
GRM7_MOUSE	0.144146	11.24	12.54	12.50	9.36	11.47	0.00	0.84
RL40_MOUSE	0.147393	19.31	18.88	18.98	18.04	18.14	19.20	0.83
RS27A_MOUSE	0.147393	19.31	18.88	18.98	18.04	18.14	19.20	0.83
UBC_MOUSE	0.147393	19.31	18.88	18.98	18.04	18.14	19.20	0.83
UBB_MOUSE	0.147393	19.31	18.88	18.98	18.04	18.14	19.20	0.83
M21_MPV15	0.148508	8.24	15.22	15.12	16.19	16.17	16.15	0.83
NAA15_MOUSE	0.150828	13.48	13.36	12.60	13.06	12.46	11.03	0.82
RS14_MOUSE	0.151861	14.26	15.89	16.28	14.67	14.50	14.57	0.82
PDIA3_MOUSE	0.153504	15.84	16.03	15.35	15.16	15.70	14.97	0.81
EF1G_MOUSE	0.156411	13.27	13.92	13.27	13.39	13.03	12.04	0.81
LEG1_MOUSE	0.1623	18.70	18.71	18.85	18.59	19.60	19.56	0.79
PGRC1_MOUSE	0.163353	14.32	14.37	13.93	14.08	13.96	13.18	0.79
ECI1_MOUSE	0.163424	11.09	12.41	11.78	11.44	10.80	11.21	0.79
YBOX1_MOUSE	0.164124	16.68	15.95	16.70	15.46	16.04	16.29	0.78
NACA_MOUSE	0.164437	14.53	14.47	14.50	14.01	13.06	14.55	0.78
NACAM_MOUSE	0.164437	14.53	14.47	14.50	14.01	13.06	14.55	0.78
CNPY2_MOUSE	0.17626	12.46	12.99	10.04	13.61	13.63	12.29	0.75
RS25_MOUSE	0.183649	14.17	15.71	15.46	15.74	16.79	15.35	0.74
ERD21_MOUSE	0.186208	13.46	13.59	12.78	14.44	13.81	13.28	0.73
PSMD7_MOUSE	0.187718	12.96	13.17	10.64	11.64	9.38	11.52	0.73
1433T_MOUSE	0.192931	14.35	17.16	17.25	17.07	17.93	17.52	0.71
AIMP2_MOUSE	0.19423	10.62	11.43	10.63	10.92	9.42	10.34	0.71
1433F_MOUSE	0.201946	10.88	16.87	16.97	16.94	17.85	17.47	0.69
SCOT1_MOUSE	0.206602	13.33	15.39	15.53	13.71	12.49	14.59	0.68
1433B_MOUSE	0.2371	15.42	17.34	17.39	17.11	17.88	17.52	0.63
NPHP3_MOUSE	0.243422	15.80	15.77	14.74	13.76	15.10	15.37	0.61
RL30_MOUSE	0.247419	17.16	15.05	17.18	15.81	15.95	15.10	0.61
ELOV5_MOUSE	0.258405	13.97	13.78	13.99	13.88	13.83	12.76	0.59
NPC2_MOUSE	0.259318	11.08	11.64	12.19	11.76	10.87	10.79	0.59
DNPEP_MOUSE	0.264156	13.89	11.40	13.97	13.04	9.50	12.33	0.58
ARSA_MOUSE	0.264454	13.47	10.82	11.47	10.86	11.57	10.60	0.58
TPM3_MOUSE	0.266344	17.81	17.98	17.64	18.06	18.04	17.89	0.57
RS5_MOUSE	0.266611	17.31	17.41	17.23	17.32	17.65	17.53	0.57

Table A2. *Cont.*

Protein Name	*p*-Value	Log2 Intensity						
		Control01	Control02	Control03	Liposome01	Liposome02	Liposome03	log(*p*-Value)
RS13_MOUSE	0.285418	13.47	14.36	13.52	12.41	13.83	13.49	0.54
PUR9_MOUSE	0.285772	10.00	8.60	10.64	9.62	8.78	8.79	0.54
PTK6_MOUSE	0.28915	0.00	0.00	0.00	0.00	0.00	9.70	0.54
K2C79_MOUSE	0.289352	18.86	19.12	18.84	18.83	19.52	19.25	0.54
K2C7_MOUSE	0.292854	15.60	15.21	15.45	15.42	15.55	16.08	0.53
GFAP_MOUSE	0.292854	15.60	15.21	15.45	15.42	15.55	16.08	0.53
KRT85_MOUSE	0.292854	15.60	15.21	15.45	15.42	15.55	16.08	0.53
VATE1_MOUSE	0.298284	12.38	12.47	12.01	12.23	12.30	10.48	0.53
UNC79_MOUSE	0.301243	9.70	11.27	0.00	10.56	10.17	10.54	0.52
TPM1_MOUSE	0.310726	17.63	17.77	17.32	17.80	17.70	17.76	0.51
ROAA_MOUSE	0.312399	17.53	17.26	17.46	17.13	17.28	17.35	0.51
TCPQ_MOUSE	0.328145	14.84	13.00	14.61	12.67	13.87	13.94	0.48
IPO4_MOUSE	0.330746	10.94	12.57	12.78	11.74	11.10	11.72	0.48
RB11A_MOUSE	0.332879	12.52	12.83	10.92	11.77	11.88	10.66	0.48
FUS_MOUSE	0.334276	14.99	14.25	14.38	14.36	14.39	14.12	0.48
RL9_MOUSE	0.335137	11.52	14.10	14.36	11.45	12.84	12.90	0.47
PSME2_MOUSE	0.345432	10.59	12.14	12.92	12.12	12.94	12.54	0.46
RL22_MOUSE	0.356258	14.62	15.45	14.44	14.52	14.64	14.47	0.45
H2AX_MOUSE	0.363435	19.34	19.36	18.97	19.20	19.02	18.95	0.44
PGK1_MOUSE	0.365098	17.57	17.47	17.76	17.39	17.38	17.60	0.44
ACTB_MOUSE	0.372193	16.23	17.58	17.39	16.60	17.10	16.23	0.43
RS15_MOUSE	0.373532	8.56	11.70	9.45	11.67	9.21	12.03	0.43
H2A2B_MOUSE	0.375984	19.31	19.32	18.93	19.14	18.99	18.94	0.42
TCTP_MOUSE	0.389556	15.70	16.11	16.01	16.15	15.85	16.41	0.41
K2C73_MOUSE	0.394735	17.48	17.28	17.24	17.18	16.86	17.43	0.40
1433G_MOUSE	0.397852	16.37	17.70	17.82	17.36	18.05	17.71	0.40
NDKB_MOUSE	0.400574	15.12	15.57	15.59	15.81	15.35	15.69	0.40
RLA1_MOUSE	0.409863	18.51	18.47	18.84	18.63	18.76	18.85	0.39
ABCE1_MOUSE	0.411465	14.21	12.17	14.39	13.79	11.70	13.11	0.39
PUR4_MOUSE	0.415174	13.27	13.77	12.35	13.18	13.17	11.32	0.38
H2AJ_MOUSE	0.419493	19.30	19.32	18.91	19.17	18.99	18.92	0.38
H2A1K_MOUSE	0.419493	19.30	19.32	18.91	19.17	18.99	18.92	0.38
H2A1H_MOUSE	0.419493	19.30	19.32	18.91	19.17	18.99	18.92	0.38
H2A1F_MOUSE	0.419493	19.30	19.32	18.91	19.17	18.99	18.92	0.38
H2A3_MOUSE	0.419493	19.30	19.32	18.91	19.17	18.99	18.92	0.38
H2A1P_MOUSE	0.419493	19.30	19.32	18.91	19.17	18.99	18.92	0.38
H2A1O_MOUSE	0.419493	19.30	19.32	18.91	19.17	18.99	18.92	0.38
H2A1N_MOUSE	0.419493	19.30	19.32	18.91	19.17	18.99	18.92	0.38
H2A1I_MOUSE	0.419493	19.30	19.32	18.91	19.17	18.99	18.92	0.38
H2A1G_MOUSE	0.419493	19.30	19.32	18.91	19.17	18.99	18.92	0.38
H2A1E_MOUSE	0.419493	19.30	19.32	18.91	19.17	18.99	18.92	0.38
H2A1D_MOUSE	0.419493	19.30	19.32	18.91	19.17	18.99	18.92	0.38
H2A1C_MOUSE	0.419493	19.30	19.32	18.91	19.17	18.99	18.92	0.38
H2A1B_MOUSE	0.419493	19.30	19.32	18.91	19.17	18.99	18.92	0.38
H2A2A_MOUSE	0.436243	19.27	19.28	18.86	19.11	18.96	18.91	0.36
H2A2C_MOUSE	0.436243	19.27	19.28	18.86	19.11	18.96	18.91	0.36
RS17_MOUSE	0.439512	13.88	10.54	13.59	13.79	12.66	14.01	0.36
LYAG_MOUSE	0.450325	12.54	9.97	11.66	10.70	10.94	10.93	0.35
THOC4_MOUSE	0.462349	14.38	14.69	14.65	14.27	14.65	14.39	0.34
ALRF2_MOUSE	0.462349	14.38	14.69	14.65	14.27	14.65	14.39	0.34
NDUA4_MOUSE	0.471139	13.69	12.04	13.71	12.39	12.33	13.40	0.33
AT5F1_MOUSE	0.482549	16.29	12.89	14.77	12.77	14.33	14.64	0.32
RM12_MOUSE	0.506745	13.70	13.65	12.79	13.52	13.33	12.46	0.30
RCN2_MOUSE	0.518016	14.14	13.87	13.66	13.93	13.85	14.32	0.29
RS10_MOUSE	0.52846	11.58	14.25	14.08	12.52	13.43	12.35	0.28
RL31_MOUSE	0.540627	16.99	16.90	15.24	15.88	16.37	15.86	0.27
RS18_MOUSE	0.556426	12.92	12.59	12.58	12.37	12.42	12.91	0.25
TCEA1_MOUSE	0.565193	12.59	14.22	13.07	13.90	10.80	13.47	0.25
TMED9_MOUSE	0.58254	12.07	12.94	9.11	13.17	11.80	11.11	0.23
GSTP2_MOUSE	0.589048	16.49	14.57	15.66	15.38	15.37	15.13	0.23
GSTP1_MOUSE	0.589048	16.49	14.57	15.66	15.38	15.37	15.13	0.23
SKP1_MOUSE	0.610602	12.96	15.48	15.74	15.01	15.06	15.35	0.21
QCR2_MOUSE	0.618647	10.32	9.89	11.16	11.05	9.97	9.53	0.21
AQP1_MOUSE	0.642568	8.72	9.75	11.31	9.64	11.31	9.98	0.19
MTPN_MOUSE	0.674227	13.79	14.86	14.62	14.51	14.60	13.64	0.17

Table A2. *Cont.*

Protein Name	p-Value	Log2 Intensity						
		Control01	Control02	Control03	Liposome01	Liposome02	Liposome03	log(p-Value)
ARL1_MOUSE	0.677382	14.41	13.58	14.00	14.95	14.22	13.41	0.17
SCRB2_MOUSE	0.678152	12.27	11.87	11.01	11.23	12.12	11.24	0.17
K2C1B_MOUSE	0.689647	18.32	18.49	18.36	18.23	18.37	18.80	0.16
PTMA_MOUSE	0.696432	21.09	21.06	21.09	21.06	21.37	20.99	0.16
1433E_MOUSE	0.697313	18.88	19.57	19.80	19.17	19.76	18.90	0.16
TPM4_MOUSE	0.7118	19.25	19.19	18.96	18.93	19.47	19.23	0.15
S10AA_MOUSE	0.715815	12.43	14.11	14.26	13.96	13.71	13.74	0.15
MARCS_MOUSE	0.726697	14.39	14.24	13.95	14.54	14.64	13.72	0.14
CAPR1_MOUSE	0.731991	12.49	13.16	13.04	12.61	12.96	13.42	0.14
SET_MOUSE	0.745103	17.88	18.09	18.00	17.96	18.02	17.86	0.13
PLP2_MOUSE	0.748147	11.74	13.38	12.77	13.13	12.22	12.02	0.13
ATPD_MOUSE	0.758636	11.61	12.28	12.48	11.64	13.01	11.20	0.12
PUSL1_MOUSE	0.767776	10.10	12.69	12.69	9.81	11.75	12.92	0.11
ACTY_MOUSE	0.782008	11.23	13.96	14.26	13.44	13.43	13.30	0.11
ACTZ_MOUSE	0.782008	11.23	13.96	14.26	13.44	13.43	13.30	0.11
DBF4A_MOUSE	0.788214	12.57	0.00	0.00	0.00	0.00	17.81	0.10
GPM6A_MOUSE	0.809664	18.65	18.79	18.22	19.09	18.01	18.80	0.09
PSME1_MOUSE	0.842895	12.30	13.00	13.15	12.88	13.04	12.70	0.07
COX41_MOUSE	0.930115	11.07	14.37	14.74	12.28	13.97	13.62	0.03
PP14B_MOUSE	0.932589	12.48	12.62	12.40	12.41	12.79	12.25	0.03
CO6A1_MOUSE	0.945765	9.84	11.39	10.75	12.41	9.66	9.73	0.02
CO1A2_MOUSE	0.94961	11.48	14.35	13.89	12.85	13.72	13.31	0.02
HARS1_MOUSE	0.960346	9.84	10.24	11.74	11.12	10.46	10.32	0.02
CALR_MOUSE	0.967791	16.10	16.32	16.31	16.07	15.89	16.80	0.01
BAP31_MOUSE	0.973337	10.98	13.81	13.19	12.35	13.30	12.25	0.01
EI3JB_MOUSE	0.97778	12.28	12.65	11.10	10.89	12.88	12.31	0.01
EI3JA_MOUSE	0.97778	12.28	12.65	11.10	10.89	12.88	12.31	0.01
MIF_MOUSE	0.98048	18.20	18.05	18.36	17.61	19.16	17.87	0.01

References

1. Udenigwe, C.C.; Aluko, R.E. Food protein-derived bioactive peptides: Production, processing, and potential health benefits. *J. Food Sci.* **2012**, *77*, R11–R24. [CrossRef] [PubMed]
2. Aoyama, T.; Fukui, K.; Takamatsu, K.; Hashimoto, Y.; Yamamoto, T. Soy protein isolate and its hydrolysate reduce body fat of dietary obese rats and genetically obese mice (yellow KK). *Nutrition* **2000**, *16*, 349–354. [CrossRef]
3. Bruno, B.J.; Miller, G.D.; Lim, C.S. Basics and recent advances in peptide and protein drug delivery. *Ther. Deliv.* **2013**, *4*, 1443–1467. [CrossRef]
4. Jo Feeney, M.; Miller, A.M.; Roupas, P. Mushrooms-biologically distinct and nutritionally unique: Exploring a "third food kingdom". *Nutr. Today* **2014**, *49*, 301–307. [CrossRef] [PubMed]
5. Mendoza, G.; Suarez-Medellin, J.; Espinoza, C.; Ramos-Ligonio, A.; Fernandez, J.J.; Norte, M.; Trigos, A. Isolation and characterization of bioactive metabolites from fruiting bodies and mycelial culture of ganoderma oerstedii (higher basidiomycetes) from Mexico. *Int. J. Med. Mushrooms* **2015**, *17*, 501–509. [CrossRef]
6. Cao, Q.Z.; Lin, Z.B. Ganoderma lucidum polysaccharides peptide inhibits the growth of vascular endothelial cell and the induction of VEGF in human lung cancer cell. *Life Sci.* **2006**, *78*, 1457–1463. [CrossRef]
7. Sanodiya, B.S.; Thakur, G.S.; Baghel, R.K.; Prasad, G.B.; Bisen, P.S. Ganoderma lucidum: A potent pharmacological macrofungus. *Curr. Pharm. Biotechnol.* **2009**, *10*, 717–742. [CrossRef]
8. Yoon, H.M.; Jang, K.J.; Han, M.S.; Jeong, J.W.; Kim, G.Y.; Lee, J.H.; Choi, Y.H. Ganoderma lucidum ethanol extract inhibits the inflammatory response by suppressing the NF-kappaB and toll-like receptor pathways in lipopolysaccharide-stimulated BV2 microglial cells. *Exp. Ther. Med.* **2013**, *5*, 957–963. [CrossRef]
9. Krobthong, S.; Choowongkomon, K.; Suphakun, P.; Kuaprasert, B.; Samutrtai, P.; Yingchutrakul, Y. The anti-oxidative effect of Lingzhi protein hydrolysates on lipopolysaccharide-stimulated A549 cells. *Food Biosci.* **2021**, *41*, 101093. [CrossRef]
10. Hayes, M. Food proteins and bioactive peptides: New and novel sources, characterisation strategies and applications. *Foods* **2018**, *7*, 38. [CrossRef]
11. Udenigwe, C.C.; Mohan, A.; Wu, S. Peptide aggregation during plastein reaction enhanced bile acid-binding capacity of enzymatic chicken meat hydrolysates. *J. Food Biochem.* **2015**, *39*, 344–348. [CrossRef]
12. Segura-Campos, M.; Chel-Guerrero, L.; Betancur-Ancona, D.; Hernandez-Escalante, V.M. Bioavailability of bioactive peptides. *Food Rev. Int.* **2011**, *27*, 213–226. [CrossRef]
13. McClements, D.J. *Nanoparticle and Microparticle-based Delivery Systems: Encapsulation, Protection and Release of Active Compounds*; Taylor & Francis: Oxfordshire, UK, 2014.

14. Allen, T.M.; Cullis, P.R. Liposomal drug delivery systems: From concept to clinical applications. *Adv. Drug. Deliv. Rev.* **2013**, *65*, 36–48. [CrossRef] [PubMed]
15. Moeller, E.H.; Holst, B.; Nielsen, L.H.; Pedersen, P.S.; Ostergaard, J. Stability, liposome interaction, and in vivo pharmacology of ghrelin in liposomal suspensions. *Int. J. Pharm.* **2010**, *390*, 13–18. [CrossRef] [PubMed]
16. Pisal, D.S.; Kosloski, M.P.; Balu-Iyer, S.V. Delivery of therapeutic proteins. *J. Pharm. Sci.* **2010**, *99*, 2557–2575. [CrossRef] [PubMed]
17. Malam, Y.; Loizidou, M.; Seifalian, A.M. Liposomes and nanoparticles: Nanosized vehicles for drug delivery in cancer. *Trends Pharmacol. Sci.* **2009**, *30*, 592–599. [CrossRef] [PubMed]
18. Sereewatthanawut, I.; Prapintip, S.; Watchiraruji, K.; Goto, M.; Sasaki, M.; Shotipruk, A. Extraction of protein and amino acids from deoiled rice bran by subcritical water hydrolysis. *Bioresour. Technol.* **2008**, *99*, 555–561. [CrossRef] [PubMed]
19. Nielsen, P.M.; Petersen, D.; Dambmann, C. Improved method for determining food protein degree of hydrolysis. *J. Food Sci.* **2001**, *66*, 642–646. [CrossRef]
20. Riss, T.L.; Moravec, R.A.; Niles, A.L.; Duellman, S.; Benink, H.A.; Worzella, T.J.; Minor, L. Cell viability assays. In *Assay Guidance Manual*; Sittampalam, G.S., Grossman, A., Brimacombe, K., Arkin, M., Auld, D., Austin, C.P., Baell, J., Bejcek, B., Caaveiro, J.M.M., Chung, T.D.Y., et al., Eds.; Assay Guidance Manual. Eli Lilly & Company and the National Center for Advancing Translational Sciences: Bethesda, MD, USA, 2004.
21. Shilov, I.V.; Seymour, S.L.; Patel, A.A.; Loboda, A.; Tang, W.H.; Keating, S.P.; Hunter, C.L.; Nuwaysir, L.M.; Schaeffer, D.A. The paragon algorithm, a next generation search engine that uses sequence temperature values and feature probabilities to identify peptides from tandem mass spectra. *Mol. Cell. Proteom.* **2007**, *6*, 1638–1655. [CrossRef]
22. Willforss, J.; Chawade, A.; Levander, F. NormalyzerDE: Online tool for improved normalization of omics expression data and high-sensitivity differential expression analysis. *J. Proteome. Res.* **2019**, *18*, 732–740. [CrossRef] [PubMed]
23. DeFelice, S.L. The nutraceutical revolution: Its impact on food industry R&D. *Trends Food Sci. Technol.* **1995**, *6*, 59–61. [CrossRef]
24. Rebeca, B.D.; Peña-Vera, M.T.; Díaz-Castañeda, M. Production of fish protein hydrolysates with bacterial proteases; yield and nutritional value. *J. Food Sci.* **1991**, *56*, 309–314. [CrossRef]
25. Ruderman, G.; Grigera, J.R. Effect of Triton X-100 on the physical properties of liposomes. *Biochim. Biophys. Acta* **1986**, *863*, 277–281. [CrossRef]
26. Gupta, K.; Jang, H.; Harlen, K.; Puri, A.; Nussinov, R.; Schneider, J.P.; Blumenthal, R. Mechanism of membrane permeation induced by synthetic beta-hairpin peptides. *Biophys. J.* **2013**, *105*, 2093–2103. [CrossRef] [PubMed]
27. Smith, M.C.; Crist, R.M.; Clogston, J.D.; McNeil, S.E. Zeta potential: A case study of cationic, anionic, and neutral liposomes. *Anal. Bioanal. Chem.* **2017**, *409*, 5779–5787. [CrossRef] [PubMed]
28. Andar, A.U.; Hood, R.R.; Vreeland, W.N.; Devoe, D.L.; Swaan, P.W. Microfluidic preparation of liposomes to determine particle size influence on cellular uptake mechanisms. *Pharm. Res.* **2014**, *31*, 401–413. [CrossRef] [PubMed]
29. Sheng, Y.; Chang, L.; Kuang, T.; Hu, J. PEG/heparin-decorated lipid–polymer hybrid nanoparticles for long-circulating drug delivery. *RSC Adv.* **2016**, *6*, 23279–23287. [CrossRef]
30. Taira, M.C.; Chiaramoni, N.S.; Pecuch, K.M.; Alonso-Romanowski, S. Stability of liposomal formulations in physiological conditions for oral drug delivery. *Drug Deliv.* **2004**, *11*, 123–128. [CrossRef]
31. Kim, J.; Jang, D.S.; Kim, H.; Kim, J.S. Anti-lipase and lipolytic activities of ursolic acid isolated from the roots of Actinidia arguta. *Arch. Pharmacal Res.* **2009**, *32*, 983–987. [CrossRef]
32. Fernandez-Galilea, M.; Perez-Matute, P.; Prieto-Hontoria, P.L.; Martinez, J.A.; Moreno-Aliaga, M.J. Effects of lipoic acid on lipolysis in 3T3-L1 adipocytes. *J. Lipid Res.* **2012**, *53*, 2296–2306. [CrossRef]
33. Valikangas, T.; Suomi, T.; Elo, L.L. A systematic evaluation of normalization methods in quantitative label-free proteomics. *Brief Bioinform.* **2018**, *19*, 1–11. [CrossRef]
34. Wang, Y.; Botolin, D.; Xu, J.; Christian, B.; Mitchell, E.; Jayaprakasam, B.; Nair, M.G.; Peters, J.M.; Busik, J.V.; Olson, L.K.; et al. Regulation of hepatic fatty acid elongase and desaturase expression in diabetes and obesity. *J. Lipid Res.* **2006**, *47*, 2028–2041. [CrossRef] [PubMed]

foods

MDPI

Article

Chemical, Nutritional and Sensory Characteristics of Six Ornamental Edible Flowers Species

Jiri Mlcek [1,*], Anna Plaskova [1], Tunde Jurikova [2], Jiri Sochor [3], Mojmir Baron [3] and Sezai Ercisli [4]

1 Department of Food Analysis and Chemistry, Faculty of Technology, Tomas Bata University in Zlin, Vavreckova 275, 760 01 Zlin, Czech Republic; plaskova@utb.cz
2 Institute for Teacher Training, Faculty of Central European Studies, Constantine the Philosopher University in Nitra, Dražovská 4, 949 74 Nitra, Slovakia; tjurikova@ukf.sk
3 Department of Viticulture and Enology, Faculty of Horticulture, Mendel University in Brno, Valtická 337, 691 44 Lednice, Czech Republic; jiri.sochor@mendelu.cz (J.S.); mojmir.baron@mendelu.cz (M.B.)
4 Department of Horticulture, Agricultural Faculty, Ataturk University, Erzurum 25240, Turkey; sercisli@gmail.com
* Correspondence: mlcek@utb.cz; Tel.: +420-576-033-030

Abstract: Ornamental edible flowers can be used as novel nutraceutical sources with valuable biological properties. The purpose of this study was to establish nutritional, chemical, and sensory characteristics, antioxidant capacity (AC), and the relationship between their bioactive components and AC. The selected flowers *Begonia* × *tuberhybrida*, *Tropaeolum majus*, *Calendula officinalis*, *Rosa*, *Hemerocallis*, and *Tagetes patula*, can be easily collected due to their larger size. Their methanolic extracts were spectrophotometrically determined for polyphenols, flavonoids, and AC. Mineral elements were analyzed by atomic-absorption spectroscopy; crude protein was quantified by the Kjeldahl method. Eventually, 30 panelists evaluated sensory properties in 11 attributes. In addition, this study may serve to popularize selected blossoms. In flowers the contents of minerals were in this order: K > Ca > P > Mg > Na > Zn > Mn > Fe > Cu > Mo. AC ranged between 4.11 and 7.94 g of ascorbic acid equivalents/kg of fresh mass. The correlation coefficients between AC-total phenolics and AC-total flavonoids were $r = 0.73^*$ and $r = 0.58^*$, respectively. It is also possible to observe a strong correlation between mineral elements and bioactive compounds. *Hemerocallis* was rated as the best and most tasteful; additionally, it exhibited the highest AC, total phenolic and flavonoid contents.

Keywords: edible flower; antioxidant; bioactive compound; phenolic; flavonoid; mineral element; sensory evaluation

Citation: Mlcek, J.; Plaskova, A.; Jurikova, T.; Sochor, J.; Baron, M.; Ercisli, S. Chemical, Nutritional and Sensory Characteristics of Six Ornamental Edible Flowers Species. *Foods* **2021**, *10*, 2053. https://doi.org/10.3390/foods10092053

Academic Editor: Antonello Santini

Received: 10 August 2021
Accepted: 28 August 2021
Published: 31 August 2021

Publisher's Note: MDPI stays neutral with regard to jurisdictional claims in published maps and institutional affiliations.

1. Introduction

In recent years, edible flowers (EFs) have gained more attention due to their potential as a functional food with health benefits. This increased interest is also because customers are increasingly demanding in terms of nutrition. Flowers used in the human diet must be edible, which means harmless and non-toxic; thus, blossoms treated with chemical pesticides are unfit for consumption [1,2].

The primary gastronomic use of blossoms stems from their alluring colour; the assortment of EFs includes several dozen species with a significant number of shapes, colours, and flavours [3,4]. According to Lu et al., the number of EFs varies depending on region, but roughly 97 families, 100 genera, and 180 species are found worldwide [5]. Popular edible ornamental flowers are begonia, borage, calendula, carnation, cornflower, daylily, chrysanthemum, hibiscus, nasturtium, pansy, rose, tulip, and others. In addition to the ornamental flowers described above, EFs include several vegetables (broccoli and cauliflower), herbs (common sage, chives, marjoram, mint, thyme, and summer savory), or the blossoms of some fruit trees (elderberry and apple).

Nowadays, EFs are used in cuisine for flavour, garnish, and improved nutritional value of food, but they also represent a new opportunity for gastronomic innovations [6,7].

Flowers can be consumed in various forms, including fresh, dehydrated, lyophilised, cooked, and candied. Common uses for them are salads, sauces, jellies, soups, meat dishes, dyes, beverages, ice cubes, syrups, desserts, and cakes [8–10].

The main component of blossoms is water; its content ranges from 70 to 95% [11]. The content of nutrients like lipids, proteins, carbohydrates, and vitamins is similar to those in vegetables [12]. However, the mineral content of EFs shows the most significant variability regarding nutrient composition, as it is affected by the micronutrients in the soil [13]. The beneficial compounds for human health include antioxidants, vitamins, phenolics, carotenoids, flavonoids, minerals, and others [14]. Moreover, bioactive substances high content represents a beneficial component of the diet because of the possible development of valuable nutraceuticals [15]. However, the recommended limits of toxic agents must be regarded to utilise flowers as food [12].

Some edible flowers are traditionally used as medicinal herbs, and several previous studies revealed their biologically active molecules with potential health effects [16]. For example, these bioactive compounds can lower the risk of cardiovascular and cancer diseases, and they additionally have many beneficial properties like anti-inflammatory, antibacterial, antidiabetic, diuretic, antifungal and others [17–25]. The positive efficacy in lessening the risk of some diseases is due to these molecules' ability to reduce the damage induced by reactive oxygen species (ROS) [26]. These bioactive molecules probably may have a prolonged effect inside the body, as significant antioxidant activities (AA) have been found after the digestion in vitro of selected blossoms [27].

For example, the importance of AA in roses, begonias, and nasturtiums was mentioned by Friedman [28]. Because of the inclusion of polyphenols and ascorbic acid, nasturtium flowers demonstrated an exceptional tendency to exhibit unstable and highly reactive free radicals [29,30]. High values of AA recorded in extracts of rose flowers show a significant inhibitive effect on ROS [31–33]. In daylily extracts, intense scavenging ROS activity and lipid peroxidation were also observed [34,35]. *C. officinalis* extract had stronger AA when it came to scavenging free radicals than the synthetic antioxidant butylhydroxytoluene, commonly used as a food additive [18].

The utilisation of flowers for the human diet is associated with higher demands on their quality, sensory and nutritional properties [1] (Osimitz, Franzosa, Maciver, & Maibach, 2008). The larger size of blossoms with a simple collection is currently preferred if they can be economically effectively used [12]. The popular ornamental EFs of our gardens, specific in their size, colour, or aroma, could be convenient for the above properties and increase edible flowers' consumption. However, the sensory properties of selected cultivars of ornamental EFs are not sufficiently described in the literature. Additionally, only a small amount of EFs have been studied, and further research is needed to use them effectively. All flower samples in this research were non-toxic, allowing them to be included in human nutrition; nevertheless, it should be noted that the daily limit of intake for some of them is yet unknown [14].

The study aimed to investigate the nutritional composition, total phenolics and flavonoids contents, antioxidant capacity and mineral elements (P, K, Ca, Mg, Na, Fe, Mn, Cu, Zn, and Mo) of selected EFs. This work was supplemented and extended by studying the relationship between bioactive compounds and antioxidant capacity to assess their potential benefits to human metabolism. Furthermore, the sensory properties (appearance, fragrance, consistency, acid, bitter, astringent, sweet, spicy, overall taste, juiciness, and overall evaluation) were evaluated and described.

2. Materials and Methods

2.1. Plant Material

During the period of 2018–2019, the examined plants were grown in an unheated greenhouse on the plots of experimental orchards belonging to the Mendel University in Brno. These grounds are situated in the south-eastern part of Czechia at an elevation of 170 m above the sea level in Lednice town. Besides, the average yearly temperature and precipitation are 9.2 °C and 516 mm, respectively. The soil type was classified as black soil; the value of pH/KCl is 6.8. The agrochemical attributes of the used soil are shown in Table 1 [36].

Table 1. Agrochemical characteristics of the soil.

Mineral Element	Content in Soil [1]	Mineral Element	Content in Soil [1]
Phosphorus	84	Iron	4980
Potassium	269	Manganese	560
Calcium	4989	Copper	18
Magnesium	293	Zinc	22
Sodium	55	Molybdenum	4.3

Note: [1] All values of mineral content are expressed in mg/kg.

The criteria for selecting suitable ornamental flowers for our experiment include well-known ornamental edible species with larger blossoms that can be easily collected. The six selected flowers varied in species and colour, namely a pink cultivar of *Begonia × tuberhybrida*, a red cultivar of *Tropaeolum majus*, an orange cultivar of *Calendula officinalis*, a light yellow to a cream colour cultivar of *Rosa*, *Hemerocallis* cultivar with two-coloured petals (yellow and red), and orange cultivar of *Tagetes patula*. Some of their characteristics as shown in Table 2.

Table 2. Species and cultivars of edible flowers used in the present experiment.

Latin Name	English Name	Cultivar
Begonia × tuberhybrida	Tuberous Begonia	Doublet Rose
Tropaeolum majus	Nasturtium	Empress of India
Calendula officinalis	Pot Marigold	Orange King
Rosa	Rose	Gloria Dei
Hemerocallis	Daylilies	Bonanza
Tagetes patula	French marigolds	Antiqua Orange

2.2. Preparation of Samples

Flowers were collected in full ripeness from five randomly chosen plants of each species (cultivar). The degree of full readiness of flowers was determined from the blossom size, opening and colouring [37]. Five flowers from each cultivar were mixed and used for analyses.

The flowers of the unique cultivar were processed promptly after harvest (within 24 h at the latest). The reaped flowers were ground in a disc type mill SJ 500 (MEZOS, Hradec Králové, Czechia). Finally, the average sample was obtained by dividing pureed samples into quarters. Each parameter was measured in five replications. The obtained data were expressed as the average of a two-year experiment (2018–2019).

2.3. Extraction of Samples

The extraction of samples was performed according to the method developed by Kim et al. [38] and modified by Barros et al. [39]. The fresh flowers (10 g) were homogenised for 10 s in methanol; the plant and solvent ratio was 1:10 (w/v). The subsequent slurries were left in a water bath at a constant temperature of +25 °C for 24 h. The exact amount of methanol (100 mL) was used twice to extract residues. Both portions of methanolic extracts were combined, and the final solution was concentrated through evaporation at +40 °C to dryness (rotary evaporator R-215, Buchi Ltd., Oldham, UK). The samples were redissolved in methanol at a 100 g/mL concentration and kept at +4 °C for further utilisation.

2.4. Total Phenolic Content Assay

Total phenolic content (TPC) was measured by the method presented by Kim et al. with some modifications [38]. The sample (500 μL of extract) was brought quantitatively into a volumetric flask and diluted with distilled water to a volume of 50 mL. Then Folin-Ciocalteu's reagent (2.5 mL) and 20% solution of sodium carbonate (7.5 mL) were added to the sample. The resulting absorbance was measured at 765 nm against a blank on UV/VIS spectrophotometer LIBRA S6 (Biochrom Ltd., Cambridge, UK). TPC in different methanolic extracts was calculated and reported as g of gallic acid equivalents (GAE) per kg of fresh mass (FM).

2.5. Antioxidant Capacity by the DPPH Test Assay

Total antioxidant capacity (TAC) assay was carried out according to Brand-Williams et al. [40,41] with some modifications to analyse the antioxidant capacity of given samples [40,41]. The determination of free radical scavenging activity of methanolic flower extracts on 2,2-diphenyl-1-picrylhydrazyl (DPPH) free radical was estimated. The preparation of the stock methanol solution of DPPH and then the working solution was performed according to the procedure described by Rop et al. [14] using a spectrophotometer (LIBRA S6). In this method, 150 µL of flower extract was allowed to react with 2.85 µL of the DPPH methanolic solution. After 1 h in the dark, the absorbance was measured at 515 nm, then the values were converted using a calibration curve and expressed as g of ascorbic acid equivalents (AAE) per kg of FM [42].

2.6. Ferric Reducing Antioxidant Power Assay

The total antioxidant potential was estimated spectrophotometrically using the ferric reducing antioxidant power (FRAP) assay determined by Benzie and Strain with a slight modification [43]. The FRAP reagent was prepared from sodium acetate buffer (300 mM/L, pH 3.6), 10 mM/L TPTZ solution in 40 mM/L HCl and 20 mM/L $FeCl_3$ solution in proportions of 10:1:1. The sample (50 µL) was added into a testing tube with 4 mL of FRAP reagent, and its absorbance was measured at 593 nm after 10 min of incubation. The standard curve was prepared using different gallic acid concentrations; the results were expressed as g GAE/kg of FM.

2.7. Total Flavonoid Content Assay

Total flavonoid content (TFC) was determined with the aluminium chloride colourimetric assay described by Singleton et al. [44]. Into a microcentrifuge tube, 0.3 mL of the flower extract, 3.4 mL of 30% ethanol, 0.1 mL of sodium nitrite (c = 0.5 mol/L) and 0.15 mL of aluminium chloride hexahydrate (c = 0.3 mol/L) were put and mixed. After 5 min of incubation, 1 mL of sodium hydroxide (c = 1 mol/L) was added to the mixture. The absorbance of samples was measured against the blank at 506 nm using the LIBRA S6 spectrophotometer. TFCs were calculated from the calibration curve using rutin as a standard and were expressed as g of rutin equivalents (RE) per kg of FM.

2.8. Dry Matter and Mineral Content Assay

The dry matter (DM) and the mineral content were measured using modified methods described by Higson and Novotny [45,46]. The plant samples were dried in a laboratory oven Venticell 111 (BMT, Brno, Czech Republic), at 105 ± 2 °C to a constant weight. The dried samples' weights were measured and expressed as a percentage of weight concentration (*w*/*w*).

The next step was homogenisation of dried flowers to a particle size of up to 1 mm using a SJ500 laboratory grinder (MEZOS, Hradec Kralove, Czech Republic). About 1 g of DM of the homogenised sample was mineralised with concentrated sulphuric acid and 30% hydrogen peroxide in digestion tubes using a Bloc-digest M 24 apparatus (JP Selecta, Abrera, Spain). The digested samples were quantitatively added into a volumetric flask and then diluted to a final volume of 250 mL with double-distilled water.

The flower mineralizate was measured using an atomic absorption spectrometer PHILIPS PU 9200× with flame atomisation (Philips, Eindhoven, The Netherlands). A spectrophotometer Libra S6 was used to analyse the amount of phosphorus in the mineralizate quantitatively. The sample was prepared in a 100 mL volumetric flask, where 10 mL of the mineralizate and the same volume of ammonium-vanadomolybdate reagent were mixed. Lastly, samples were diluted up to a total volume of 100 mL with redistilled water and measured at 410 nm wavelength. For preparing the stock standard solution, potassium dihydrogen phosphate was used. The average contents of mineral elements were expressed as mg/kg of FM.

The content of total nitrogen was established according to the Kjeldahl method (ISO 1871:2009) using the analyser unit Kjeltec™ 2300 (Foss, Hillerod, Denmark). This apparatus provides automatic distillation and approves colourimetric titration. The crude protein in g/kg of FM was estimated by multiplying the determined nitrogen content by the standard default conversion factor of 6.25 [47].

2.9. Sensory Analysis

The sensory evaluation of edible flowers was performed by 30 panellists (trained students thanks to the course Sensory analysis and trained academic staff). They were acquainted with the monitored materials and instructed on the principles of analysis in advance. The course of sensory evaluation and equipment of the room for sensory analysis met precisely defined conditions according to the international standard ISO 6658. The sensor room at Tomas Bata University in Zlín was equipped with 12 separate evaluation boxes, placed next to each other and modified to limit contact with other evaluators. The room temperature was approximately 21 °C and was lit by artificial lighting. The assessment took place at 10:00 am, approximately 1 h (6 samples). It was recommended to take a break of approximately two minutes between the evaluations of the individual samples. Individual samples (each 3 flowers from one species) were administered in order *Rosa, Hemerocalis, Calendula officinalis, Begonia × tuberhybrida, Tagetes patula* and *Tropaeolum majus*. Before tasting, the samples were stored in reusable plastic containers at 7 °C for 12 h. Between individual samples, participants could neutralise the taste in their mouths with common water and white bread. The following sensory attributes were evaluated: appearance, fragrance, consistency, acidity, bitterness, astringency, sweetness, spiciness, overall taste, juiciness, and overall flower evaluation. The panellists assessed each blossom attribute using a 9-point hedonic scale; 1 = dislike extremely, 2, 3, 4 = subjective sense of dislike (very much/moderately/slightly), 5 = neutral, 6, 7, 8 = like slightly/moderately/very much, 9 = like extremely for overall taste and overall evaluation. They also determined the perceived intensity of each taste (acidity, bitterness, astringency, sweetness, and spiciness); 1 = very strong, maximum, 5 = slightly, moderate, 9 = without the taste. The scales for the remaining descriptors were as follows: 1 = unacceptable, 5 = neutral, 9 = ideal for appearance (suitability for food decoration); 1 = very intense and unpleasant, 5 = odourless, 9 = very intense pleasant for fragrance; 1 = very stiff, 5 = ideally crispy, 9 = flowable for consistency; 1 = dry, 5 = moderately juicy, 9 = watery for juiciness. The results were expressed graphically as the mean values of all ratings for each component and the overall score.

2.10. Statistical Analysis

Microsoft Office-Excel 2013 (Microsoft Corporation, Redmond, WA, USA) and STATISTICA CZ version 12 (StatSoft, Inc., Tulsa, OK, USA) were used for data analysis. The results were expressed by mean ± standard deviation (M ± SD). To establish statistically significant differences between individual species, Shapiro-Wilk test of normality and Levene's test of homogeneity of variances was performed. Since the conditions for the calculation by ANOVA analysis were not complied in any of the monitored data sets, a non-parametric Kruskal-Walllis test ($\alpha = 0.05$) were performed. Correlation functions were calculated using statistic software Unistat 5.1 (Unistat Ltd., London, UK) and Microsoft Office-Excel 2010 (Microsoft Corporation, Redmond, WA, USA).

3. Results and Discussion

The results of the chemical analyses are shown in Tables 3–8. The results were expressed as an average of a two-year experiment (2018–2019) since there was no statistically significant difference between the years in any parameter researched.

Table 3. Total phenolic content (g of GAE/kg of FM), total antioxidant capacity (g of AAE/kg of FM–DPPH or as g GAE/kg of FM-FRAP) and total flavonoid content (g of RE/kg of FM) in 6 species of edible flowers.

Species	TPC	TFC	TAC-DPPH	TAC-FRAP
Begonia × tuberhybrida	4.82 ± 0.27 [a]	2.32 ± 0.21 [a]	5.85 ± 0.31 [a]	5.15 ± 0.57 [a]
Tropaeolum majus	3.23 ± 0.35 [b]	1.29 ± 0.32 [b]	6.23 ± 0.48 [b]	4.98 ± 0.56 [a]
Calendula officinalis	3.65 ± 0.19 [b]	2.05 ± 0.20 [a]	4.11 ± 0.30 [c]	3.44 ± 0.29 [b]
Rosa	4.45 ± 0.20 [a]	2.43 ± 0.18 [a]	6.61 ± 0.41 [b]	4.57 ± 0.31 [a]
Hemerocallis	6.59 ± 0.23 [c]	3.76 ± 0.23 [c]	7.94 ± 0.26 [d]	4.71 ± 0.25 [a]
Tagetes patula	4.78 ± 0.44 [a]	2.02 ± 0.17 [a]	6.64 ± 0.38 [b]	5.62 ± 0.28 [c]

Note: All values are expressed as the mean ± standard deviation (SD) (n = 10). Values in a column that do not share the same superscript letters ([a,b,c,d]) are significantly different at $p < 0.05$. TPC: total phenolic content; TFC: total flavonoid content; TAC: total antioxidant capacity; GAE: gallic acid equivalents; FM: fresh mass; RE: rutin equivalents.

Table 4. Content of macroelements in 6 species of edible flowers.

Species	Phosphorus	Potassium	Calcium	Magnesium	Sodium
Begonia × tuberhybrida	208.59 ± 11.71 [a]	2008.55 ± 91.02 [a]	385.75 ± 12.33 [a]	131.18 ± 7.66 [a]	70.34 ± 5.28 [a]
Tropaeolum majus	452.02 ± 8.23 [b]	2353.30 ± 105.71 [b]	317.54 ± 14.47 [b]	152.89 ± 8.23 [b]	85.21 ± 4.55 [b]
Calendula officinalis	268.98 ± 7.89 [a]	2988.64 ± 90.73 [c]	294.08 ± 10.18 [b]	189.56 ± 9.20 [c]	86.31 ± 2.10 [b]
Rosa	245.15 ± 12.32 [a]	2033.44 ± 89.36 [a]	285.58 ± 15.15 [b]	142.45 ± 6.65 [ab]	79.23 ± 4.59 [ab]
Hemerocallis	235.05 ± 10.18 [a]	2759.22 ± 90.65 [d]	490.82 ± 9.68 [c]	284.15 ± 4.29 [d]	96.08 ± 3.50 [c]
Tagetes patula	397.08 ± 8.94 [c]	3623.78 ± 100.05 [e]	362.95 ± 7.26 [a]	203.14 ± 4.84 [c]	121.00 ± 1.58 [d]

Note: All values are expressed as the mean ± SD (n = 10). Values in the column that do not share the same superscript letters ([a,b,c,d,e]) indicate significant differences at $p < 0.05$. The content of macroelements is expressed as mg/kg of FM.

Table 5. Content of microelements in 6 species of edible flowers.

Species	Iron	Manganese	Copper	Zinc	Molybdenum
Begonia × tuberhybrida	3.12 ± 0.34 [a]	5.77 ± 0.19 [a]	1.28 ± 0.10 [a]	4.78 ± 0.93 [a]	0.76 ± 0.08 [a]
Tropaeolum majus	6.52 ± 0.58 [b]	5.79 ± 0.22 [a]	1.22 ± 0.04 [a]	8.89 ± 1.02 [b]	0.32 ± 0.01 [b]
Calendula officinalis	4.62 ± 0.21 [c]	7.33 ± 0.21 [b]	2.14 ± 0.08 [b]	10.79 ± 0.75 [c]	0.59 ± 0.08 [c]
Rosa	4.02 ± 0.10 [c]	3.41 ± 0.24 [c]	2.31 ± 0.02 [b]	4.62 ± 0.42 [a]	0.70 ± 0.04 [a]
Hemerocallis	8.70 ± 0.26 [d]	8.75 ± 0.17 [d]	2.93 ± 0.10 [c]	11.45 ± 0.55 [cd]	0.98 ± 0.09 [d]
Tagetes patula	8.20 ± 0.23 [d]	7.64 ± 0.20 [b]	1.24 ± 0.01 [a]	14.91 ± 1.21 [e]	0.43 ± 0.01 [e]

Note: All values are expressed as the mean ± SD (n = 10). Values in the column that do not share the same superscript letters ([a,b,c,d,e]) are significantly different at $p < 0.05$. The content of microelements is expressed as mg/kg of FM.

Table 6. Dry matter and the content of crude protein in 6 species of edible flowers.

Species	Dry Matter	Crude Protein
Begonia × tuberhybrida	11.34 ± 0.09 [a]	4.51 ± 0.32 [a]
Tropaeolum majus	7.38 ± 0.12 [b]	4.56 ± 0.35 [a]
Calendula officinalis	8.98 ± 0.09 [c]	3.48 ± 0.22 [b]
Rosa	14.39 ± 0.15 [d]	2.89 ± 0.19 [c]
Hemerocallis	11.24 ± 0.12 [a]	3.54 ± 0.25 [b]
Tagetes patula	9.34 ± 0.15 [c]	3.01 ± 0.31 [b]

Note: All values are expressed as the mean ± SD (n = 10). Values in a column that do not share the same superscript letters ([a,b,c,d]) are significantly different at $p < 0.05$. The content of dry matter is expressed as% w/w, and the content of crude protein is expressed as g/kg of FM.

Table 7. Correlation analysis between TPC, TAC, TFC and phosphorus (P), potassium (K), calcium (Ca), magnesium (Mg), natrium (Na), iron (Fe), manganese (Mn), copper (Cu), zinc (Zn) and molybdenum (Mo) of six edible flowers grown for two years.

	P	K	Ca	Mg	Na	Fe	Mn	Cu	Zn	Mo	TAC	TPC	TFC
P	1												
K	0.40	1											
Ca	−0.29	0.15	1										
Mg	−0.09	0.59 *	0.73*	1									
Na	0.51	0.92 **	0.24	0.57 *	1								
Fe	0.47	0.68 *	0.57	0.81 *	0.82 *	1							
Mn	0.07	0.72 *	0.67*	0.83 *	0.59 *	0.71 *	1						
Cu	−0.59 *	−0.05	0.36	0.64 *	−0.1	0.19	0.21	1					
Zn	0.49	0.96 **	0.29	0.69 *	0.92 **	0.83 *	0.81 *	0.00	1				
Mo	−0.88 **	−0.28	0.65 *	0.46	−0.3	−0.04	0.20	0.76 *	−0.26	1			
TAC	0.01	−0.03	0.68 *	0.48	0.31	0.61 *	0.14	0.31	0.11	0.39	1		
TPC	−0.52	0.12	0.88 *	0.71 *	0.24	0.45	0.47	0.58	0.18	0.82 *	0.73 *	1	
TFC	−0.69 *	0.00	0.77 *	0.72 *	0.06	0.30	0.40	0.81 *	0.06	0.93 **	0.57 *	0.94 **	1

Note: Mean values were used in the analyses of chemical parameters at levels of statistical significance (* $p < 0.05$; ** $p < 0.01$).

Table 8. Selected correlation coefficients between TAC-DPPH and mineral content, TPC and TFC.

Variables	Coefficient	Variables	Coefficient
TAC and phosphorus	$r = 0.0141$	TAC and potassium	$r = 0.0316$
TAC and calcium	$r = 0.6790$	TAC and magnesium	$r = 0.4848$
TAC and sodium	$r = 0.3119$	TAC and iron	$r = 0.6093$
TAC and manganese	$r = 0.1407$	TAC and copper	$r = 0.3061$
TAC and zinc	$r = 0.1091$	TAC and molybdenum	$r = 0.3882$
TAC and TPC	$r = 0.7288$	TAC and TFC	$r = 0.5786$

3.1. Total Content of Phenolic Substances, Total Antioxidant Capacity and Total Flavonoid Content

Table 3 provides TPC values of six edible flowers. These TPC values varied from 3.23 g GAE/kg in *T. majus* to 6.59 g GAE/kg in *Hemerocallis*, with almost double the difference. The results of *Hemerocallis* showed up to six times higher TPC than in some cultivars of daylilies presented by Muchová [48].

The TPC values for tested pink *Begonia* × *tuberhybrida* were 4.82 g GAE/kg of FM. They were similar for the white cultivar of Begonia (*B. semperflorens* Link et Otto) [49] and double higher when compared to another pink cultivar, 'Chanson' [48].

TPC in *T. patula* (French marigolds) was 4.78 g GAE/kg of FM, and this content is slightly higher than that reported by Rop et al. [14] but lower than in different orange cultivars of French marigold flowers [50].

According to Ashraf et al., for *C. officinalis*, the TPC values were slightly lower (by 0.83 g) than the values measured by us (3.65 g GAE/kg FM), which could be caused by the fact that other parts of the plants (leaves) were used or different growing conditions [51]. Ferreira et al. quantified TPC values of 2.30 g GAE/kg DW in calendula flowers [52]. This research found an aqueous methanol solution (70:30, Me-OH:H_2O) more effective for extracting phenolic compounds than pure methanol; the values were probably higher because the solvent was highly polar and had the highest dielectric constant [52].

The content of total phenolic substances in *Rosa* was determined to be 4.45 g GAE/kg FM, which is approximately 1.12 times lower than that measured by Rop et al. [14]. Yang and Shin found the difference between TPC in ethanolic samples of roses, and their values ranged from 7.99 to 29.79 g/kg FM [1]. They also reported that yellow and pink roses had lower TPC than red roses. Despite different flower varieties and conditions of extraction, these values are lower than those reported by Gonçalves et al., where the orange rose cultivar had a slightly higher TPC content (17.60 g GAE/kg FM) than *Tagetes erecta L.*

(17.47 g GAE/kg FM) [11]. The considerable variation could indicate that the composition of substances differs within the genus.

As mentioned above, the red cultivar of *T. majus* exhibited the lowest TPC value (3.23), which is significantly lower than the value for the orange cultivar exhibited by other authors [29,53]. Nevertheless, our TPC results are comparable to those reported by Rop et al. [14] and higher than those reported by Muchová [48]. The differences could be caused by using a different variety, growing conditions, the flower's measured parts, or its adjustment before analysis.

The total phenolic range level is comparable with some berries *Vaccinium L. hybrids* and *Rubus L. species* [54,55], both of which are considered to be great sources of antioxidants [29]. This content is higher when compared to different vegetables like Ceylon spinach, white and red onions [56], lamb's lettuce [57], or other green leafy vegetables [58]. Flowers may be ideal for making salads more appealing to consumers, either adding colour or increasing the phenolic content of the food.

In this study, total flavonoids were another parameter studied in flowers, as seen in Table 3. Flavonoids are likely the most significant natural phenolics, and also they are one of the most varied and widespread natural chemicals [59]. The TFC in the flowers ranged from 1.29 to 3.76 g RE/kg FM. The highest TFC was determined in *Hemerocallis* (daylily), and the lowest was observed in the red variety of *T. majus*. For other flowers, the TFC was above 2 g of RE/kg of FM. In the Rosa' Gloria Dei', the total flavonoid contents were 2.43 g of RE/kg of FM, which is slightly higher than that observed in Rosa odorata [14]. According to a study by Yang and Shin, the TFC of edible roses ranged between 0.79 to 5.32 g/kg FM; our analysed cultivar is in this range [1]. The flowers reached significantly higher TFC values than some vegetables and fruit, such as tomatoes (0.133 to 0.474 g RE/kg FM) [60], watermelons (0.09 to 0.27 g RE/kg FM) [61], or fruit studied by Mirzaei et al., such as blackberry and black grape, whose TFC values ranged from 0.05 to 1.03 g RE/kg FM [62]. According to studies on 12 cultivars of EFs, the yellow blossoms had a higher content of flavonoids and suggested that they have a stronger antioxidant potential than other colours [59]. This correlates with our results because the flowers with the highest TFC content were *Hemerocallis* and *Rosa* with yellow colour petals. However, Garzón et al. analysed the *T. majus* flowers depending on their colour. The yellow cultivar had lower TPC values and antioxidant activity than the orange and red cultivars due to the low content of primary anthocyanins [29].

Further research could involve using high-performance liquid chromatography (HPLC) to identify and accurately quantify phenolic compounds in the sample. In the case of including HPLC analysis in this study, its length and complexity would exceed the proposed research framework.

The antioxidant potentials of flower extracts were estimated using two different colourimetric assays in vitro based on electron-transfer reactions. The first DPPH method was measured antioxidants' capacity to scavenge an organic radical; the results were expressed as ascorbic acid equivalents. The second FRAP method was estimated antioxidants' ability to reduce ferric to the ferrous ion, and findings were reported as reducing power per gallic acid equivalent. Combining these two approaches based on distinct mechanisms may provide more reliable and complex data for antioxidant capacity. A single spectrophotometric assay may not provide satisfactory results because of its deficiency of specificity and sensitivity [63]. Both methods are widely used because of their simplicity, speed, high reproducibility, and ability to measuring by simple instrumentation [41,64,65]. Each of them has some advantages and limitations. For example, the DPPH assay can detect weak antioxidants and thermally unstable compounds; however, DPPH might react with other radicals in the sample and is sensitive to light [43,64,65]. The FRAP method result may not positively correlate with the total antioxidant activity of the sample; because this assay is non-specific and has limitations in measuring slow-reacting polyphenolic antioxidants and thiols [43,66].

The total antioxidant capacity of samples ranged from 4.11 g AAE/kg FM in *C. officinalis* to 7.94 g AAE/kg FM in daylilies. TCA values above 6 g AAE/kg FM were measured in *T. patula, Rosa*, and *T. majus*. The *Hemerocallis* (daylily) achieved a higher antioxidant capacity than the edible flowers in the Rop et al. study; the TAC of edible flowers ranged from 4.21 to 6.96 g AAE/kg FM [14]. In addition, the strong antioxidant activities of daylilies extracts (aqueous and ethanolic) were observed by Que et al. [35]. These results exhibited lower AA than synthetic antioxidant (butylated hydroxyanisole) but higher than α-tocopherol. According to Fu et al., the highest antioxidant capacity and the highest proportion of phenolic substances is in the opening stage of daylilies [17]. Mao et al. found that the use of lyophilised daylily flowers is more suitable for obtaining an extract with higher AA and a higher TPC than the use of flowers dried with hot air [67]. The limitation of using daylily is that each flower only lasts one day. The flower extracts from *T. majus* are active, reducing agents, which indicates a good ability to scavenge radicals [68]. According to Pavithra et al., the methanol extracts of flowers have scavenging abilities dependent on their concentration (25 mg/mL and higher) but lesser than ascorbic acid [69].

Comparing the results obtained from TAC-FRAP with TAC-DPPH, it is evident that the extracts' ability to reduce Fe^{3+} has a different order than the ability to quench the $DPPH^{\bullet}$ radical. Additionally, the values obtained by FRAP assay show that the highest antioxidant capacity corresponded to *T. patula* (5.62 g GAE/kg FM), followed by *Begonia* × *tuberhybrida* (5.15 g GAE/kg FM) and *T. majus* (4.98 g GAE/kg FM). The lowest AC was that of *C. officinalis* (3.44 g GAE/kg FM). The FRAP values displayed a 1.6-fold difference.

People generally do not consume as many edible flowers as carrots, radishes, cucumbers, tomatoes, and other vegetables. Because some edible flowers have a pungent or strong aroma, it is advisable to use them sparingly to encourage food flavour [70]. The ornamental EFs evaluated in the study were non-toxic; nevertheless, it should also be considered that the daily limit for their ingestion is not determined for all samples, and no international authority has published the official list of EFs [12,71,72]. Consumption and culinary use of some EFs were documented in history before May 1997; consequently, these flowers are not defined as novel foods [73,74]. For example, none of the species analysed in our research was featured on official lists like the Novel Food Catalogue [72]. On the other hand, other blossoms that cannot prove their significantly large consumption by people before 15 May 1997 must be submitted to the European Food Safety Authority for their safety application as novel foods [73,75].

Lucarini et al. [75] examined available information in databases and bibliographies about the same plant genera as our study, and they discovered no scientific proof that these plants constitute potential allergens.

Even the most favourable herbs can have unpredictable effects, for example, the consumption of more than 39.5 g of fresh *T. majus* flowers surpassing the daily erucic acid allowance [70,76]. The number of blossoms consumed may be the limiting factor because of allergic and toxic reactions by sensitive persons to some of the flower unidentified compounds [14]. In addition, pollen from specific blossoms might induce an allergic response in humans with allergies or asthma [75]. Thus, it is important to study the toxicity of EFs with high antioxidant activity to establish their safety as food additives. Moreover, identification of the plant is critical because some toxic flowers could be readily confused with EFs, such as daylilies with true lilies, and confusing them might be dangerous [77].

3.2. The Content of Mineral Elements

Five macroelements (P, K, Na, Ca, Mg) and five microelements (Fe, Mn, Cu, Zn, Mo) were determined and quantified in the petals of diverse species of ornamental edible plants. These mineral elements are essential for the human body's vital functioning, but the available literature contains scant data about their content in EFs. The contents of minerals, expressed on a FM basis, are shown in Tables 4 and 5, and were in this order: K > Ca > P > Mg > Na > Zn > Mn > Fe > Cu > Mo. The macroelements amount ranged from 121 to

3623 mg/kg FM (Table 4), and the content of microelements then from 0.98 to 14.91 mg/kg FM (Table 5).

Hemerocallis had the greatest Ca, Mg, Fe, Mn, Cu, and Mo concentrations, whereas *T. patula* contained the highest amount of K, Na and Zn, and the highest P was detected in *T. majus*. In contrast, the lowest P, K, Mg, Na and Fe content was observed in *Begonia* × *tuberhybrida*; *Rosa* had the least amount of Ca, Mn and Zn, the lowest quantity of Cu and Mo was found in *T. majus*.

The content of mineral elements is comparable to the flower mineral concentration mentioned by Rop et al. [14]. When compared to ordinary fruit and vegetables, EFs are a good source of minerals. This is evidenced by the higher K content than vegetables and fruit, which have an average K content of 1500–2100 mg/kg FM (Table 4) [78–80]. Several researchers observed a similar trend in which potassium content was highest in flowers [14,81–83]. Potassium content in flowers was higher than in leaves, roots, and stem of *Chrysanthemum indicum* L. [82]. According to Grzeszczuk et al., in other Hemerocallis species, the most abundant macroelement was K, which correlates with our results, but P content was higher than that of Mg [83]. However, Navarro-González et al. reported that *T. majus* and *Tagetes erecta* blossoms contain more zinc, iron, and manganese than potassium [53]. Flowers (100 g fresh weight) provided only 10.0–18.1% of the daily recommended K intake of 2000 mg for adults [47]. Potassium content is an important source for maintaining acid-base balance in blood and tissues and preventing cardiovascular or oncogenic diseases [84].

The content of other elements in flowers is comparable to vegetables [80], but some selected leafy vegetables had a higher content of sodium than potassium [58]. Compared to fruit, a two-fold increase in Ca and Mg contents and a fourfold rise in Na content can be observed [78,85,86]. In addition, the content of mineral elements in flowers can be compared with published minerals data about wild-growing and cultivated mushrooms. Calcium and sodium contents are two to four times higher than that of fungi, the content of other elements is approximately comparable, but the phosphorus one is twice lower [87,88].

Mineral elements perform several functions: as components of enzymes, regulation of cellular energy transduction, gas transport, antioxidant defense, membrane receptor functions, second-messenger systems, and integration of several physiological functions [89–91]. Furthermore, they can strengthen the immune system [92,93], form building blocks of the human skeleton [91,94] and are associated with anti-inflammatory [24,95], antibacterial [93,96], antifungal [97] and antiviral effects [98]. A few published research papers deal with the content of mineral elements in EFs regarding their relevance for human health [14,83,99,100].

Previous research has shown that iron concentrations in ornamental flowers are highly varied, compared to our results, for example, *Begonia boliviensis* with lower content of 2.65 mg/kg FM [14], *T. majus* with slightly lower content from 5.51 to 6.47 mg/kg FM [14,53], and *T. erecta* with slightly higher amount of 10.26 mg/kg FM [53]. Different species probably caused variations in mineral elements content between the flower samples because they were grown in the same location and with identical agricultural practices.

All analysed flower species have high molybdenum levels based on recommended daily intakes for adults (50 µg) since 100 g of fresh blossoms provides 64–196% of this value [47]. The concentration of Mo affects ascorbic acid level; for example, its deficiency can cause a decrease in AA content in some vegetables [101]. Tolerable upper intake level of Mo range from 0.1 to 0.6 mg/day [102]; therefore, consuming a slight amount of flowers is unlikely to be a risk for human health. *Hemerocallis* can be considered as a possible source of Cu (0.29 mg/100 g FM), Mn (0.88 mg/100 g FM), and Zn (1.15 mg/100 g FM), and these mineral elements can contribute to daily recommended dietary allowances for adults. For example, 100 g fresh *Hemerocallis* can provide 29.3% copper, 43.8% manganese and 11.5% zinc for dietary reference intakes [47]. EFs should not be overlooked as a source of mineral elements in the human diet; however, it is unlikely that somebody would eat

100 g of flowers in a single day. Edible flowers will most likely be used as a garnish to add colour and flavour to the food.

3.3. Dry Matter and Content of Crude Protein

The dry matter and the content of crude protein of the edible flowers are shown in Table 6. As can be seen from the results, the DM of these edible flowers ranged from 7.38 to 14.39%, and this amount is slightly higher than the average content in fruit and vegetables [103]. On the other hand, according to Montañés Millán et al. [104], the DM percentage in blossoms from the fruit tree was higher. When comparing our DM results to previous research for the same plant genus, *Begonia nelumbiifolia* ranged from 5.31 to 6.15%, which is lower than Begonias results in our experiment [105]. However, Rop et al. determined *Begonia boliviensis* (14.20%) with a higher DM [14]. In addition to the last-mentioned research, they determined a higher DM for *T. patula* (9.68%) and *T. majus* (11.27%) and lower DM for *Rosa odorata* lower DM (10.09%) [14]. De Lima Franzen et al. observed a higher DM percentage for rose (*Rosa* × *grandiflora*) and *C. officinalis* of 15.44% and 10.66%, respectively [106].

The CP content of EFs samples was estimated by the Kjeldahl method, and the results ranged between 2.89 to 4.56 g/kg of FM (Table 6).

The highest values were reached for *T. majus* (4.56) and *Begonia* × *tuberhybrida* (4.51). Comparing these CP values with results obtained by Rop et al. [14], *T. majus* had slightly lower values (4.74 g/kg FM), and Begonia had one and a half times higher than another cultivar. However, the CP contents of *T. majus* and other varieties of Begonia and roses cultivated in Japan were significantly higher than in our research [9]. The difference may be caused by different cultivars, place and growth conditions. A similar CP was observed in EF *Fernaldia pandurate* with 3.0 g/kg FM [107]. This protein content can be comparable to some fruit and vegetables but not to cereals because they have an order of magnitude higher content [79]. Similar proportion content was observed in some fruit, for example, plum with 3.9 g/kg FM [108] or red banana (*Musa acuminata*) [109]. The CP content in fresh vegetables was higher than in our experiment, for example, radishes with 5 to 15.5 g/kg, beetroot with 13.22–14.43 g/kg [110], celery with 6.91 g/kg, carrot with 5.64 g/kg, and turnip with 4.88 g/kg [108]. Thus, flower petals could not be regarded as good protein sources because of their low CP levels [106]; also, people consume fewer EFs than radishes, carrots and other popular types of vegetables.

3.4. Correlation Analysis between Mineral Elements and Bioactive Compounds

The correlation coefficients of mineral elements and bioactive compounds in edible flowers are shown in Tables 7 and 8. Significantly strong positive correlations were observed between some mineral elements contents; for example, the correlation of Na-K (r = 0.92 **), Zn-K (r = 0.96 **), Zn-Na (r = 0.92 **) and Zn-Fe (r = 0.83 **). Furthermore, considerable high positive correlations between TFC-Mo (r = 0.93 **) and TFC-Cu (r = 0.81 *) were found. From a different point of view, negative relationships were noticed between the contents of M and P (r = −0.88 **), between TFC and P (r = −0.69 *), and also between Cu and P (r = −0.59 *).

Table 8 shows the selected correlation coefficients between TAC-DPPH and mineral elements, TPC or TFC. These relationships are studied to assess if these components contribute to the TAC of the flowers and if they have any potential benefits for human metabolism.

In accordance with some research studies [111–114], significant correlations between TAC, TPC and TFC were commonly achieved in our results as well, from r = 0.57 * to 0.94 **. The results imply that blossoms with a higher amount of polyphenols have a stronger antioxidant activity, and flavonoids comprise an important group of phenolic compounds. Some authors also found a strong positive correlation between TPC and FRAP assay values [43,71,115]. The antioxidant activity could be attributed to some mineral elements like copper, iron and manganese [116]. In our case, AC correlates with the Ca (r = 0.68 *) and Fe (r = 0.61 *), which means a moderate positive correlation; some authors also described these relationships [117,118]. Their articles state the importance of nutrition by given elements on the content of bioactive substances. On the other hand, the correlations between TAC and the remaining mineral elements were weak or negligible. Aside from polyphenols, the antioxidant activity of floral extracts may be affected by other biological compounds, including vitamins, pigments such as carotenoids, mineral elements, nitrogenous compounds, and other metabolites. [17,119–121].

3.5. Sensory Evaluation

Necessary criteria for evaluating the quality of edible flowers for gastronomy are organoleptic performance, flavour, and overall impression. The overall evaluation of the flowers makes a positive impression on consumers. In our research, the best-rated flowers were *Hemerocallis* with 8.2 points, followed by *T. majus* (7.4), *Begonia* × *tuberhybrida* (7.3), *C. officinalis* (7.2), *Rosa* (6.7), and *T. patula* (6.2).

As can be seen in Figure 1, the appearance of all analysed flowers was evaluated as suitable for decorating dishes; blossoms of *C. officinalis* (8.5) appeared to be the most acceptable. The difference between the species was 1.1 points. According to Kelley et al., the colour and composition of flowers are important characteristics influencing consumer preferences [3]. For example, the nasturtium mixture containing flowers of darker colours such as orange and crimson was evaluated as more appealing [3,122]. In addition, the colour of EFs can influence sales because of the appeal to the appetite of consumers; for example, red flowers may increase appetite, the yellow one can elicit happiness, and the orange expresses affordability [122]. Furthermore, the customers can associate the colour of EFs with the taste of food in the same hue [122].

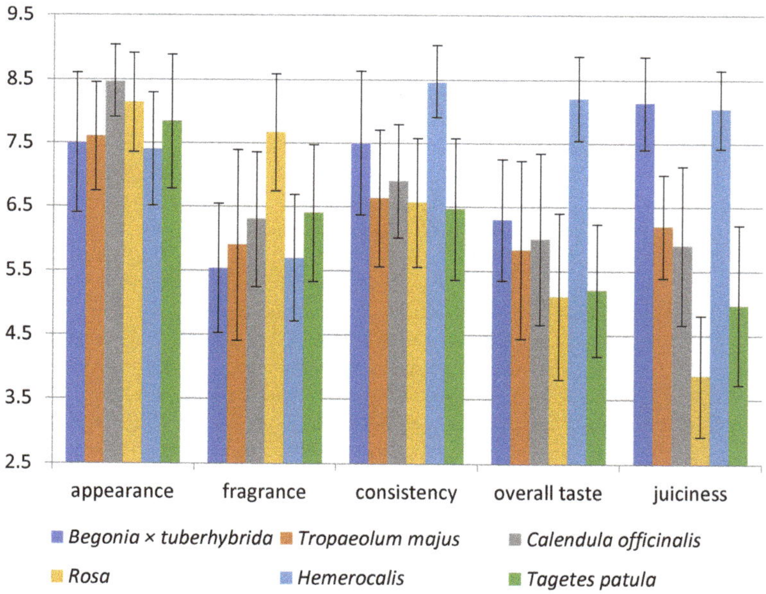

Figure 1. Sensory analysis diagram for six edible flowers; *Begonia* × *tuberhybrida*; *T. majus*; *C. officinalis*; *Rosa*; *Hemerocallis*; *T. patula*.

The fragrance of blooms was evaluated in all samples as pleasant with various levels of intensity. *Rosa* 'Gloria Dei' was identified as the flower with the most pleasant scent (7.7) because the scale ranged from very intense pleasant (=9) through odourless (=5) to very unpleasant fragrance (=1). The other blossoms were rated from 5.5 to 6.4 points, corresponding to a lower intensity scent. The fragrance may attract consumers by arousing their interest in the flowers, but the buds are generally odourless; thus, only fully ripeness blossoms were collected [123]. Therefore, the petals are the main source of aromatic compounds in many flowers; for example, the petals of *Rosa damascena* are used as a source of aroma and natural scents [107].

In evaluating consistency, the crispness is related to the water content because the cells or cavities are exhibited this property when they are filled with air instead of water [124]. The remaining *Hemerocallis* (8.5) and *Begonia* × *tuberhybrida* (7.5) were evaluated as more flowable. Another parameter related to consistency is probably juiciness due to evaluating these two EFs as watery and more flowable. The level of juiciness significantly varied. For example, *Rosa* (3.9) was evaluated as drier, and the remaining flowers showed a subjective degree of juiciness or watery. The sensory properties of the rose, such as its dryness and crispiness, can be influenced by the high DM content (14.39% w/w).

The evaluation of the overall taste of flowers is important for their acceptance and valuation as food. *Hemerocallis* flowers have an 8.4-point gain, which means delightful taste. On the contrary, *Rosa* (5.1) and *T. patula* (5.2) flowers were evaluated neutrally; for instance, they can be more suitable as garnish. The remaining flowers had a slightly pleasant overall taste.

Statistically significant differences were found in: appearance (*C. officinalis* vs. *Begonia* × *tuberhybrida* and *T. majus* and *Hemerocallis*), fragrance (*Rosa* vs. all species), consistency (*Hemerocallis* vs. all species, *Begonia* × *tuberhybrida* vs. *T. patula and Rosa)*, overall taste (*Hemerocallis* vs. all species, *Begonia* × *tuberhybrida* vs. *Rosa* and *T. patula*), juiciness (*T. majus* vs. *C. officinalis*, *Hemerocallis* vs. *Begonia* × *tuberhybrida*, *T. patula* vs. *Rosa* and *C. officinalis* and *T. majus*) ($p < 0.05$).

Figure 2 shows the sensory analysis results on the five various taste qualities-sweet, acid, bitter, astringent, and spicy. If the blossom received 9 points in the sensory analysis of taste, it did not contain the evaluated taste; for example, it was not sweet at all. The sourness intensity ratings were slightly acidic in all flowers; the range of gained points was from 5.1 to 6.9. Besides, *Begonia* × *tuberhybrida* was evaluated as the least acidic (6.9).

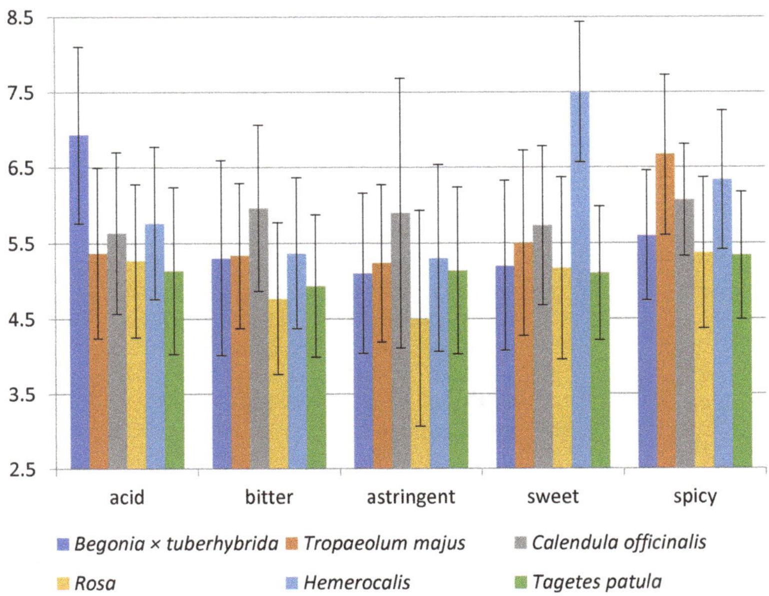

Figure 2. Sensory analysis diagram for the taste of 6 edible flowers; *Begonia* × *tuberhybrida*; *T. majus*; *C. officinalis*; *Rosa*; *Hemerocallis*; *T. patula*. The scale of taste intensity ranged from very strong, maximum (=1) through slightly moderate (=5) to without taste (=9).

Further, all blossoms were rated as slightly bitter, with bitterness perceived most intensely in rose (4.8), and *C. officinalis* with 6 points was rated as less bitter. According to Mlcek and Rop, the taste of *C. officinalis* should be slightly bitter, which corresponds with our results [12]. However, the difference is in the taste of *T. patula* because it should be bitterish or with bitter undertones [12,125].

Panellists described a similar intensity of slightly astringent taste for the evaluated flowers; for example, *Rosa* (4.5) was more astringent than *C. officinalis* (5.9).

The next evaluated taste quality was sweetness; all flowers except *Hemerocallis* were rated similarly to slightly sweet with a point range from 5.1 to 5.7. In contrast to other flowers, *Hemerocallis* was described as unsweetened at all (7.5). Mlcek and Rop described the tastes of rose as sweet and daylily as slightly sweet, which does not correspond to the evaluators' opinions [12].

Finally, the flowers' spiciness was evaluated as slightly spicy, with different intensity. The blossoms of *T. majus* (6.7) tasted the least spicy; in comparison, spicier flowers were *T. patula* (5.3) and *Rosa* (5.4).

According to sensory evaluation, *Hemerocallis* had the highest score in the sensory analysis and the most acceptable overall taste. The flower buds seem to be the most widely consumed part of the daylily [12]. The *Hemerocallis* have a mild taste, albeit with a sour, astringent, and spicy touch. These flowers were not evaluated particularly as sweet, and therefore could be used as an alternative to mustard due to their peppery, radishes, and spicy taste [12,126]. However, *Hemerocallis* 'Bonanza' was statistically assessed as one of two non-preferred cultivars (15 daylilies) according to the taste preferences [127]. According to Grosvenor, the red (dark) varieties have more bitterness, and the tested cultivar has a yellow flower with a red centre [128].

Statistically significant differences were found in: acid (*Begonia* × *tuberhybrida* vs. all species), sweet (*Hemerocallis* vs. all species), bitter (*C. officinalis* vs. *Rosa* and *T. patula*), astringent (*C. officinalis* vs. *Rosa*), spicy (*T. majus* vs. *Begonia* × *tuberhybrida* and Rosa and *T. patula*) ($p < 0.05$).

4. Conclusions

This study evaluated selected Czech flowers' suitability as nutritional food with the health benefits in terms of the content of bioactive substances, mineral elements, and sensory analysis. The individual flowers are not usually consumed in large quantities but rather as a garnish or ingredient for dishes. In terms of the current popularisation of nutraceutical, edible flowers can represent a significant natural source of bioactive substances, containing a higher concentration of these than ordinary fruit or vegetables. In this research, we determined the total phenolic, flavonoids and antioxidant capacity of six ornamental flowers. Investigation of correlations confirms the findings of earlier research. The correlation study suggests that polyphenolic and flavonoids form an important part of the antioxidant compounds of these flowers. This study confirms that the amount of bioactive substances in edible flowers is affected by various factors, including the plant's external and internal environment during growth, the time of collection, post-harvest technologies; however, optimized cultivating and harvesting protocols have the potential to standardize the produce. In the future, edible flowers can serve as a natural source for food supplements that contain these substances. Besides, the flowers presented in this study will certainly be the food industry's future with their taste characteristics, size, and ease of collection. Many pieces of research on this topic are likely to be carried out.

Author Contributions: Conceptualisation, J.M. and A.P.; methodology, J.M.; software, J.S.; validation, J.M. and T.J.; formal analysis, J.M.; investigation, J.M.; resources, J.S.; data curation, J.M.; writing—original draft preparation, J.M. and A.P.; writing—review and editing, A.P. and J.M.; visualisation, A.P.; supervision, S.E.; project administration, M.B.; funding acquisition, J.M. All authors have read and agreed to the published version of the manuscript.

Funding: This work was supported by the internal grant of TBU in Zlin (No. IGA/FT/2021/008).

Data Availability Statement: New research data were presented in this contribution.

Acknowledgments: We would like to thank Anezka Lengalova for English language editing and Jarmila Neugebauerova for samples of edible flowers.

Conflicts of Interest: The authors declare no conflict of interest.

References

1. Yang, H.; Shin, Y. Antioxidant compounds and activities of edible roses (*Rosa hybrida* spp.) from different cultivars grown in Korea. *Appl. Biol. Chem.* **2017**, *60*, 129–136. [CrossRef]
2. Guiné, R.P.F.; Florença, S.G.; Ferrão, A.C.; Bizjak, M.; Vombergar, B.; Simoni, N.; Vieira, V. Factors affecting eating habits and knowledge of edible flowers in different countries. *Open Agric.* **2021**, *6*, 67–81. [CrossRef]
3. Kelley, K.M.; Behe, B.K.; Biernbaum, J.A.; Poff, K.L. Consumer Preference for Edible-flower Color, Container Size, and Price. *HortScience* **2001**, *36*, 801–804. [CrossRef]
4. Benvenuti, S.; Bortolotti, E.; Maggini, R. Antioxidant power, anthocyanin content and organoleptic performance of edible flowers. *Sci. Hortic.* **2016**, *199*, 170–177. [CrossRef]
5. Lu, B.; Li, M.; Yin, R. Phytochemical Content, Health Benefits, and Toxicology of Common Edible Flowers: A Review (2000–2015). *Crit. Rev. Food Sci. Nutr.* **2016**, *56*, S130–S148. [CrossRef]
6. Benvenuti, S.; Mazzoncini, M. The Biodiversity of Edible Flowers: Discovering New Tastes and New Health Benefits. *Front. Plant Sci.* **2021**, *11*, 569499. [CrossRef] [PubMed]
7. Tanji, A.; Nassif, F. Edible Weeds in Morocco. *Weed Technol.* **1995**, *9*, 612–620. [CrossRef]
8. Koike, A.; Barreira, J.C.M.; Barros, L.; Santos-Buelga, C.; Villavicencio, A.L.C.H.; Ferreira, I.C.F.R. Edible flowers of *Viola tricolor* L. as a new functional food. *Food Chem.* **2015**, *179*, 6–14. [CrossRef]
9. Chensom, S.; Okumura, H.; Mishima, T. Primary Screening of Antioxidant Activity, Total Polyphenol Content, Carotenoid Content, and Nutritional Composition of 13 Edible Flowers from Japan. *Prev. Nutr. Food Sci.* **2019**, *24*, 171–178. [CrossRef]
10. Zhao, L.; Fan, H.; Zhang, M.; Chitrakar, B.; Bhandari, B.; Wang, B. Edible flowers: Review of flower processing and extraction of bioactive compounds by novel technologies. *Food Res. Int.* **2019**, *126*, 14. [CrossRef]
11. Gonçalves, F.; Gonçalves, J.C.; Ferrão, A.C.; Correia, P.; Guiné, R.P.F. Evaluation of phenolic compounds and antioxidant activity in some edible flowers. *Open Agric.* **2020**, *5*, 857–870. [CrossRef]
12. Mlcek, J.; Rop, O. Fresh edible flowers of ornamental plants—A new source of nutraceutical foods. *Trends Food Sci. Technol.* **2011**, *22*, 561–569. [CrossRef]
13. Rivas-García, L.; Navarro-Hortal, M.D.; Romero-Márquez, J.M.; Forbes-Hernández, T.Y.; Varela-López, A.; Llopis, J.; Sánchez-González, C.; Quiles, J.L. Edible flowers as a health promoter: An evidence-based review. *Trends Food Sci. Technol.* **2020**, in press. [CrossRef]
14. Rop, O.; Mlcek, J.; Jurikova, T.; Neugebauerova, J.; Vabkova, J. Edible Flowers—A New Promising Source of Mineral Elements in Human Nutrition. *Molecules* **2012**, *17*, 6672–6683. [CrossRef] [PubMed]
15. Bazylko, A.; Granica, S.; Filipek, A.; Piwowarski, J.; Stefańska, J.; Osińska, E.; Kiss, A. Comparison of antioxidant, anti-inflammatory, antimicrobial activity and chemical composition of aqueous and hydroethanolic extracts of the herb of *Tropaeolum majus* L. *Ind. Crop. Prod.* **2013**, *50*, 88–94. [CrossRef]
16. Roberts, M.; Green, P. *Edible & Medicinal Flowers*, 1st ed.; The Spearhead Press: Claremont, CA, USA, 2000; p. 166.
17. Fu, M.; He, Z.; Zhao, Y.; Yang, J.; Mao, L. Antioxidant properties and involved compounds of daylily flowers in relation to maturity. *Food Chem.* **2009**, *114*, 1192–1197. [CrossRef]
18. Rigane, G.; Ben Younes, S.; Ghazghazi, H.; Ben Salem, R. Investigation into the biological activities and chemical composition of *Calendula officinalis* L. growing in Tunisia. *Int. Food Res. J.* **2013**, *20*, 3001–3007.
19. Kim, G.; Kim, J.; Kim, G.; Choi, S. Anti-adipogenic effects of *Tropaeolum majus* (nasturtium) ethanol extract on 3T3-L1 cells. *Food Nutr. Res.* **2017**, *61*, 8. [CrossRef] [PubMed]
20. Fernandes, L.; Casal, S.; Pereira, J.A.; Saraiva, J.A.; Ramalhosa, E. Edible flowers: A review of the nutritional, antioxidant, antimicrobial properties and effects on human health. *J. Food Compos. Anal.* **2017**, *60*, 38–50. [CrossRef]
21. Jurca, T.; Baldea, I.; Filip, G.A.; Olteanu, D.; Clichici, S.; Pallag, A.; Vicas, L.G.; Marian, E.; Micle, O.; Muresan, M. The effect of *Tropaeolum majus* L. on bacterial infections and in vitro efficacy on apoptosis and DNA lesions in hyperosmotic stress. *J. Physiol. Pharmacol. Off. J. Pol. Physiol. Soc.* **2018**, *69*, 391–401.
22. Pires, T.C.; Dias, M.I.; Barros, L.; Calhelha, R.C.; Alves, M.J.; Oliveira, B.; Santos-Buelga, C.; Ferreira, I.C. Edible flowers as sources of phenolic compounds with bioactive potential. *Food Res. Int.* **2018**, *105*, 580–588. [CrossRef] [PubMed]
23. Verma, P.K.; Raina, R.; Agarwal, S.; Kour, H. Phytochemical ingredients and pharmacological potential of *Calendula officinalis* Linn. *Pharm. Biomed. Res.* **2018**, *4*, 17. [CrossRef]
24. Choi, E.-M.; Hwang, J.-K. Investigations of anti-inflammatory and antinociceptive activities of *Piper cubeba*, *Physalis angulata* and *Rosa hybrida*. *J. Ethnopharmacol.* **2003**, *89*, 171–175. [CrossRef]
25. Kumari, P.; Ujala; Bhargava, B. Phytochemicals from edible flowers: Opening a new arena for healthy lifestyle. *J. Funct. Foods* **2021**, *78*, 18. [CrossRef]

26. Arru, L.; Mussi, F.; Forti, L.; Buschini, A. Biological Effect of Different Spinach Extracts in Comparison with the Individual Components of the Phytocomplex. *Foods* **2021**, *10*, 382. [CrossRef]
27. Chen, G.-L.; Chen, S.-G.; Xie, Y.-Q.; Chen, F.; Zhao, Y.-Y.; Luo, C.-X.; Gao, Y.-Q. Total phenolic, flavonoid and antioxidant activity of 23 edible flowers subjected to in vitro digestion. *J. Funct. Foods* **2015**, *17*, 243–259. [CrossRef]
28. Friedman, H.; Rot, I.; Agami, O.; Vinokur, Y.; Rodov, V.; Reznick, N.; Umiel, N.; Dori, I.; Ganot, L.; Shmuel, D.; et al. Edible Flowers: New Crops with Potential Health Benefits. *Acta Hortic.* **2007**, *755*, 283–290. [CrossRef]
29. Garzón, G.; Wrolstad, R. Major anthocyanins and antioxidant activity of Nasturtium flowers (*Tropaeolum majus*). *Food Chem.* **2009**, *114*, 44–49. [CrossRef]
30. Barros, R.G.C.; Andrade, J.K.S.; Pereira, U.C.; de Oliveira, C.S.; Rafaella Ribeiro Santos Rezende, Y.; Oliveira Matos Silva, T.; Pedreira Nogueira, J.; Carvalho Gualberto, N.; Caroline Santos Araujo, H.; Narain, N. Phytochemicals screening, antioxidant capacity and chemometric characterization of four edible flowers from Brazil. *Food Res. Int.* **2020**, *130*, 10. [CrossRef]
31. Ng, T.B.; He, J.S.; Niu, S.M.; Pi, Z.F.; Shao, W.; Liu, F.; Zhao, L. A gallic acid derivative and polysaccharides with antioxidative activity from rose (*Rosa rugosa*) flowers. *J. Pharm. Pharmacol.* **2004**, *56*, 537–545. [CrossRef]
32. Friedman, H.; Agami, O.; Vinokur, Y.; Droby, S.; Cohen, L.; Refaeli, G.; Resnick, N.; Umiel, N. Characterization of yield, sensitivity to *Botrytis cinerea* and antioxidant content of several rose species suitable for edible flowers. *Sci. Hortic.* **2010**, *123*, 395–401. [CrossRef]
33. Cho, E.J.; Yokozawa, T.; Rhyu, D.Y.; Kim, H.Y.; Shibahara, N.; Park, J.C. The Inhibitory Effects of 12 Medicinal Plants and Their Component Compounds on Lipid Peroxidation. *Am. J. Chin. Med.* **2003**, *31*, 907–917. [CrossRef]
34. Chen, H.-Y.; Bor, J.-Y.; Huang, W.-H.; Yen, G.-C. Effect of sulfite-treated daylily (*Hemerocallis fulva* L.) flower on the production of nitric oxide and DNA damage in macrophages. *J. Food Drug Anal.* **2007**, *15*, 63–70.
35. Que, F.; Mao, L.; Zheng, X. In vitro and vivo antioxidant activities of daylily flowers and the involvement of phenolic compounds. *Asia Pac. J. Clin. Nutr.* **2007**, *16*, 196–203.
36. UKZUZ. *UKZUZ Data from Central Institute for Supervising and Testing in Agriculture*; UKZUZ: Brno, Czech Republic, 2019.
37. Hertle, B.; Nickig, M.; Kiermeier, P. *Gartenblumen*, 1st ed.; Gräfe and Unzer: München, Germany, 1993; pp. 30–200.
38. Kim, D.-O.; Jeong, S.W.; Lee, C.Y. Antioxidant capacity of phenolic phytochemicals from various cultivars of plums. *Food Chem.* **2003**, *81*, 321–326. [CrossRef]
39. Barros, L.; Baptista, P.; Ferreira, I.C.F. R Effect of lactarius piperatus fruiting body maturity stage on antioxidant activity measured by several biochemical assays. *Food Chem. Toxicol.* **2007**, *45*, 1731–1737. [CrossRef] [PubMed]
40. Brand-Williams, W.; Cuvelier, M.; Berset, C. Use of a free radical method to evaluate antioxidant activity. *LWT* **1995**, *28*, 25–30. [CrossRef]
41. Thaipong, K.; Boonprakob, U.; Crosby, K.; Cisneros-Zevallos, L.; Byrne, D.H. Comparison of ABTS, DPPH, FRAP, and ORAC assays for estimating antioxidant activity from guava fruit extracts. *J. Food Compos. Anal.* **2006**, *19*, 669–675. [CrossRef]
42. Rupasinghe, H.V.; Jayasankar, S.; Lay, W. Variation in total phenolics and antioxidant capacity among European plum genotypes. *Sci. Hortic.* **2006**, *108*, 243–246. [CrossRef]
43. Benzie, I.; Strain, J. The Ferric Reducing Ability of Plasma (FRAP) as a Measure of "Antioxidant Power": The FRAP Assay. *Anal. Biochem.* **1996**, *239*, 70–76. [CrossRef] [PubMed]
44. Singleton, V.L.; Orthofer, R.; Lamuela-Raventós, R.M. Analysis of total phenols and other oxidation substrates and antioxidants by means of folin-ciocalteu reagent. *Meth. Enzymol.* **1999**, *299*, 152–178.
45. Higson, S.P.J. *Analytical Chemistry*, 1st ed.; Oxford University Press: Oxford, UK, 2004; pp. 30–420.
46. Novotny, F. *Methodologies of Chemical Analyzes for Quality Evaluation of Species [In Czech: Metodiky Chemických Rozborů pro Hodnocení Kvality Odrůd]*, 1st ed.; UKZUZ: Brno, Czech Republic, 2000; pp. 15–120.
47. Publication Office of European Union Regulation (EU) No 1168/2011 of the European Parliament and of the Council of 25 October 2011 Amending Council Regulation (EC). No 2007/2004 Establishing a European Agency for the Management of Operational Cooperation at the External Borders of the Member States of the European Union. *Off. J. Eur. Union* **2011**, *54*, 18–63. Available online: https://eur-lex.europa.eu/legal-content/EN/TXT/PDF/?uri=OJ:L:2011:304:FULL&from=EN (accessed on 10 June 2021). [CrossRef]
48. Muchová, T. Use of the Edible Flowers, Medicinal and Spice Plants in Gastronomy [Využití Jedlých Květů, Léčivých a Kořenových Rostlin pro Potřeby Gastronomie]. Master's Thesis, Mendel University in Brno, Lednice, Czech Republic, 2017. (In Czech).
49. Grzeszczuk, M.; Stefaniak, A.; Pachlowska, A. Biological value of various edible flower species. *Acta Sci. Pol. Hortorum Cultus* **2016**, *15*, 109–119.
50. Anisha, R. Variations in Lycopene Content and Antioxidant Activity of Indian Marigold. Ph.D. Thesis, Central University of Jharkhand, Jharkhand, India, 2018.
51. Ashraf, A.; Riaz, M.; Nasrullah, M.; Hanif, M.; Ali, S.; Javaid, B.; Ali, S. Phytochemical, antioxidant and cytotoxicity studies of *Calendula officinalis* L. (Pot Marigold) leaves extracts. *Oxid. Commun* **2016**, *40*, 120–130.
52. Ferreira, C.S.; Pereyra, A.; Patriarca, A.; Mazzobre, M.F.; Abram, V.; Buera, M.P.; Ulrih, N.P. Phenolic Compounds in Extracts from *Eucalyptus globulus* Leaves and *Calendula officinalis* Flowers. *J. Nat. Prod. Resour.* **2016**, *2*, 53–57.
53. Navarro-González, I.; González-Barrio, R.; García-Valverde, V.; Bautista-Ortín, A.; Periago, M. Nutritional composition and antioxidant capacity in edible flowers: Characterization of phenolic compounds by HPLC-DAD-ESI/MSn. *Int. J. Mol. Sci.* **2015**, *16*, 805–822. [CrossRef] [PubMed]

54. Moyer, R.A.; Hummer, K.E.; Finn, C.E.; Frei, B.; Wrolstad, R.E. Anthocyanins, Phenolics, and Antioxidant Capacity in Diverse Small Fruits: Vaccinium, Rubus, and Ribes. *J. Agric. Food Chem.* **2002**, *50*, 519–525. [CrossRef] [PubMed]
55. You, Q.; Wang, B.; Chen, F.; Huang, Z.; Wang, X.; Luo, P.G. Comparison of anthocyanins and phenolics in organically and conventionally grown blueberries in selected cultivars. *Food Chem.* **2011**, *125*, 201–208. [CrossRef]
56. Lin, J.-Y.; Tang, C.-Y. Determination of total phenolic and flavonoid contents in selected fruits and vegetables, as well as their stimulatory effects on mouse splenocyte proliferation. *Food Chem.* **2007**, *101*, 140–147. [CrossRef]
57. Kołton, A.; Wojciechowska, R.; Długosz-Grochowska, O.; Grzesiak, W. The Storage Ability of Lamb'S Lettuce Cultivated in The Greenhouse under Led or Hps Lamps. *J. Hortic. Res.* **2014**, *22*, 159–165. [CrossRef]
58. Arasaretnam, S.; Kiruthika, A.; Mahendran, T. Nutritional and mineral composition of selected green leafy vegetables. *Ceylon J. Sci.* **2018**, *47*, 35. [CrossRef]
59. Kaisoon, O.; Siriamornpun, S.; Weerapreeyakul, N.; Meeso, N. Phenolic compounds and antioxidant activities of edible flowers from Thailand. *J. Funct. Foods* **2011**, *3*, 88–99. [CrossRef]
60. Ilahy, R.; Hdider, C.; Lenucci, M.S.; Tlili, I.; Dalessandro, G. Antioxidant activity and bioactive compound changes during fruit ripening of high-lycopene tomato cultivars. *J. Food Compos. Anal.* **2011**, *24*, 588–595. [CrossRef]
61. Park, H.; Kim, Y.-J.; Shin, Y. Estimation of daily intake of lycopene, antioxidant contents and activities from tomatoes, watermelons, and their processed products in Korea. *Appl. Biol. Chem.* **2020**, *63*, 11. [CrossRef]
62. Mirzaei, A.; Mirzaei, N.; Salehpour, Z.; Khosravani, S.A.; Amouei, M. Phenolic, ascorbic Contents and Antioxidant activities of 21 Iranian Fruits. *Life Sci. J.* **2013**, *10*, 1240–1245.
63. Zheng, J.; Yu, X.; Maninder, M.; Xu, B. Total phenolics and antioxidants profiles of commonly consumed edible flowers in China. *Int. J. Food Prop.* **2018**, *21*, 1524–1540. [CrossRef]
64. Sadeer, N.B.; Montesano, D.; Albrizio, S.; Zengin, G.; Mahomoodally, M.F. The Versatility of Antioxidant Assays in Food Science and Safety—Chemistry, Applications, Strengths, and Limitations. *Antioxidants* **2020**, *9*, 709. [CrossRef]
65. Kedare, S.B.; Singh, R.P. Genesis and development of DPPH method of antioxidant assay. *J. Food Sci. Technol.* **2011**, *48*, 412–422. [CrossRef]
66. Ou, B.; Huang, D.; Hampsch-Woodill, M.; Flanagan, J.A.; Deemer, E.K. Analysis of Antioxidant Activities of Common Vegetables Employing Oxygen Radical Absorbance Capacity (ORAC) and Ferric Reducing Antioxidant Power (FRAP) Assays: A Comparative Study. *J. Agric. Food Chem.* **2002**, *50*, 3122–3128. [CrossRef]
67. Mao, L.-C.; Pan, X.; Que, F.; Fang, X.-H. Antioxidant properties of water and ethanol extracts from hot air-dried and freeze-dried daylily flowers. *Eur. Food Res. Technol.* **2006**, *222*, 241. [CrossRef]
68. Koike, A.; Barreira, J.C.M.; Barros, L.; Santos-Buelga, C.; Villavicencio, A.L.C.H.; Ferreira, I.C.F.R. Irradiation as a novel approach to improve quality of *Tropaeolum majus* L. flowers. *Innov. Food Sci. Emerg. Technol.* **2015**, *30*, 138–144. [CrossRef]
69. Pavithra, G.M.; Saba, S.; Abhishiktha, S.N.; Prashith Kekuda, T.R.; Vinayaka, K.S. Antioxidant and antimicrobial activity of flowers of *Wendlandia thyrsoidea, Olea dioica, Lagerstroemia speciosa* and *Bombax malabaricum*. *J. Appl. Pharm. Sci.* **2013**, *3*, 114–120.
70. Clevely, A.; Richmondová, K. *The Complete Book of Herbs [Velká Kniha Bylinek]*, 1st ed.; Svojtka & Co., s. r. o.: Prague, Czech Republic, 2007; p. 28. (In Czech)
71. Demasi, S.; Caser, M.; Donno, D.; Enri, S.R.; Lonati, M.; Scariot, V. Exploring wild edible flowers as a source of bioactive compounds: New perspectives in horticulture. *Folia Hortic.* **2021**, *33*, 1–22.
72. EU Novel Food Catalogue (v.1.2). Available online: https://ec.europa.eu/food/safety/novel_food/catalogue/search/public/ (accessed on 10 August 2021).
73. Council of the European Union. Regulation (EC) No 258/97 of the European Parliament and of the Council of 27 January 1997 Concerning Novel Foods and Novel Food Ingredients. *Off. J. Eur. Union* **1997**, *L43*, 1–6. Available online: https://eur-lex.europa.eu/legal-content/EN/TXT/PDF/?uri=CELEX:31997R0258&from=EN (accessed on 10 August 2021).
74. Kirker, C.L.; Newman, M. *Edible Flowers: A Global History*, 1st ed.; Reaktion Books: London, UK, 2016; p. 144.
75. Lucarini, M.; Copetta, A.; Durazzo, A.; Gabrielli, P.; Lombardi-Boccia, G.; Lupotto, E.; Santini, A.; Ruffoni, B. A Snapshot on Food Allergies: A Case Study on Edible Flowers. *Sustainability* **2020**, *12*, 8709. [CrossRef]
76. Egebjerg, M.M.; Olesen, P.T.; Eriksen, F.D.; Ravn-Haren, G.; Bredsdorff, L.; Pilegaard, K. Are wild and cultivated flowers served in restaurants or sold by local producers in Denmark safe for the consumer? *Food Chem. Toxicol.* **2018**, *120*, 129–142. [CrossRef]
77. Guiné, R.P.F.; Santos, E.; Correia, P.M.R. Edible Flowers: Knowledge and Consumption Habits. *Int. J. Nutr. Health Sci.* **2017**, *1*, 18–22.
78. Mayer, A. Historical changes in the mineral content of fruits and vegetables. *Br. Food J.* **1997**, *99*, 207–211. [CrossRef]
79. Kovacikova, E.; Vojtassakova, A.; Holcikova, K.; Simonova, E. *Fruit and Vegetables: Food Tables [Ovocie a Zelenina: Potravinové Tabuľky]*, 1st ed.; VUP: Bratislava, Slovak, 1997; p. 208. (In Slovak)
80. Kopec, K. *Vegetable in Human Nutrition [Zelenina ve Výživě Člověka]*, 1st ed.; Grada: Praha, Czech Republic, 2010; p. 168. (In Czech)
81. Pavithra, G.M.; Rakesh, K.N.; Dileep, N.; Junaid, S.; Kumar, R.K.A.; Kekuda, P.T.R. Elemental analysis, antimicrobial and radical scavenging activity of *Lagerstroemia speciosa* (L.) flower. *J. Chem. Pharm. Res.* **2013**, *5*, 215–222.
82. Cui, X.-S.; Guo, Y.-H. [Analysis of mineral elements in different organs of *Chrysanthemum indicum* L. based on ICP-AES]. *Guang Pu Xue Yu Guang Pu Fen Xi* **2012**, *32*, 2828–2830.
83. Grzeszczuk, M.; Meller, E.; Stefaniak, A.; Wysocka, G. Mineral composition of some edible flowers. *J. Elementol.* **2018**, *23*, 151–162. [CrossRef]

84. Kader, A.A. Flavor quality of fruits and vegetables. *J. Sci. Food Agric.* **2008**, *88*, 1863–1868. [CrossRef]
85. Sanchez-Castillo, C.; Dewey, P.J.S.; Aguirre, A.; Lara, J.J.; Vaca, R.; de la Barra, P.L.; Ortiz, M.; Escamilla, I.; James, W.P. The Mineral Content of Mexican Fruits and Vegetables. *J. Food Compos. Anal.* **1998**, *11*, 340–356. [CrossRef]
86. Smanalieva, J.; Iskakova, J.; Oskonbaeva, Z.; Wichern, F.; Darr, D. Investigation of nutritional characteristics and free radical scavenging activity of wild apple, pear, rosehip, and barberry from the walnut-fruit forests of Kyrgyzstan. *Eur. Food Res. Technol.* **2020**, *246*, 1095–1104. [CrossRef]
87. Kalač, P. Chemical composition and nutritional value of European species of wild growing mushrooms. *Food Chem.* **2009**, *113*, 9–16. [CrossRef]
88. Mattila, P.; Könkö, K.; Eurola, M.; Pihlava, J.-M.; Astola, J.; Vahteristo, L.; Hietaniemi, V.; Kumpulainen, J.; Valtonen, M.; Piironen, V. Contents of Vitamins, Mineral Elements, and Some Phenolic Compounds in Cultivated Mushrooms. *J. Agric. Food Chem.* **2001**, *49*, 2343–2348. [CrossRef]
89. Hanć, A.; Komorowicz, I.; Iskra, M.; Majewski, W.; Barałkiewicz, D. Application of spectroscopic techniques: ICP-OES, LA-ICP-MS and chemometric methods for studying the relationships between trace elements in clinical samples from patients with atherosclerosis obliterans. *Anal. Bioanal. Chem.* **2011**, *399*, 3221–3231. [CrossRef]
90. Hicsonmez, U.; Ozdemir, C.; Cam, S.; Ozdemir, A.; Erees, F.S. Major-minor element analysis in some plant seeds consumed as feed in Turkey. *Nat. Sci.* **2012**, *4*, 298–303. [CrossRef]
91. Institute of Medicine; Otten, J.J.; Hellwig, J.P.; Meyers, L.D. *Dietary Reference Intakes: The Essential Guide to Nutrient Requirements*, 1st ed.; National Academies Press: Washington, DC, USA, 2006; p. 1344.
92. Campbell, N.A.; Reece, J.B. *Biology [Biologie]*, 1st ed.; Computer Press: Brno, Czech Republic, 2006; p. 1332. (In Czech)
93. Gombart, A.F.; Pierre, A.; Maggini, S. A Review of Micronutrients and the Immune System–Working in Harmony to Reduce the Risk of Infection. *Nutrients* **2020**, *12*, 236. [CrossRef] [PubMed]
94. Velisek, J. *Food Chemistry [Chemie Potravin]*, 1st ed.; OSSIS: Tabor, Czech Republic, 2002; pp. 252–324. (In Czech)
95. Jarosz, M.; Olbert, M.; Wyszogrodzka, G.; Mlyniec, K.; Librowski, T. Antioxidant and anti-inflammatory effects of zinc. Zinc-dependent NF-κB signaling. *Inflammopharmacology* **2017**, *25*, 11–24. [CrossRef]
96. Schreiner, M.; Krumbein, A.; Mewis, I.; Ulrichs, C.; Huyskens-Keil, S. Short-term and moderate UV-B radiation effects on secondary plant metabolism in different organs of nasturtium (*Tropaeolum majus* L.). *Innov. Food Sci. Emerg. Technol.* **2009**, *10*, 93–96. [CrossRef]
97. Shafaghat, A.; Larijani, K.; Salimi, F. Composition and Antibacterial Activity of the Essential Oil of *Chrysanthemum parthenium* Flower from Iran. *J. Essent. Oil Bear. Plants* **2009**, *12*, 708–713. [CrossRef]
98. Mahmood, N.; Piacente, S.; Pizza, C.; Burke, A.; Khan, A.I.; Hay, A.J. The Anti-HIV Activity and Mechanisms of Action of Pure Compounds Isolated from *Rosa damascena*. *Biochem. Biophys. Res. Commun.* **1996**, *229*, 73–79. [CrossRef] [PubMed]
99. dos Santos, A.M.P.; Silva, E.F.R.; dos Santos, W.N.L.; da Silva, E.G.P.; dos Santos, L.O.; da Santos, B.R.; da Sauthier, M.C.; dos Santos, W.P.C. Evaluation of minerals, toxic elements and bioactive compounds in rose petals (*Rosa* spp.) using chemometric tools and artificial neural networks. *Microchem. J.* **2018**, *138*, 98–108. [CrossRef]
100. Drava, G.; Iobbi, V.; Govaerts, R.; Minganti, V.; Copetta, A.; Ruffoni, B.; Bisio, A. Trace Elements in Edible Flowers from Italy: Further Insights into Health Benefits and Risks to Consumers. *Molecules* **2020**, *25*, 2891. [CrossRef]
101. Moncada, A.; Miceli, A.; Sabatino, L.; Iapichino, G.; D'Anna, F.; Vetrano, F. Effect of Molybdenum Rate on Yield and Quality of Lettuce, Escarole, and Curly Endive Grown in a Floating System. *Agronomy* **2018**, *8*, 171. [CrossRef]
102. European Food Safety Authority. *EFSA Tolerable Upper Intake Levels for Vitamins and Minerals*; European Food Safety Authority: Parma, Italy, 2006; p. 480.
103. Sularz, O.; Smoleń, S.; Koronowicz, A.; Kowalska, I.; Leszczyńska, T. Chemical Composition of Lettuce (*Lactuca sativa* L.) Biofortified with Iodine by KIO3, 5-Iodo-, and 3.5-Diiodosalicylic Acid in a Hydroponic Cultivation. *Agronomy* **2020**, *10*, 1022. [CrossRef]
104. Montañés Millán, L.; Val, J.; Betrán, J.; Monge, E.; Moreno, M.A.; Montañés, L. Floral analysis: Fresh and dry weight of flowers from different fruit tree species. *Acta Hortic.* **1997**, *448*, 233–240. [CrossRef]
105. Villa-Ruano, N.; Pacheco-Hernández, Y.; Cruz-Durán, R.; Lozoya-Gloria, E.; Betancourt-Jiménez, M.G. Seasonal variation in phytochemicals and nutraceutical potential of *Begonia nelumbiifolia* consumed in Puebla, México. *J. Food Sci. Technol.* **2017**, *54*, 1484–1490. [CrossRef] [PubMed]
106. De Lima Franzen, F.; Rodrigues de Oliveira, M.S.; Lidório, H.F.; Farias Menegaes, J.; Martins Fries, L.L. Chemical composition of rose, calendula flower petals for human food use. *Cienc. Tecnol. Agropecu.* **2019**, *20*, 159–168.
107. Lara-Cortés, E.; Osorio-Díaz, P.; Jiménez-Aparicio, A.; Bautista-Baños, S. Nutritional content, functional properties and conservation of edible flowers. Review [Contenido Nutricional, Propiedades Funcionales Y Conservación De Flores Comestibles. Revisión]. *Arch. Latinoam. Nutr.* **2013**, *63*, 197–208. (In Spanish) [PubMed]
108. Marino, C.T.; Hector, B.; Rodrigues, P.H.M.; Borgatti, L.M.O.; Meyer, P.M.; Alves da Silva, E.J.; Ørskov, E.R. Characterization of vegetables and fruits potential as ruminant feed by in vitro gas production technique. *Livestock Res. Rural. Dev.* **2010**, *22*, 168.
109. Punchay, K.; Inta, A.; Tiansawat, P.; Balslev, H.; Wangpakapattanawong, P. Nutrient and Mineral Compositions of Wild Leafy Vegetables of the Karen and Lawa Communities in Thailand. *Foods* **2020**, *9*, 1748. [CrossRef]

110. Petek, M.; Herak Ćustić, M.; Toth, N.; Slunjski, S.; Čoga, L.; Pavlović, I.; Karažija, T.; Lazarević, B.; Cvetković, S. Nitrogen and Crude Proteins in Beetroot (*Beta vulgaris* var. conditiva) under Different Fertilization Treatments. *Not. Bot. Horti Agrobot. Cluj-Napoca* **2012**, *40*, 215–219. [CrossRef]
111. Cai, Y.; Luo, Q.; Sun, M.; Corke, H. Antioxidant activity and phenolic compounds of 112 traditional Chinese medicinal plants associated with anticancer. *Life Sci.* **2004**, *74*, 2157–2184. [CrossRef] [PubMed]
112. Rigane, G.; Ben Salem, R.; Sayadi, S.; Bouaziz, M. Phenolic Composition, Isolation, and Structure of a New Deoxyloganic Acid Derivative from Dhokar and Gemri-Dhokar Olive Cultivars. *J. Food Sci.* **2011**, *76*, 965–973. [CrossRef]
113. Ulewicz-Magulska, B.; Wesolowski, M. Total Phenolic Contents and Antioxidant Potential of Herbs Used for Medical and Culinary Purposes. *Plant. Foods Hum. Nutr.* **2019**, *74*, 61–67. [CrossRef]
114. Yin, D.-D.; Yuan, R.-Y.; Wu, Q.; Li, S.-S.; Shao, S.; Xu, Y.-J.; Hao, X.-H.; Wang, L.-S. Assessment of flavonoids and volatile compounds in tea infusions of water lily flowers and their antioxidant activities. *Food Chem.* **2015**, *187*, 20–28. [CrossRef]
115. Kaisoon, O.; Konczak, I.; Siriamornpun, S. Potential health enhancing properties of edible flowers from Thailand. *Food Res. Int.* **2012**, *46*, 563–571. [CrossRef]
116. Grela, E.R.; Samolińska, W.; Kiczorowska, B.; Klebaniuk, R.; Kiczorowski, P. Content of Minerals and Fatty Acids and Their Correlation with Phytochemical Compounds and Antioxidant Activity of Leguminous Seeds. *Biol. Trace Element Res.* **2017**, *180*, 338–348. [CrossRef]
117. Aghdam, M.S.; Dokhanieh, A.Y.; Hassanpour, H.; Fard, J.R. Enhancement of antioxidant capacity of cornelian cherry (*Cornus mas*) fruit by postharvest calcium treatment. *Sci. Hortic.* **2013**, *161*, 160–164. [CrossRef]
118. Chaturvedi, N.; Ahmed, J.; Dhal, N.K. Effects of iron ore tailings on growth and physiological activities of *Tagetes patula* L. *J. Soils Sediments* **2014**, *14*, 721–730. [CrossRef]
119. Hou, W.; Pan, Y.; Zhang, Y. Changes in quality, antioxidant compounds and DPPH radical-scavenging activity of *Rosa hybrida* flowers during development. *N. Z. J. Crop. Hortic. Sci.* **2014**, *42*, 31–37. [CrossRef]
120. Gonzalez-Barrio, R.; Periago, M.J.; Luna-Recio, C.; Garcia-Alonso, F.J.; Navarro-González, I. Chemical composition of the edible flowers, pansy (*Viola wittrockiana*) and snapdragon (*Antirrhinum majus*) as new sources of bioactive compounds. *Food Chem.* **2018**, *252*, 373–380. [CrossRef]
121. Tai, Z.; Cai, L.; Dai, L.; Dong, L.; Wang, M.; Yang, Y.; Cao, Q.; Ding, Z. Antioxidant activity and chemical constituents of edible flower of *Sophora viciifolia*. *Food Chem.* **2011**, *126*, 1648–1654. [CrossRef]
122. Kelley, K.M.; Behe, B.K.; Biernbaum, J.A.; Poff, K.L. Combinations of Colors and Species of Containerized Edible Flowers: Effect on Consumer Preferences. *HortScience* **2002**, *37*, 218–221. [CrossRef]
123. Takahashi, J.A.; Rezende, F.A.G.G.; Moura, M.A.F.; Dominguete, L.C.B.; Sande, D. Edible flowers: Bioactive profile and its potential to be used in food development. *Food Res. Int.* **2020**, *129*, 108868. [CrossRef]
124. Tunick, M.H.; Onwulata, C.I.; Thomas, A.E.; Phillips, J.G.; Mukhopadhyay, S.; Sheen, S.; Liu, C.-K.; Latona, N.; Pimentel, M.R.; Cooke, P.H. Critical Evaluation of Crispy and Crunchy Textures: A Review. *Int. J. Food Prop.* **2013**, *16*, 949–963. [CrossRef]
125. Burdock, G.A.; Fenaroli, G. *Fenaroli's Handbook of Flavor Ingredients*, 6th ed.; CRC Press Taylor & Francis Group: Boca Raton, FL, USA, 2010; p. 1861.
126. Barbano, P. Be Certain That You Are Eating True Edible Daylily Plants. Available online: https://www.capegazette.com/article/be-certain-you-are-eating-true-edible-daylily-plants/111652 (accessed on 11 March 2021).
127. Pollard, A.N.; Coggins, P.C.; Knight, P.R. Sensory Evaluation of Edible Daylilies (*Hemerocallis* sp.). *HortScience* **2004**, *39*, 783. [CrossRef]
128. Grosvenor, G. *Daylilies for the Garden*; Timber Press: Portland, OR, USA, 1999; p. 176.

foods

MDPI

Article

Sugar-Free Milk Chocolate as a Carrier of Omega-3 Polyunsaturated Fatty Acids and Probiotics: A Potential Functional Food for the Diabetic Population

Andrea R. Gómez-Fernández [1], Paulinna Faccinetto-Beltrán [1], Norma E. Orozco-Sánchez [2], Esther Pérez-Carrillo [3], Luis Martín Marín-Obispo [3], Carmen Hernández-Brenes [3], Arlette Santacruz [3] and Daniel A. Jacobo-Velázquez [1,*]

1 Tecnologico de Monterrey, Escuela de Ingeniería y Ciencias, Av. General Ramón Corona 2514, Zapopan 45201, Mexico; andrea.rebeca.gf@gmail.com (A.R.G.-F.); paulinnafb1995@gmail.com (P.F.-B.)

2 Escuela Mexicana de Confitería y Chocolatería, Melchor Ocampo 926, San Luis Potosi 78280, Mexico; ccinvestigacionc@gmail.com

3 Tecnologico de Monterrey, Escuela de Ingeniería y Ciencias, Av. Eugenio Garza Sada 2501, Monterrey 64849, Mexico; perez.carrillo@tec.mx (E.P.-C.); lmmarin@live.com.mx (L.M.M.-O.); chbrenes@tec.mx (C.H.-B.); asantacruz@tec.mx (A.S.)

* Correspondence: djacobov@tec.mx; Tel.: +52-312-119-1650

Abstract: Chocolate is an adequate matrix to deliver bioactive ingredients. However, it contains high sugar levels, one of the leading causes of chronic degenerative diseases. This work aimed to evaluate the effects of milk chocolate reformulation with alternative sugar sweeteners (Sw; isomalt + stevia), probiotics (Prob), and ω-3 polyunsaturated fatty acids (PUFAs) on its physicochemical properties and consumers' acceptability. *Lactobacillus plantarum* 299v (L. p299v) and *Lactobacillus acidophilus* La3 (DSMZ 17742) were added as Prob strains, and fish oil (FO) was added as the source of ω-3 PUFAs. Prob addition resulted in chocolates with >2 $\times$ 10^7 colony forming unit (CFU) per serving size (12 g). Except for Prob, a_w values of all treatments were <0.46. Sw and Sw + Prob presented the nearest values to the control in hardness, whereas Sw without FO increased fracturability. FO, Sw + FO, and Sw + Prob + FO contained 107.4 $\pm$ 12.84, 142.9 $\pm$ 17.9, and 133.78 $\pm$ 8.76 mg of ω-3 PUFAs per chocolate, respectively. Prob + FO increased the resistance of chocolate to shear stress, while Sw + FO showed a similar flow behavior to the control. The consumers' acceptability of Sw + Prob chocolate was adequate, while Sw + Prob + FO had higher acceptability than Prob + FO. Health benefits of reformulated milk chocolates requires further assessment by in vitro, in vivo and clinical studies.

Keywords: sugar-free; sweeteners; isomalt; stevia; milk chocolate; functional foods

Citation: Gómez-Fernández, A.R.; Faccinetto-Beltrán, P.; Orozco-Sánchez, N.E.; Pérez-Carrillo, E.; Marín-Obispo, L.M.; Hernández-Brenes, C.; Santacruz, A.; Jacobo-Velázquez, D.A. Sugar-Free Milk Chocolate as a Carrier of Omega-3 Polyunsaturated Fatty Acids and Probiotics: A Potential Functional Food for the Diabetic Population. *Foods* **2021**, *10*, 1866. https://doi.org/10.3390/foods10081866

Academic Editor: Antonello Santini

Received: 4 July 2021
Accepted: 10 August 2021
Published: 12 August 2021

1. Introduction

Consumption of foods high in sugar is associated with the development of metabolic syndrome, which is defined as a collection of physiological, biochemical, and clinical factors, and is one of the leading causes of death worldwide [1]. Therefore, there is a need to develop new sugar-free products. Sweeteners are sugar substitutes, with natural sweeteners being more accepted in the market [2]. Additionally, there is an interest in the addition of bioactive ingredients to food formulations, in order to obtain food products that provide health benefits, including the prevention and treatment of diseases related to metabolic syndrome [3]. The term nutraceutical was coined in 1989 by Stephen DeFelice from the words "nutrition" and "pharmaceutical", and he defined it as a food or part of a food that provides health benefits, including the prevention and treatment of disease beyond basic nutritional functions [4]. Recently, the term nutraceutical was revisited to separate the concept of food supplements and nutraceuticals [5]. Food supplements are food-derived products that compensate the lack of specific components (i.e., vitamins and minerals) in the daily diet and/or can exert a beneficial effect on health without any proven biological effect. On the other hand, nutraceuticals should have a proven beneficial pharmacological effect as a requirement [5]. In practical terms, as stated by Santini and Novellino, nutraceuticals should go beyond the diet, before the drug [5].

One of the food products experiencing more dynamic changes through this healthy demand is chocolate, since it represents 60% of the world's confectionery market and is liked by adults and children due to its sweet taste and pleasant mouthfeel [3,6]. Sugar-free chocolates usually use a combination of sweeteners with high sweet power, such as stevia (Stev), and sweeteners as bulking agents, such as isomalt (Iso) [7]. Both sweeteners (Stev and Iso) are considered prebiotics [8,9]. Prebiotics are defined as non-digestible food ingredients that are metabolized by gut microbiota, improving host health [10]. Additionally, Stev is reported to exert beneficial effects on type 2 diabetes since this molecule interacts with intestinal and pancreatic cells, improving glucose uptake and helping to maintain glucose homeostasis [2,11].

The consumption of fish oil (FO) has been related to decreasing the risk of type 2 diabetes and other coronary diseases due to its high content of ω-3 PUFAs, such as eicosapentaenoic acid (EPA) and docosahexaenoic acid (DHA) [12,13]. The World Health Organization [14] recommend a consumption of 250–500 mg per day of combined EPA and DHA for healthy adults. Furthermore, several studies suggest that dietary ω-3 PUFAs from FO improve insulin sensitivity or reduce the incidence of type 2 diabetes thorough inhibition of adipose tissue inflammation [15].

Another bioactive ingredient that can be used to improve the health of the diabetic and non-diabetic populations are probiotics, since their consumption modulates gut microbiota [16]. Probiotics are defined as live microorganisms that confer health-promoting properties when administrated in adequate amounts to the host [17]. In this context, *Lactobacillus platarum* and *Lactobacillus acidophilus* have demonstrated to improve the health of type 2 diabetes patients by balancing the gut microbiota [18,19].

Chocolate could be an adequate vehicle for the delivery of probiotics and ω-3 PUFAs due to its main ingredients (cocoa butter, cocoa paste, soy lecithin, and milk) that generate a food matrix with low water activity, low oxygen tension, and low moisture permeability [20]. In addition, microencapsulation of probiotics provides double protection due to the covalently or ionically crosslinked polymer networks that enclose bacterial cells [21]. However, there are few reports in the literature on the development of functional sugar-free chocolates that could be consumed by the diabetic population.

The milk chocolate system comprises solid particles (cocoa, sugar, and milk powder) dispersed in the fat phase (cocoa butter). The composition of these ingredients affects the final sensory properties and rheological behavior as a fluid mass. To obtain high-quality products, the determination of these properties in chocolate manufacture must be well-defined to obtain the right palatable products and fulfill consumers' preferences [22]. Rheological properties affect the final texture of chocolates, which plays a crucial role in the confectionery industry's elaboration process [23]. For instance, if chocolate viscosity is too low, the texture would not be optimal, and if it is too high, bubbles may appear in the molded tablet. In addition to modifying texture, viscosity also affects the flavor of chocolate because the taste depends on the order and rate of contact, which is dependent on viscosity and melt rate. Chocolate rheology is usually determined by yield stress and apparent viscosity parameters. Yield stress provides information related to the transition behavior from elastic to viscous deformation. Furthermore, sensory evaluation is also a key element to evaluate the elaboration process of chocolate and ensure high-quality products that reach consumers' preferences [24].

The objective of this study was to evaluate the effect of sugar substitution, probiotics and ω-3 PUFAs addition on the physicochemical properties and consumers' acceptability of milk chocolate. Sugar was replaced by isomalt (Iso) and stevia (Stev), whereas the probiotics (Prob) strains added were *Lactobacillus plantarum* 299v (L. p299v) and *Lactobacillus acidophilus* La3 (DSMZ 17742). Furthermore, fish oil (FO) was used as a source of ω-3 PUFAs.

2. Materials and Methods

2.1. Bacterial Strains and Chemicals

Probiotic strains *Lactobacillus plantarum* 299v (L. p299v) and *Lactobacillus acidophilus* La3 (DSMZ 17742) were obtained from the German Collection of Microorganisms and Cell Cultures (DSMZ, Braunschweig, Germany) and the American Type Culture Collection (ATCC, Manassas, VA, USA), respectively. Sodium alginate was purchased from Deiman (Guadalajara, JAL, Mexico) and food-grade maltodextrin was obtained from Best Ingredients (Monterrey, NL, Mexico). Alkalinized cocoa paste, alkalinized cocoa, cocoa butter, whey powder, soy lecithin, polyglycerol polyricinoleate (PGPR), NaCl, vanilla, and sugar were obtained from Escuela Mexicana de Confitería y Chocolatería (San Luis Potosí, SLP, Mexico). Isomalt low moisture powder fine (LMPF) was obtained from Palsgaard Industry de México S de RL de CV (San Luis Potosí, SLP, Mexico). Stevia was obtained from Grupo Químico Amillán S.A. de C.V. (Zapopan, JAL, Mexico). Fish oil (Omega Pure®) was purchased from America Alimentos S.A. de C.V. (Zapopan, JAL, Mexico). For the fatty acid methyl esters profile determination, toluene-hexane mixture (1:1 v/v), undecanoic acid (100 ppm), and external standard fatty acid mixtures GLC 566 (39 fatty acid methyl esters) were purchased from Nu Chek Prep Inc (Elysian, MN, USA). Finally, for microbiological determinations, reconstituted skim milk (Svelty, Nestlé®) was obtained from a local market, whereas Violet Red Bile Agar (VRB agar), potato Dextrose Peptone Agar (DP agar), Xylose Lysine Deoxycholate Agar (XLD agar), Salmonella Shigella Agar (SS agar), Tetrathionate Broth Base, Rappaport Vassiliadis Broth, VRBA agar, and MRS agar were obtained from Sigma-Aldrich® (St. Louis, MO, USA).

2.2. Bacterial Strains' Propagation, Microencapsulation, and Viability Assessment

Bacteria were propagated by inoculating an aliquot (100 µL) from a stock of *Lactobacillus plantarum* 299v (L. p299v) and a stock of *Lactobacillus acidophilus* La3 (DSMZ 17742) in 10 mL of MRS broth, which was incubated at 37 °C in a Shel lab 1535 incubator (VWR, Randor, PA, USA) for 16 h under aerobic conditions. Then, propagation was scaled-up to a final volume of 800 mL under the same incubation conditions. Bacteria cells were harvested by centrifugation (at 10,000× g, 25 °C for 15 min). Cell pellets were washed in peptone water (0.1% peptone, 0.85% NaCl, pH 7) and resuspended in a final volume of 30 mL in peptone water.

Suspended cells were added to 750 mL of microencapsulation mix (10% w/v maltodextrin, and w/v 2% food-grade alginate) and spray-dried (ADL 311S, Yamato Scientific Co., Ltd., Santa Clara, CA, USA) at 130 °C inlet, 60 °C outlet, and 0.13 MPa. The viability of probiotics was determined by homogenizing the powder with microencapsulated probiotics (0.1 g) or the chocolates with added probiotics (1 g), with 90 mL of peptone water preheated at 37 °C in a stomacher (IUL Instruments, Barcelona, Spain) for 90 s. Proper dilutions (10^4, 10^6, and 10^8) of each replicate were plated twice on MRS agar and incubated at 37 °C for 48 h, aerobically.

2.3. Chocolate Preparation

Milk chocolates were prepared in a confectionery pilot plant factory as previously described [25]. Chocolates were formulated to develop a sugar-free product rich in *Lactobacillus plantarum* 299v (L. p299v), *Lactobacillus acidophilus* La3 (DSMZ 17742), and ω-3 PUFAs (EPA and DHA). Eight milk chocolate formulations were tested using the same base (alkalized cocoa paste 12%, natural cocoa powder 3%, cocoa butter 26%, whole milk powder 13%, skim milk powder 10%, soy lecithin 0.3%, PGPR 0.2%, NaCl 0.08%, and vanilla 0.003% w/w). Sugar was replaced with a mixture of Iso and Stev as sweeteners (Table 1). Likewise, FO and probiotics (*L. plantarum* 299v and *L. acidophilus* La3) were added as indicated in the formulations shown in Table 1.

Table 1. Milk chocolate formulations added with probiotics and fish oil.

Ingredients	% Percentage in Each Formulation (*w/w*)							
	Control	Prob	FO	Prob + FO	Sw	Sw + Prob	Sw + FO	Sw + Prob + FO
Alkalinized cocoa paste	12.46	12.43	11.64	11.61	13.00	12.97	12.12	12.12
Natural cocoa	3.00	3.00	2.80	2.80	3.00	3.00	2.80	2.80
Cocoa butter	26.15	26.10	24.43	24.39	23.24	23.19	21.67	21.67
Whole milk powder	13.42	13.39	12.54	12.51	14.00	13.97	13.05	13.05
Skim milk powder	10.54	10.51	9.85	9.83	11.00	10.98	10.25	10.25
Soy lecithin	0.384	0.38	0.36	0.36	0.30	0.30	0.30	0.30
PGPR	0.192	0.19	0.18	0.18	0.20	0.20	0.19	0.19
NaCl	0.08	0.08	0.07	0.07	0.08	0.08	0.08	0.08
Vanilla	0.03	0.03	0.03	0.03	0.03	0.03	0.03	0.03
Sugar	33.75	33.68	31.53	31.47	-	-	-	-
Isomalt LMPF	-	-	-	-	35.12	35.05	32.95	32.74
Stevia	-	-	-	-	0.03	0.03	0.03	0.03
Probiotic	-	0.21	-	0.21	-	0.21	-	0.21
Fish oil	-	-	6.57	6.55	-	-	6.54	6.54

Abbreviations: Prob, probiotics; FO, fish oil; Sw, sweeteners; PGPR, polyglycerol polyricin; LMPF: low moisture powder fine.

Each chocolate formulation was produced by the following procedure: (1) melting and heating, (2) coaching, (3) refining, (4) tempering, and (5) molding. In the melting step, a water bath at 40 °C was used; for the coaching and refining steps, the temperature was 25 °C, and the duration was 24 h using a chocolate refiner (Premier, Diamond Custom Machines Corp., Hillsborough Township, NJ, USA). The tempering step followed three changes of temperature. The first stage of tempering was maintained at 45 °C to melt fat crystals (3–5 min); then, in the second stage, chocolate was cooled at 27 °C under manual agitation using a spatula (3–5 min), and finally, chocolate was reheated to 29 °C. Chocolate formulations were molded at 14 °C for 1 h and stored at 11 °C until analysis. FO and microencapsulated probiotics were added to chocolate after tempering at 29 °C at a ratio of 1×10^{13} UFC/g, resulting in chocolates with 2×10^7 CFU per serving size (12 g).

2.4. Water Activity, Color, Texture, and Rheological Determinations

Water activity a_w of chocolate samples was measured using a water activity meter (Aqualab CX-2, Decagon Divices Inc., Pullman, WA, USA) at 25 °C using 3.0 g of the samples previously homogenized with a grinder (80350R, Hamilton Beach, Glen Allen, VA, USA). The color was determined with a spectrophotometer cm-600d (Konica Minolta Inc., Tokyo, Japan). Colorimetric parameters obtained (CIE L^*, a^*, and b^*) were used to calculate the whiteness index (WI^*) value, as indicated in Equation (1):

$$WI^* = 100 - \left[(100 - L^*)^2 + a^{*2} + b^{*2}\right]^{1/2} \quad (1)$$

Treatments: Control = milk chocolate formulation, Prob = milk chocolate + probiotics, FO = milk chocolate + fish oil, Prob + FO = milk chocolate + probiotics + fish oil, Sw = sugar-free chocolate formulation (with added isomalt + stevia as sweeteners), Sw + Prob = sugar-free chocolate + probiotics, Sw + FO = sugar-free chocolate + fish oil, Sw + Prob + FO = sugar-free chocolate + probiotics + fish oil.

Hardness and fracturability (N) of the samples were determined using a texture analyzer (TVT 6700, Perten Instruments, Sydney, NSW, Australia) equipped with a cylinder probe (height 45 mm, diameter 3 mm). The conditions used were: sample height: 8 mm; starting distance from sample: 5 mm; compression: 2 mm; initial speed: 0.5 mm/s; test speed: 0.5 mm/s; retract speed: 10 mm/s; trigger force: 5 g; data rate: 500 pps, at 25 °C [25–27]. Five replicates of each treatment were evaluated.

Rheological experiments (flow behavior, stress sweep, and frequency sweep test) were carried out with a previously reported protocol [25,28]. A controlled stress rheometer (Physica MCR 101, Anton Paar, Ostfildern, Germany) fitted with a parallel plate geometry

(PP25/S, 24.973 mm diameter, 1.0 mm gap) was used. Chocolate samples were melted in a water bath at 35 °C and poured on the bottom plate based on the methodology previously reported [25,28].

2.5. Fatty Acid Methyl Esters (FAMEs) Profile

Chocolate fat was extracted following the AOAC 948.22 Soxhlet method, using ethyl ether as the extraction solvent [29]. For each formulation, fat extraction was performed in triplicate from the chocolate bars (12 g). A sample of extracted fat (5 mg) was dissolved in a toluene-hexane mixture (0.6 mL, 1:1 v/v). Undecanoic acid (100 ppm) was added to samples as an internal standard for quantification. Subsequently, samples were derivatized using methanol-sulfuric acid (1 mL, 93:7 v/v) in capped vials placed in a water-bath (80 °C, 60 min). Thereafter, the samples were chilled, and the FAMEs were extracted with hexane and volume-adjusted (2 mL) for chromatographic analysis.

FAMEs profile was determined on a GC Agilent 6850A gas chromatograph coupled with a flame ionization detector (GC-FID, Agilent Technologies Inc., Santa Clara, CA, USA). The chromatography column employed was a fused-silica SP-2380 capillary column (100 m × 0.25 mm i.d., 0.2 μm film thickness, Supelco, Bellefonte, PA, USA). The chromatographic setup and FAMEs' identification and quantification were performed as previously reported by Faccinetto-Beltrán et al. [25]. Quantification for each compound and the total amount of fatty acids (FAs) were calculated by the AOAC method 996.06. Concentration of FAs were expressed as mg of each individual FA per 100 g of product on a fresh weight (FW) basis.

2.6. Sensory Evaluation

A sensory acceptability test was performed using the 9-point hedonic scale to assess the consumers' acceptability of chocolate formulation. A total of 223 students and staff (59% male and 41% female) from Tecnológico de Monterrey (Monterrey, NL, México) that consumed chocolate at least once a week were selected for the study, with ages ranging between 17 and 21 years old. Each chocolate sample was provided with a different random three-digit number. The samples were provided in different orders. Participants were asked to eat the chocolate samples one at a time, drink water, and eat a cookie with a plain flavor before the evaluation and between testing different samples. For each chocolate, the participants were requested to evaluate the attributes of appearance, flavor, texture, and overall acceptability using a 9-point hedonic scale ranging from 1 to 9: 1 = "dislike extremely," 2 = "dislike very much," 3 = "dislike moderately," 4 = "dislike slightly," 5 = "neither like nor dislike," 6 = "like slightly," 7 = "like moderately," 8 = "like very much," and 9 = "like extremely."

To determine the microbial safety of chocolate samples before sensory evaluation, chocolate formulations were analyzed for total coliforms, yeast, molds, and *Salmonella* spp., according to methods previously reported in the literature [30–32]. Briefly, 10 g of each chocolate sample was introduced into a sample bag (Whirl-Pak®, Nasco, Fort Atkinson, WI, USA), diluted with sterile peptone water (0.1 % peptone, 0.85% NaCl, pH 7), and homogenized for 2 min in a stomacher. Triplicate counts were performed for all dilutions. Total coliforms were determined using violet red bile agar and incubated at 37 °C for 24 h. Fungi and molds were grown in potato dextrose peptone agar and incubated at 25 °C for 5 days. All chocolates presented <10 CFU/mL for total coliforms, fungi, and molds.

For *Salmonella* spp. analysis, 25 g of chocolate sample was placed in 225 mL of reconstituted skim sterilized milk for 60 min at 25 °C. Then, 1 mL of each sample was put in 10 mL of Vassiliadis-Rappaport and in 10 mL of tetrathionate for 24 h. *Salmonella* spp. counts were performed in XLD agar and SS agar. Chocolate formulations were free of *Salmonella* spp., and thus, all chocolates were safe for human consumption and suitable for sensory evaluations (Ethics ID: CSERMBIGDL-002).

2.7. Statistical Analysis

Results are expressed as mean ± standard error of three independent measurements unless otherwise indicated. Data were analyzed with full factorial analysis of variance to evaluate main effects and interactions, followed by the LSD test to determine significant differences among groups ($p < 0.05$), using JMP software version 14.3.0 (SAS Inst. Inc., Cary, NC, USA).

3. Results and Discussion

3.1. Probiotics Viability

Microencapsulation by spray-drying is a common technology to protect the viability of probiotics [33]. In the present study, maltodextrin (10%, *w*/*v*) and sodium alginate (2%, *w*/*v*) were used as bacteria-protecting ingredients to generate powders with microencapsulated probiotics. Spray-drying microencapsulation resulted in powders with 7×10^{13} CFU/g and 1×10^{14} CFU/g for *Lactobacillus plantarum* L299v and *Lactobacillus acidophilus* La 3, respectively. These results agree with previous reports that evaluated microencapsulation of probiotics with sodium alginate, demonstrating that it can be used as a heat protector agent for different probiotic strains, such as *L. rhamnosus*, *B. longum*, *L. salivarius*, *L. plantarum*, *L. acidophilus*, *L. paracasei*, *B. lactis* B1-O4, *B. lactis* Bi-07 [34], and *L. casei* [35]. Furthermore, the use of prebiotic agents such as maltodextrin, in addition to alginate, is recommended to generate a physical barrier with a symbiotic relationship [36]. In this tenor, previous reports have demonstrated that maltodextrin can be used as an effective microencapsulating protective agent for probiotics, reducing the caking and stickiness to the spray-dryer's wall, increasing the free-flowing nature of the spray-dried powder [37], and exerting heat protection [38].

For all chocolate formulations, the addition of microencapsulated probiotics resulted in a product with $\geq 2 \times 10^7$ CFU per portion (12 g). This value is in the range of the minimum count of probiotic bacteria intake ($\geq 1 \times 10^6$ CFU) recommended to have a beneficial effect [39,40]. Prior reports have shown that chocolate ingredients are suitable as a vehicle for probiotics [20,25,27]. For instance, the high total solids in milk chocolate, including fat and protein, generate a protective matrix for probiotics [40]. Furthermore, the low water activity (a_w) and fat concentration in chocolate aid in preserving the viability of probiotic bacteria in an inactive state.

3.2. Physicochemical Properties of Sugar-Free Milk Chocolate Formulations with Added Probiotics and Fish Oil

3.2.1. Water Activity (a_w)

Water activity has an important role in the safety, quality, processing, shelf-life, texture, and sensory characteristics of confectionary products [3]. The a_w values for milk chocolate formulations are shown in Table 2. The control showed $a_w = 0.46$, which is in the threshold for a_w values of pathogenic microbial growth in foods. Sweetener addition (Sw), FO addition, and their combination (Sw*FO) showed a significant reduction in a_w, whereas Prob addition impeded this effect. Water activity reduction by isomalt addition has been previously reported for sugar-free milk chocolate formulations, which has been attributed to its hygroscopic property [41]. FO addition generated chocolate formulations with lower a_w values. This phenomenon could be attributed to the degree of unsaturation in fatty acid, generating electric charges that affect the molecular interaction with water molecules [42].

Table 2. Water activity (a_w), whiteness index (WI), and texture parameters' (hardness and fracturability) values of sugar-free milk chocolate formulations with added probiotics and fish oil.

Sample	a_w [a]	WI	Hardness [b] (N)	Fracturability [b] (N)
Control	0.46 ± 0.02 a	19.08 ± 0.99 de	3072.8 ± 93.6 a	2824.2 ± 117.5 b
Prob	0.47 ± 0.01 a	27.21 ± 0.29 a	2170.6 ± 198.3 c	2676.6 ± 129.8 b
FO	0.45 ± 0.01 a	20.73 ± 0.49 de	1644.6 ± 103.9 d	2834.2 ± 202.9 b
Prob + FO	0.45 ± 0.01 a	26.01 ± 0.26 ab	1719.8 ± 176.3 d	2823.8 ± 294.1 b
Sw	0.41 ± 0.01 b	24.68 ± 1.29 bc	2709.2 ± 140.5 b	3606.2 ± 96.8 a
Sw + Prob	0.40 ± 0.01 b	14.69 ± 1.41 f	2599.0 ± 103.6 b	3300.6 ± 101.9 a
Sw + FO	0.45 ± 0.01 a	18.95 ± 0.49 e	1241.2 ± 47.7 e	2031.6 ± 121.6 c
Sw + Prob + FO	0.46 ± 0.01 a	21.45 ± 1.47 cd	1545.2 ± 44.5 de	2454.2 ± 63.9 bc
Significance [c]				
Sw	**	***	NS	NS
FO	*	NS	***	***
Prob	NS	*	NS	NS
Sw*FO	***	NS	NS	***
Sw*Prob	NS	***	*	NS
Prob*FO	NS	**	***	NS
Sw*FO*Prob	NS	***	NS	NS

W.I., white index; a_w, water activity. Treatments: Control = milk chocolate formulation, Prob = milk chocolate + probiotics, FO = milk chocolate + fish oil, Prob + FO = milk chocolate + probiotics + fish oil, Sw = isomalt + stevia, Sw + Prob = isomalt + stevia + probiotics, Sw + FO = isomalt + stevia + fish oil, Sw + Prob + FO = isomalt + stevia + probiotics + fish oil. Values with different letters within the same column indicate a statistically significant difference by the LSD test ($p < 0.05$). [a] Values represent the mean of 3 replicates with their standard error. [b] Values represent the mean of 5 replicates with their standard error. [c] Asterisks indicate significant difference from a full factorial analysis of variance showing the main effects and interactions of the variables evaluated: * $p < 0.05$, ** $p < 0.01$, *** $p < 0.001$. Sw, sweetener; FO, fish oil; Prob, probiotic; NS, non-significant.

3.2.2. Whiteness Index (WI)

The WI indicates fat bloom formation [43]. Fat blooming plays a crucial role in the final structure, mechanical properties, appearance, quality, and marketability of chocolate products [44]. The effects of Sw, FO, and Prob addition as well as their interactions in the WI values of chocolate are shown in Table 2. Sw and Prob added alone (without sugar replacement) showed a significant increase on the WI value, whereas FO added alone did not affect the WI value. However, when FO was added to sugar-free chocolate (Sw + FO and Sw + Prob + FO), the individual effect of Sw and Prob on the increase in WI value was impeded.

The lower WI values observed in Sw treatments indicate that sugar replacement by sweeteners generates darker chocolates less prone to fat blooming. This result is in agreement with previous reports, where sucrose replacement with polyols, such as maltitol, xylitol, isomalt, and stevia, generated darker chocolates compared to their reference chocolate [45,46]. Particle size and distribution play an important role in instrumental color measurements. The tempering process of Sw and Sw + Prob chocolates could be responsible for the development of appropriate cocoa butter nucleation, generating more stable microparticle interaction due to the generation of adequate amounts and sizes of β V polymorphic form crystals [47].

On the other hand, Prob addition increased fat blooming predisposition in the formulation. These agreed with a report where the incorporation of *L. paracasei* to white chocolate formulation generated brighter chocolates as compared to the control [48]. The color of microencapsulated probiotic powder can explain this increase since probiotics could affect the particle size distribution in the chocolate matrix [49]. Since microcapsules are composed of carbohydrates such as maltodextrin, sugar bloom and fat bloom could be occurring. Sugar bloom is caused by absorption of moisture solubilizing sugar and then re-crystallized at the surface as a thin film of sugar crystals [50]. The fat bloom is distinguished from loss of gloss caused by larger crystals' growth, causing scattering of the light, and the surface appears dull, due to an incorrect tempering [51]. Similar results from chocolate with added probiotics were obtained by Silva et al. [52]. The authors attributed this phenomenon to the addition of probiotics during the tempering process, which influences the recrystallization of lipids. However, the interaction of Sw and Prob shows a decrease in WI values, which could be related to the microstructure interaction between sweeteners, Prob, and other ingredients in the chocolate formulation.

3.2.3. Texture

Hardness and fracturability are two texture parameters that have a direct correlation with the acceptability by consumers. Hardness represents the physical rigidity, whereas fracturability is associated with the maximum force for penetration [6,7]. Prob, FO, and sugar replacement (Sw) either evaluated alone or combined significantly reduced the hardness value of chocolate. The hardness decrease by FO addition could be attributed to the increase of PUFAs in the chocolate matrix, yielding a softer product that melts easier [53]. Lipids in chocolate represent the continuous phase in the chocolate emulsion, which governs the physical and the textural properties. The hardness of chocolate is affected by the extent and nature of the crystalline lipid phase, linked to the control of the proper polymorphic form controlled by tempering [47].

Probiotics' addition also decreased the hardness value of chocolate. This result is in agreement with a previous report where the hardness of chocolate was evaluated in dark chocolates with and without probiotics [54]. The authors attributed the lower hardness values to the effect that microencapsulated probiotics' addition could have on crystal formation during the tempering process. Furthermore, sugar replacement (Sw) also generated chocolates with lower hardness values. Polyols sweeteners such as isomalt and high-intensity sweeteners such as stevia affect the texture of chocolate due to their hydroxyl sites, which interact with intermolecular bonds between particles in chocolate [45,46].

Fracturability of chocolate was increased only when Prob and FO were added either alone or combined in sugar-free chocolates. For instance, treatments without sugar and without FO showed higher fracturability values as compared with the control. Interestingly, FO addition in sugar-free chocolates decreased fracturability, showing the lowest values among treatments (Table 2). Previous studies on physical analyses of chocolate formulations with added EPA/DHA in the triglyceride form resulted in a softer product as compared with the control, attributed to the high content of PUFAs, which contributes to the generation of a softer product with lower fracturability when sugar is replaced [55]. Other authors have suggested that process and product optimization could improve the texture of chocolates when the formulation has added FO or EPA/DHA as microencapsulated oil and powder, overcoming undesirable textural and physiological effects of FO addition [3,55].

Texture values presented herein are influenced by the tempering process, since properly tempered chocolate contains numerous β V polymorph crystals of cocoa butter that form a tight crystalline matrix, giving a high degree of hardness and fracturability. Besides, in milk chocolates, it is important to consider the effect of milk fat on cocoa butter crystallization since it can influence the modification of β V crystals to β'polymorph, which foment disorder in the emulsion matrix [6,47]. Therefore, the fatty acid composition affects liquid fat solidification, and thus the texture properties. The addition of isomalt in chocolate has been reported to increase hardness and fracturability [56]. However, the interaction between Sw and FO treatments decreased fracturability values, likely due to the increased concentration of PUFAs [3,55].

3.3. *Rheological Analysis: Shear Stress, Apparent Viscosity, and Frequency Sweep Test*

3.3.1. Flow Behavior

Rheological characteristics of chocolate are directly related to the quality attributes of the product [23]. Viscosity plays an important role in texture, flavor, and mouthfeel. Likewise, flow properties can be perceived by consumers in flavor and mouthfeel, since the perceived taste depends on the melting rate [6].

The variations of shear stress versus shear rate as well as apparent viscosity versus shear stress of milk chocolate formulations are shown in Figure 1. Probiotics' addition induced a significant increase in shear stress and apparent viscosity values, whereas sugar replacement (Sw) and FO addition evaluated individually did not affect shear stress values or apparent viscosity values. However, the Sw*FO interaction significantly modified the rheological behavior of chocolate. Chocolates with added Prob showed a plunge more stable than the control (Figure 1A). Likewise, FO combined with Sw significantly modified the shear stress. For instance, Sw + FO treatment showed a similar flow behavior as compared with the control. Additionally, Prob showed the highest apparent viscosity values compared with the control and the other sucrose milk chocolate formulations (Figure 1B). Nevertheless, FO addition affected the apparent viscosity as well as the use of Sw. As the apparent viscosity decreased, the shear rate increased, which agrees with the pseudoplastic or shear-thinning nature of chocolate [22].

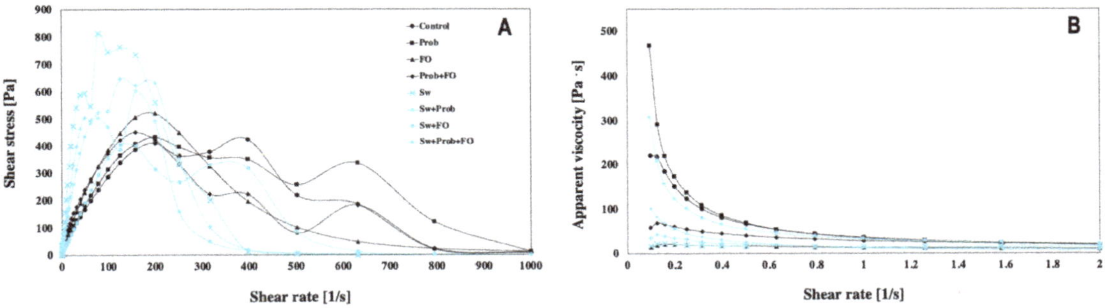

C

Significance	Shear stress [Pa]	Apparent viscosity [Pa · s]
Sw	NS	NS
FO	NS	NS
Prob	*	*
Sw*FO	***	***
Sw*Prob	NS	NS
Prob*FO	NS	NS
Sw*FO*Prob	NS	NS

Figure 1. (**A**) Shear stresses were at a range of steady-shear rate 0.1 to 100 s^{-1} and temperature 35 °C. (**B**) Apparent viscosity was at a range of steady-shear rate 0.1 to 100 s^{-1} and temperature 35 °C. (**C**) Full factorial analysis of variance showing the main effects and interactions of the variables evaluated. Values represent the mean of 3 replicates. Asterisks indicate significant difference from a full factorial analysis of variance showing the main effects and interactions of the variables evaluated: * $p < 0.05$, *** $p < 0.001$. Sw, sweetener; FO, fish oil; Prob, probiotic; NS, non-significant. Treatments: Control = milk chocolate formulation, Prob = milk chocolate + probiotics, FO = milk chocolate + fish oil, Prob + FO = milk chocolate + probiotics + fish oil, Sw = isomalt + stevia, Sw + Prob = isomalt + stevia + probiotics, Sw + FO = isomalt + stevia + fish oil, Sw + Prob + FO = isomalt + stevia + probiotics + fish oil.

The higher apparent viscosity observed in the Prob treatment could be attributed to an increase in the size and number of solid particles in the chocolate formulation. A study conducted by Afoakwa et al. [49] showed that the increase of an average particle size resulted in a decrease of Casson plastic viscosity, shear stress, yield stress, and apparent viscosity. Furthermore, previous reports have demonstrated that the addition of lyophilized probiotics increased rheological parameters and negatively affected chocolate flow properties [57,58].

As described earlier, the content and type of ingredients, such as the incorporation of PUFAs, have a critical role in chocolate viscosity. For instance, FO addition in chocolates induced a decrease in the shear stress since the fat in chocolate recovers solid particles, allowing an easy flow [23]. Similar observations were reported by Konar et al. [3], who evaluated the addition of different sources of DHA/EPA, and the authors reported a decrease in shear stress.

On the other hand, sweeteners induced an increase in shear stress, indicating that sugar-free chocolate formulations did not reach a steady condition in their rheology (Figure 1A). Previous authors studied the rheology of chocolates with different added bulk sweeteners, including isomalt, and observed that the shear-thinning index changes between the control (chocolate with sucrose) and the different bulk sweeteners. Likewise, the authors concluded that each sweetener's structure interacts with other particles in the chocolate matrix in each void space [28,47]. Void spaces between cocoa particles and cocoa butter allow optimal rheology. When the void space is too tight, the shear-thinning index is increased, and the viscosity is reduced. Similar behavior occurs when adding isomalt. However, the special molecular conformation of isomalt allows more void spaces, reducing the shear-thinning index [28].

3.3.2. Mechanical Properties

The mechanical spectra of chocolate samples are shown in Figure 2. G′ is an index of a sample's elastic behavior and represents the deformation energy stored in the sample during the shear process. On the other hand, the G″ value measures the viscous component of a sample and compares the energy lost during the shear process [59]. The addition of Prob increased the storage modulus G′ over the loss modulus G″. On the other hand, FO addition and Iso + Stev showed a contrast effect on G′ and G″ at the frequency range of 0.1 to approximately 70 Hz, indicating a liquid-like behavior of a weakly structured system.

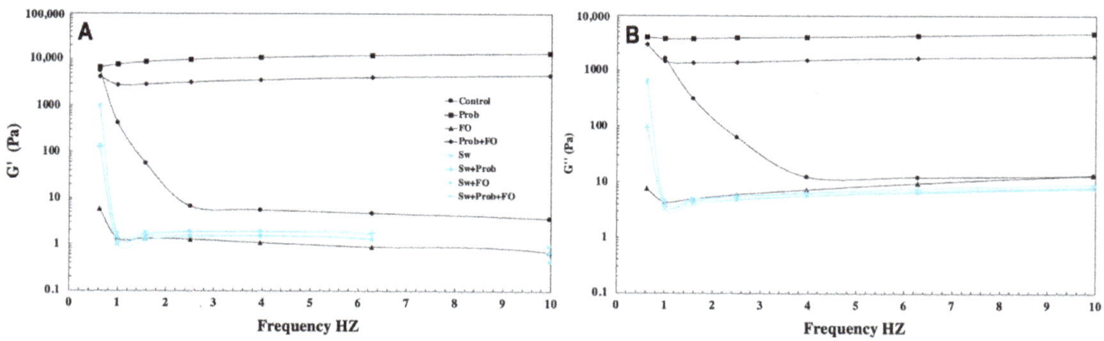

C

Significance	G' [Pa]	G" [Pa]
Sw	***	***
FO	***	***
Prob	***	***
Sw*FO	***	***
Sw*Prob	***	***
Prob*FO	***	***
Sw*FO*Prob	***	***

Figure 2. Frequency sweep test of chocolate at 35 °C with a linear viscoelastic region of 6 Pa. (**A**) Changes in storage modulus G′. (**B**) Changes in loss modulus G″. (**C**) Full factorial analysis of variance showing the main effects and interactions of the variables evaluated. Values represent the mean of 3 replicates. Asterisks indicate significant difference from a full factorial analysis of variance showing the main effects and interactions of the variables evaluated: *** $p < 0.001$. Treatments: Control = milk chocolate formulation, Prob = milk chocolate + probiotics, FO = milk chocolate + fish oil, Prob + FO = milk chocolate + probiotics + fish oil, Sw = isomalt + stevia, Sw + Prob = isomalt + stevia + probiotics, Sw + FO = isomalt + stevia + fish oil, Sw + Prob + FO = isomalt + stevia + probiotics + fish oil.

Prob addition generated a structured system to the chocolate emulsion due to the results of G′ over G″ presented herein. This behavior has been previously reported for milk chocolate [22,59]. Additionally, the combination of Prob with FO (Prob + FO) showed a similar result to Prob, meaning that Prob as an ingredient increased the stability of chocolate's mechanical properties. Although the addition of Prob generated a strong matrix, when Prob was mixed with Sw (Sw + Prob) or Sw + FO (Sw + Prob + FO), a solid-like and more elastic formulation was observed mainly due to a higher solid fraction. On the other hand, the addition of FO generated a liquid-like behavior when it was combined with Sw. These results indicate that FO addition dominates the mechanical properties in sugar-free chocolates, such as Sw + FO and Sw + Prob + FO, because FO increased the number of fatty acids in the chocolate's emulsion [53]. The fat content of chocolate determines the mass fraction of particulates, which governs the proximity of those particles to each other. Thus, if fat content increases, the distance between particles increases, resulting in a lower viscosity [47]. These observations are in agreement with the results obtained for chocolates with added FO. Furthermore, results of the substitution of sucrose by Sw in chocolates showed an unstable chocolate matrix, generating a deep increase of G' and G'', as observed in Figure 2. This behavior is attributed to the higher solid volume fraction and lower density of isomalt, resulting in more flexible chocolates [28].

3.4. Fatty Acid Methyl Esters (FAMEs) Profile

The most widely available dietary source of EPA and DHA is cold-water oily fish or fish oil offered to consumers as a dietary supplement [55]. The chocolates formulated herein had 790 mg of FO added per serving size (12 g) of chocolate, expecting to obtain 200 mg of ω-3 PUFAs.

The fatty acid composition of chocolates with and without added FO are shown in Table 3. Likewise, the fatty acid composition of FO used as an ingredient for chocolate formulations is shown in the Supplementary Material (Table S1). FO contains a high amount of PUFAs (38,746.6 $\pm$ 45.8 mg per 100 g fish oil), of which ω-3 were the most abundant (34,712.6 mg $\pm$ 0.06 g per 100 g fish oil), with DHA ($C_{22:6}$, 14,122.2 $\pm$ 27.0 mg/100 g) and EPA ($C_{20:5}$, 12,862.1 $\pm$ 17.8 mg/100 g) being the major ω-3 PUFAs.

Chocolates without FO showed a low concentration of ω-3 PUFAs, mainly due to the presence of alfa linolenic acid in cocoa butter [53]. Additional fatty acids detected in the control and Prob treatment included linoleic acid, palmitic acid, stearic acid, and oleic acid, which are the primary fatty acids of cocoa butter [6,53,60]. FO addition resulted in a chocolate formulation with 107.4 $\pm$ 12.84 mg of ω-3 PUFAs per serving size (12 g). Interestingly, higher ω-3 PUFAs content was quantified when FO was added in sugar-free chocolate formulations (Sw + FO and Sw + Prob + FO) as compared with FO added alone, showing ω-3 PUFAs levels of 141.9 $\pm$ 17.9 mg and 133.8 $\pm$ 8.76 mg per 12 g of Sw + FO and Sw + Prob + FO formulations, respectively.

FO was added to the chocolate formulation to obtain 200 mg of ω-3 PUFAs per portion (12 g). However, results indicate that lower amounts were detected, indicating that ω-3 PUFAs were degraded during the chocolate-making process. Fatty acid degradation during the chocolate-making process can be attributed to lipid oxidation induced by low water activity and thermal treatment [3], which degrades EPA and DHA by breaking down the double bonds by oxidation [61,62].

Table 3. Fatty acid profile content (mg fatty acid per 100 g sample FW^{-1}) of fish oil source, milk chocolate with added probiotics and fish oil samples, and sugar-free milk chocolates with added probiotics and fish oil samples.

Fatty Acid	Chocolate Samples							
	Control	Prob	FO	Prob + FO	Sw	Sw + Prob	Sw + FO	Sw + Prob + FO
Octanoic acid (C8:0)	46.55 ± 4.61 b	55.47 ± 2.67 ab	58.027 ± 7.67 ab	56.19 ± 4.20 ab	53.46 ± 4.78 ab	62.021 ± 5.08 ab	65.36 ± 8.81 a	613.28 ± 6.05 ab
Decanoic acid (C10:0)	10.48 ± 1.70 b	12.84 ± 0.93 ab	16.55 ± 3.33 ab	14.20 ± 1.78 ab	13.89 ± 1.61 ab	16.02 ± 2.69 ab	10.4792 ± 1.71 a	104.79 ± 1.70 a
Lauric acid (C12:0)	8.02 ± 0.94 c	9.91 ± 0.75 bc	14.33 ± 1.23 abc	16.33 ± 3.34 ab	12.41 ± 1.12 abc	13.89 ± 3.61 abc	17.80 ± 2.41 a	165.47 ± 2.92 a
Myristic acid (C14:0)	33.48 ± 0.76 c	39.85 ± 0.31 c	261.24 ± 14.08 b	265.52 ± 30.74 b	46.89 ± 1.38 c	78.45 ± 32.45 c	354.92 ± 26.85 a	3208.28 ± 18.87 ab
Pentadecanoic acid (C15:0)	9.22 ± 0.44 c	10.53 ± 0.25 c	28.57 ± 1.48 b	29.23 ± 3.04 b	11.79 ± 0.84 c	13.51 ± 2.85 c	38.37 ± 3.15 a	365.43 ± 2.9 a
Palmitic acid (C16:0)	5412.38 ± 61.76 bc	6390.28 ± 63.55 a	5061.02 ± 242.55 c	4739.24 ± 326.71 c	6117.75 ± 198.24 ab	6237.88 ± 389.55 ab	6098.85 ± 332.69 ab	64,200.62 ± 440.60 a
Heptadecanoic acid (C17:0)	43.37 ± 0.45 c	49.97 ± 0.16 bc	54.58 ± 3.12 b	51.79 ± 3.55 bc	48.62 ± 1.71 bc	49.84 ± 5.11 bc	67.13 ± 4.22 a	674.77 ± 5.82 a
Stearic acid (C18:0)	6684.21 ± 2.10 b	7846.70 ± 79.76 a	5617.02 ± 225.79 c	5190.71 ± 335.91 c	7476.05 ± 246.78 ab	7566.89 ± 421.72 ab	6785.18 ± 354.72 b	72,410.12 ± 507.05 ab
Arachidic acid (C20:0)	220.47 ± 2.10 b	257.99 ± 2.45 a	181.37 ± 7.51 c	164.50 ± 10.85 c	249.25 ± 8.49 ab	251.35 ± 13.82 ab	227.57 ± 11.71 ab	2423.21 ± 16.91 ab
Behenic acid (C22:0)	40.01 ± 0.16 bc	46.55 ± 0.27 a	35.98 ± 1.43 cd	31.59 ± 2.16 d	45.18 ± 1.22 ab	46.05 ± 2.48 ab	45.35 ± 2.64 ab	471.32 ± 4.05 a
Lignoceric acid (C24:0)	25.69 ± 0.30 bcd	29.93 ± 3.52 abc	23.82 ± 1.83 cd	22.92 ± 1.19 c	31.81 ± 2.09 ab	30.36 ± 1.49 abc	33.14 ± 2.44 a	317.14 ± 3.39 ab
Myristoleic acid (C14:1)	N.D.	N.D.	N.D.	N.D.	N.D.	N.D.	N.D.	N.D.
Palmitoleic acid (C16:1)	50.39 ± 1.30 c	59.17 ± 0.00 c	336.49 ± 18.50 b	330.66 ± 36.51 b	57.69 ± 1.96 c	77.91 ± 23.23 c	453.92 ± 30.49 a	419.48 ± 24.94 a
Oleic Acid (C18:1)	6329.61 ± 52.79 c	7403.02 ± 0.07 a	4967.76 ± 194.57 d	4563.29 ± 265.40 d	7118.32 ± 246.68 abc	7211.03 ± 373.29 ab	6473.06 ± 341.50 bc	6854.78 ± 433.29 abc
Vaccenic acid (C18:1)	65.78 ± 1.12 c	76.13 ± 0.00 c	129.99 ± 65.57 b	124.2 ± 12.02 b	72.46 ± 2.90 c	77.29 ± 9.70 c	168.56 ± 10.77 a	165.24 ± 11.53 a
Eicosenoic acid (C20:1)	190.24 ± 0.31 c	13.11 ± 0.00 c	18.511 ± 1.10 b	17.89 ± 1.79 b	12.57 ± 0.44 c	14.46 ± 2.20 bc	26.20 ± 2.11 a	25.61 ± 2.05 a
Nervonic acid (C24:1)	N.D.	N.D.	18.511 ± 1.10 a	15.93 ± 1.48 a	N.D.	N.D.	18.51 ± 3.11 a	11.09 ± 1.88 b
Linoleic acid (C18:2)	654.58 ± 5.59 b	767.95 ± 8.11 a	407.71 ± 19.88 c	391.90 ± 19.86 c	733.25 ± 21.76 ab	730.29 ± 28.96 ab	687.41 ± 40.22 ab	736.66 ± 49.90 ab
Gamma Linolenic acid (C18:3)	N.D.	N.D.	9.55 ± 1.49 a	8.65 ± 0.75 a	N.D.	N.D.	10.83 ± 1.45 a	11.46 ± 2.28 a
Alpha Linolenic acid (C18:3)	48.60 ± 0.66 c	58.40 ± 0.94 c	79.41 ± 4.60 b	73.60 ± 6.67 b	53.22 ± 1.51 c	55.44 ± 3.47 c	114.22 ± 7.76 a	112.20 ± 6.96 a
Stearidionic acid (C18:4)	N.D.	N.D.	90.37 ± 6.48 b	87.67 ± 11.05 b	N.D.	N.D.	114.66 ± 6.62 a	109.23 ± 7.67 a
Eicosadienoic acid (C20:2)	N.D.	N.D.	28.21 ± 2.38 a	26.51 ± 2.89 a	N.D.	N.D.	17.69 ± 2.24 b	18.40 ± 2.98 b
Homo-gamma-linolenic acid (C20:3)	N.D.	N.D.	9.012 ± 0.82 b	8.86 ± 0.55 b	N.D.	N.D.	10.63 ± 1.44 ab	12.04 ± 1.39 a
Dihomogamma linolenic acid (C20:3)	N.D.	N.D.	13.99 ± 1.85 a	11.07 ± 1.20 a	N.D.	N.D.	12.57 ± 1.47 a	13.87 ± 2.02 a

Table 3. *Cont.*

Fatty Acid	Chocolate Samples							
	Control	Prob	FO	Prob + FO	Sw	Sw + Prob	Sw + FO	Sw + Prob + FO
Arachidonic acid (C20:4)	N.D.	N.D.	22.92 ± 2.66 b	22.75 ± 2.02 b	N.D.	N.D.	31.18 ± 2.73 a	28.57 ± 2.26 a
Eicosapentaenoic acid (C20:5)	N.D.	N.D.	316.75 ± 20.57 b	305.6 ± 35.34 b	N.D.	N.D.	421.21 ± 32.36 a	309.91 ± 22.28 a
Docosapentaenoic acid n-6 (C22:5)	N.D.	N.D.	8.37 ± 20.57 b	9.02 ± 0.66 b	N.D.	N.D.	14.98 ± 2.19 a	14.84 ± 0.87 a
Docosapentaenoic acid n-3 (C22:5)	N.D.	N.D.	42.39 ± 6.66 b	46.33 ± 5.10 b	N.D.	N.D.	64.63 ± 5.93 a	52.013 ± 7.22 ab
Docosahexaenoic acid (C22:6)	N.D.	N.D.	351.83 ± 22.20 b	341.07 ± 43.00 b	N.D.	N.D.	463.481 ± 31.99 a	436.57 ± 27.53 a
Total ω-3	48.61 ± 0.66 c	58.40 ± 0.94 c	894.76 ± 61.76 b	865.35 ± 102.23 b	53.23 ± 1.51 c	55.44 ± 3.47 c	1190.80 ± 86.05 a	1114.80 ± 73.02 a
Total ω-6	654.58 ± 5.60 b	767.95 ± 8.11 a	548.78 ± 28.01 c	467.70 ± 25.30 c	733.25 ± 21.76 ab	730.29 ± 28.96 ab	772.74 ± 49.35 a	821.99 ± 58.79 a
Saturated fatty acids (SFA)	12,533.89 ± 130.45 bc	14,750.05 ± 152.21 a	11,353.24 ± 505.99 cd	10,582.23 ± 716 d	14,107.12 ± 460.99 ab	14,366.42 ± 879.43 ab	13,752.27 ± 746.68 ab	145,029.57 ± 1010.38 a
Monounsaturated fatty acids (MUFA)	6456.17 ± 54.99 b	7551.44 ± 78.89 a	5469.07 ± 221.59 c	5051.99 ± 315.66 c	7261.04 ± 251.89 ab	7380.69 ± 407.77 ab	7140.27 ± 387.76 ab	7476.23 ± 473.488 a
Polyunsaturated fatty acids (PUFA)	703.19 ± 6.25 c	826.35 ± 8.75 c	1443.54 ± 89.76 b	1333.05 ± 125.75 b	786.48 ± 23.26 c	785.74 ± 32.41 c	1963.54 ± 133.79 a	1936.79 ± 131.78 a
Total fatty acids	19,693.26 ± 191.51 bc	23,127.85 ± 239.28 a	18,255.87 ± 816.67 c	16,967.28 ± 1155.56 c	22,154.64 ± 735.78 ab	22,532.86 ± 1319.25 ab	22,856.09 ± 1266.02 a	23,915.98 ± 1615.57 a

N.D., not detected. Moisture: Control 2.00%, Prob 2.00%, FO 0.77%, Prob + FO 0.99%, Sw 1.70%, Sw + Prob 1.70%, Sw + FO 1.82%, Sw + Prob + FO 1.13%. Treatments: Control = milk chocolate formulation, Prob = milk chocolate + probiotics, FO = milk chocolate + fish oil, Prob + FO = milk chocolate + probiotics + fish oil, Sw = isomalt + stevia, Sw + Prob = isomalt + stevia + probiotics, Sw + FO = isomalt + stevia + fish oil, Sw + Prob + FO = isomalt + stevia + probiotics + fish oil. Values with different letters within the same row indicate statically significant difference by the LDS test ($p < 0.05$). Values represent the mean of 3 replicates with their standard error. Sw, sweetener; FO, fish oil; Prob, probiotic; NS, non-significant.

3.5. Consumers' Acceptability

A consumer acceptability test was performed to evaluate the appearance, taste, texture, and overall acceptability of chocolates, using a 9-point hedonic scale (Table 4).

Table 4. Sensory acceptability values of milk chocolate and sugar-free milk chocolate formulations with added probiotics and fish oil.

Sample	Appearance [a]	Flavor [a]	Texture [a]	Overall Acceptability [a]
Control	8.25 ± 0.07 a	7.07 ± 0.11 a	7.49 ± 0.09 ab	7.21 ± 0.09 a
Prob	7.22 ± 0.14 bcd	6.93 ± 0.18 a	7.63 ± 0.46 a	7.035 ± 0.15 ab
FO	6.96 ± 0.17 d	4.60 ± 0.21 d	5.409 ± 0.20 d	4.75 ± 0.21 e
Prob + FO	7.02 ± 0.16 cd	4.63 ± 0.20 d	5.48 ± 0.20 d	5.035 ± 0.21 e
Sw	7.34 ± 0.13 bc	6.39 ± 0.17 b	6.97 ± 0.14 bc	6.56 ± 0.17 c
Sw + Prob	7.49 ± 0.13 b	6.24 ± 0.17 b	6.86 ± 0.14 c	6.57 ± 0.15 bc
Sw + FO	7.15 ± 0.15 bcd	5.07 ± 0.21 cd	6.66 ± 0.18 c	5.69 ± 0.19 d
Sw + Prob + FO	7.15 ± 0.15 bcd	5.07 ± 0.20 c	6.59 ± 0.17 c	5.67 ± 0.19 d
Significance [b]				
Sw	NS	NS	NS	NS
FO	**	***	***	***
Prob	NS	NS	NS	NS
Sw*FO	NS	***	***	***
Sw*Prob	NS	NS	NS	NS
Prob*FO	NS	NS	NS	NS
Sw*FO*Prob	NS	NS	NS	NS

Treatments: Control = milk chocolate formulation, Prob = milk chocolate + probiotics, FO = milk chocolate + fish oil, Prob + FO = milk chocolate + probiotics + fish oil, Sw = isomalt + stevia, Sw + Prob = isomalt + stevia + probiotics, Sw + FO = isomalt + stevia + fish oil, Sw + Prob + FO = isomalt + stevia + probiotics + fish oil. [a] Values with different letters within the same column indicate statically significant difference by the LSD test ($p < 0.05$). [b] Asterisks indicate significant difference from a full factorial analysis of variance showing the main effects and interactions of the variables evaluated: ** $p < 0.01$, *** $p < 0.001$. Sw, sweetener; FO, fish oil; Prob, probiotic; NS, non-significant.

FO addition significantly reduced the acceptability of chocolate for all the parameters evaluated. On the other hand, Prob did not affect the acceptability by consumers. These results are in agreement with previous reports, where probiotics' addition did not affect the acceptability of chocolate [25,27,49]. FO addition mainly affected the acceptability of flavor and texture in FO and Prob + FO chocolates. This may be related to the fish odor present in fish oil. Interestingly, when FO was added to sugar-free formulations (Sw + FO and Sw + Prob + FO), chocolates showed higher acceptability as compared with formulations containing sugar and FO (FO and Prob + FO). This behavior can be explained by the fact that sugar can enhance flavors [63], and Sw chocolates have antioxidant properties (due to isomalt) that could protect FO from lipid oxidation [64].

Sugar-free chocolates showed lower flavor, texture, and overall acceptability values as compared with the control. The lower acceptability scores could be attributed to stevia's bitter taste and to the changes in rheological and mechanical properties induced by Sw addition [65,66]. It is important to point out that sugar-free chocolates, and sugar-free chocolates with added probiotics, showed values in the acceptable range, indicating that they could be excellent candidates for commercialization.

4. Conclusions

In the present study, it was demonstrated that it is possible to formulate sugar-free milk chocolate formulations with added ω-3 PUFAs and probiotics, showing adequate acceptability by consumers. One of the drawbacks of the formulations evaluated was the decrease in acceptability by consumers when FO was added as an ingredient. Therefore, further studies should consider using lower concentrations of FO, or adding the ω-3 PUFAs from other sources, such as microalgae. The results presented herein support the idea that chocolate could be used as a good delivery system of bioactive ingredients, and thus further studies should evaluate the effect on these new chocolate formulations on the prevention of diseases through the evaluation of their efficacy by in vitro, in vivo and clinical studies.

Supplementary Materials: The following are available online at https://www.mdpi.com/2304-8158/10/8/1866/s1, Table S1: Fatty acid profile (mg fatty acid per 100 g sample FW^{-1}) of fish oil used as a source of ω-3 polyunsaturated fatty acids.

Author Contributions: Conceptualization, A.R.G.-F., N.E.O.-S., A.S. and D.A.J.-V.; methodology, A.R.G.-F., P.F.-B., N.E.O.-S., L.M.M.-O., A.S., E.P.-C., C.H.-B. and D.A.J.-V.; formal analysis, A.R.G.-F., P.F.-B., N.E.O.-S., L.M.M.-O. and E.P.-C.; investigation, A.R.G.-F., P.F.-B., N.E.O.-S., A.S., E.P.-C. and D.A.J.-V.; resources, A.S., E.P.-C., D.A.J.-V. and C.H.-B.; writing—original draft preparation, A.R.G.-F. and D.A.J.-V.; writing—review and editing, E.P.-C., A.S., D.A.J.-V. and C.H.-B.; supervision, D.A.J.-V. and A.S.; funding acquisition, D.A.J.-V., C.H.-B. and A.S. All authors have read and agreed to the published version of the manuscript.

Funding: This study was based upon research supported by Tecnologico de Monterrey—Bioprocess Research Group, NutriOmics Research Group, Translational Omics Research Group, and VitaWard Research Foundation. Authors A.R.G.-H., P.F.-B. and L.M.M.-O. also acknowledge their scholarships #966535, #923811, and #445941 respectively, from CONACYT.

Institutional Review Board Statement: The study was conducted according to the guidelines of the Declaration of Helsinki and was approved by the Institutional Ethics Committee for Sensory Evaluation Studies of the Department of Bioengineering, School of Engineering and Sciences of Tecnologico de Monterrey (Ethics ID: CSERMBIGDL-002, approved on 31 May 2019).

Informed Consent Statement: Informed consent was obtained from all subjects involved in the study.

Data Availability Statement: The data presented in this study are available on request from the corresponding author.

Conflicts of Interest: The authors declare no conflict of interest.

References

1. Stanhope, K.L. Sugar consumption, metabolic disease and obesity: The state of the controversy. *Crit. Rev. Clin. Lab. Sci.* **2016**, *53*, 52–67. [CrossRef]
2. Gómez-Fernández, A.R.; Santacruz, A.; Jacobo-Velázquez, D.A. The complex relationship between metabolic syndrome and sweeteners. *J. Food Sci.* **2021**, *86*, 1511–1531. [CrossRef]
3. Konar, N.; Toker, O.S.; Pirouzian, H.R.; Oba, S.; Polat, D.G.; Palabiyik, İ.; Polyrazoglu, E.S.; Sagdic, O. Enrichment of milk chocolate by using EPA and DHA originated from various origins: Effects on product quality. *Sugar Tech* **2018**, *20*, 745–755. [CrossRef]
4. De Felice, S.L. The nutraceutical revolution: Its impact on food industry R&D. *Trends Food Sci. Technol.* **1995**, *6*, 59–61.
5. Santini, A.; Novellino, E. Nutraceuticals-shedding light on the grey area between pharmaceuticals and foods. *Expert Rev. Clin. Pharmacol.* **2018**, *11*, 545–547. [CrossRef] [PubMed]
6. Skytte, U.P.; Kaylegian, K.E. Ingredients from milk. In *Beckett's Industrial Chocolate Manufacture and Use*; Beckett, S.T., Fowler, M.S., Ziegler, G.R., Eds.; Wiley-Blackwell: Hoboken, NJ, USA, 2017; pp. 102–134.
7. Shah, A.B.; Jones, G.P.; Vasiljevic, T. Sucrose-free chocolate sweetened with *Stevia rebaudiana* extract and containing different bulking agents–effects on physicochemical and sensory properties. *Int. J. Food Sci. Technol.* **2010**, *45*, 1426–1435. [CrossRef]
8. Carocho, M.; Morales, P.; Ferreira, I.C.F.R. Sweeteners as food additives in the XXI century: A review of what is known, and what is to come. *Food Chem. Toxicol.* **2017**, *107*, 302–317. [CrossRef] [PubMed]
9. Ruiz-Ojeda, F.J.; Plaza-Díaz, J.; Sáez-Lara, M.J.; Gil, A. Effects of sweeteners on the gut microbiota: A review of experimental studies and clinical trials. *Adv. Nutr.* **2019**, *10*, S31–S48. [CrossRef]
10. Gibson, G.R.; Roberfroid, M.B. Dietary modulation of the human colonic microbiota: Introducing the concept of prebiotics. *J. Nutr.* **1995**, *125*, 1401–1412. [CrossRef]
11. Philippaert, K.; Pironet, A.; Mesuere, M.; Sones, W.; Vermeiren, L.; Kerselaers, S.; Pinto, S.; Antoine, N.; Gysemans, C.; Laureys, J.; et al. Steviol glycosides enhance pancreatic beta-cell function and taste sensation by potentiation of TRPM5 channel activity. *Nat. Commun.* **2017**, *8*, 14733. [CrossRef] [PubMed]
12. Rahmawaty, S.; Lyons-Wall, P.; Batterham, M.; Charlton, K.; Meyer, B.J. Food patterns of Australian children ages 9 to 13 y in relation to ω-3 long chain polyunsaturated intake. *Nutrition* **2014**, *30*, 169–176. [CrossRef] [PubMed]
13. Gao, H.; Geng, T.; Huang, T.; Zhao, Q. Fish oil supplementation and insulin sensitivity: A systematic review and meta-analysis. *Lipids Health Dis.* **2017**, *16*, 131. [CrossRef] [PubMed]

14. Interim Summary of Conclusions and Dietary Recommendations on Total Fat and Fatty Acids. Available online: https://www.who.int/nutrition/topics/FFA_summary_rec_conclusion.pdf?ua=1 (accessed on 28 June 2021).
15. Albert, B.B.; Derraik, J.G.B.; Brennan, C.M.; Biggs, J.B.; Smith, G.C.; Garg, M.L.; Cameron-Smith, D.; Hofman, P.L.; Cutfield, W.S. Higher omega-3 index is associated with increased insulin sensitivity and more favorable metabolic profile in middle-aged overweight men. *Sci. Rep.* **2014**, *4*, 6697. [CrossRef] [PubMed]
16. El Hage, R.; Hernandez-Sanabria, E.; Van de Wiele, T. Emerging trends in "smart probiotics": Functional consideration for the development of novel health and industrial applications. *Front. Microbiol.* **2017**, *8*, 1889. [CrossRef] [PubMed]
17. Guidelines for the Evaluation of Probiotics in Food. Available online: https://www.who.int/foodsafety/fs_management/en/probiotic_guidelines.pdf (accessed on 28 June 2021).
18. Sohag, M.S.U.; Paul, M.; Al-Bari, M.A.A.; Wahed, M.I.I.; Khan, M.R.I. Potential antidiabetic activities of probiotic strains, *L. acidophilus* and *L. bulgaricus* against fructose-fed hyperglycemic rats. *Food Nutr. Sci.* **2019**, *10*, 1419–1432.
19. Stevenson, C.; Blaauw, R.; Fredericks, E.; Visser, J.; Roux, S. Randomized clinical trial: Effect of *lactobacillus plantarum* 299 v on symptoms of irritable bowel syndrome. *Nutrition* **2014**, *30*, 1151–1157. [CrossRef]
20. Marcial-Coba, M.S.; Saaby, L.; Knøchel, S.; Nielsen, D.S. Dark chocolate as a stable carrier of microencapsulated *Akkermansia muciniphila* and *Lactobacillus casei*. *FEMS Microbiol. Lett.* **2019**, *366*, fny290. [CrossRef]
21. Mirković, M.; Seratlić, S.; Kilcawley, K.; Mannion, D.; Mirković, N.; Radulović, Z. The sensory quality and volatile profile of dark chocolate enriched with encapsulated probiotic *Lactobacillus plantarum* bacteria. *Sensors* **2018**, *18*, 2570. [CrossRef] [PubMed]
22. Glicerina, V.; Balestra, F.; Rosa, M.D.; Romani, S. Effect of manufacturing process on the microstructural and rheological properties of milk chocolate. *J. Food Eng.* **2015**, *145*, 45–50. [CrossRef]
23. Gonçalves, E.V.; Lannes, S.C.d.S. Chocolate rheology. *Food Sci. Technol.* **2010**, *30*, 845–851. [CrossRef]
24. Muñoz, Y.; Gutiérrez, T. Evaluation of the sensory quality of chocolate. In *Chocolate-Cocoa Byproducts Technology, Rheology, Styling, and Nutrition*; Sira, E.P., Ed.; Nova: New York, NY, USA, 2015; pp. 167–189.
25. Faccinetto-Beltrán, P.; Gómez-Fernández, A.R.; Orozco-Sánchez, N.E.; Pérez-Carrillo, E.; Marín-Obispo, L.M.; Hernández-Brenes, C.; Santacruz, A.; Jacobo-Velázquez, D.A. Physicochemical properties and sensory acceptability of a next-generation functional chocolate added with omega-3 polyunsaturated fatty acids and probiotics. *Foods* **2021**, *10*, 333. [CrossRef] [PubMed]
26. Cikrikci, S.; Yucekutlu, M.; Mert, B.; Oztop, M.H. Physical characterization of low-calorie chocolate formulations. *J. Food Meas. Charact.* **2017**, *11*, 41–49. [CrossRef]
27. Gómez-Fernández, A.; Faccinetto-Beltrán, P.; Orozco-Sánchez, N.; Pérez-Carrillo, E.; Santacruz, A.; Jacobo-Velázquez, D.A. Physicochemical properties and sensory acceptability of sugar-free dark chocolate formulations added with probiotics. *Rev. Mex. Ing. Química* **2021**, *20*, 697–709. [CrossRef]
28. Kiumarsi, M.; Rafe, A.; Yeganehzad, S. Effect of different bulk sweeteners on the dynamic oscillatory and shear rheology of chocolate. *Appl. Rheol.* **2017**, *27*, 11–19.
29. Torres-Moreno, M.; Torrescasana, E.; Salas-Salvadó, J.; Blanch, C. Nutritional composition and fatty acids profile in cocoa beans and chocolates with different geographical origin and processing conditions. *Food Chem.* **2015**, *166*, 125–132. [CrossRef]
30. BAM. Chapter 5: Salmonella. Available online: https://www.fda.gov/food/laboratory-methods-food/bam-chapter-6-shigella (accessed on 28 June 2021).
31. BAM. Chapter 4: Enumeration of Escherichia coli and the Coliform Bacteria. Available online: https://www.fda.gov/food/laboratory-methods-food/bam-chapter-4-enumeration-escherichia-coli-and-coliform-bacteria (accessed on 28 June 2021).
32. NORMA Oficial Mexicana. NOM-186-SSA1/SCFI-2013, Cacao, Chocolate y Pruebas Similares, y Derivados del Cacao. *Especificaciones Sanitarias. Denominación Comercial. Métodos de Prueba.* Available online: https://www.dof.gob.mx/nota_detalle.php?codigo=5332832&fecha=17/02/2014 (accessed on 28 June 2021).
33. Yao, M.; Wu, J.; Li, B.; Xiao, H.; McClements, D.J.; Li, L. Microencapsulation of *Lactobacillus salivarious* Li01 for enhanced storage viability and targeted delivery to gut microbiota. *Food Hydrocoll.* **2017**, *72*, 228–236. [CrossRef]
34. Ding, W.K.; Shah, N.P. Acid, bile, and heat tolerance of free and microencapsulated probiotic bacteria. *J. Food Sci.* **2007**, *72*, M446–M450. [CrossRef] [PubMed]
35. Hernández-Carranza, P.; López-Malo, A.; Jiménez, T. Microencapsulation quality and efficiency of *Lactobacillus casei* by spray drying using maltodextrin and vegetable extracts. *J. Food Res.* **2013**, *3*, 61–69. [CrossRef]
36. Tárrega, A.; Rocafull, A.; Costell, E. Effect of blends of short and long-chain inulin on the rheological and sensory properties of prebiotic low-fat custards. *LWT Food Sci. Technol.* **2010**, *43*, 556–562. [CrossRef]
37. Anekella, K.; Orsat, V. Optimization of microencapsulation of probiotics in raspberry juice by spray drying. *LWT Food Sci. Technol.* **2013**, *50*, 17–24. [CrossRef]
38. Boza, Y.; Barbin, D.; Scamparini, A.R.P. Survival of *Beijerinckia* sp. microencapsulated in carbohydrates by spray-drying. *J. Microencapsul.* **2004**, *21*, 15–24. [CrossRef] [PubMed]
39. Shah, N.P. Functional cultures and health benefits. *Int. Dairy J.* **2007**, *17*, 1262–1277. [CrossRef]
40. Mandal, S.; Hati, S.; Puniya, A.K.; Singh, R.; Singh, K. Development of synbiotic milk chocolate using encapsulated *Lactobacillus casei* ncdc 298. *J. Food Process. Preserv.* **2013**, *37*, 1031–1037. [CrossRef]
41. Konar, N.; Oba, S.; Toker, O.S.; Palabiyik, I.; Goktas, H.; Artik, N.; Sagdic, O. Rapid tempering of sucrose-free milk chocolates by β V seeding: Textural, rheological and melting properties. *Eur. Food Res. Technol.* **2017**, *243*, 1849–1860. [CrossRef]

42. Silva, N.D.S.E.; Hernández, E.J.G.P.; Araújo, C.D.S.; Joele, M.R.S.P.; Lourenço, L.D.F.H. Development and optimization of biodegradable fish gelatin composite film added with buritic oil. *CyTA J. Food* **2018**, *16*, 340–349. [CrossRef]
43. Toker, O.S.; Oba, S.; Palabiyik, I.; Pirouzian, H.R.; Konar, N.; Artik, N.; Sagdic, O. Alternative tempering of sugar-free dark chocolates by βv seeding: Sensorial, micro-structural and some physical properties and volatile profile. *Int. J. Food Eng.* **2019**, *15*, 20180067. [CrossRef]
44. Hartel, R.W. Chocolate: Fat bloom during storage, the influence of structural elements. *Manuf. Confect.* **1999**, *79*, 89–99.
45. Aidoo, R.P.; Afoakwa, E.O.; Dewettinck, K. Rheological properties, melting behaviours and physical quality characteristics of sugar-free chocolates processed using inulin/polydextrose bulking mixtures sweetened with stevia and thaumatin extracts. *LWT Food Sci. Technol.* **2015**, *62*, 592–597. [CrossRef]
46. Rad, A.H.; Pirouzian, H.R.; Toker, O.S.; Konar, N. Application of simplex lattice mixture design for optimization of sucrose-free milk chocolate produced in a ball mill. *LWT Food Sci. Technol.* **2019**, *115*, 108435.
47. Harter, R.W.; von Ele, J.H.; Hoberger, R. Fats, oils and emulsifiers. In *Confectionary Science and Technology*; Harter, R.W., von Ele, J.H., Hoberger, R., Eds.; Confectionery Science and Technology: Cham, Switzerland, 2018; pp. 85–124.
48. Konar, N.; Palabiyik, I.; Toker, O.S.; Polat, D.G.; Kelleci, E.; Pirouzian, H.R.; Akcicek, A.; Sagdic, O. Conventional and sugar-free probiotic white chocolate: Effect of inulin DP on various quality properties and viability of probiotics. *J. Funct. Foods* **2018**, *43*, 206–213. [CrossRef]
49. Afoakwa, E.O.; Paterson, A.; Fowler, M. Effects of particle size distribution and composition on dark chocolate's rheological properties. *Eur. Food Res. Technol.* **2008**, *226*, 1259–1268. [CrossRef]
50. Timms, R.E. Interactions between fats, bloom and rancidity. In *Confectionery Fats Handbook*, 1st ed.; Timms, R.E., Ed.; Woodhead Publishing: Cambridge, UK, 2012; pp. 255–294.
51. Timms, R.E. Phase behavior of fats and their mixtures. *Prog. Lipid Res.* **1984**, *23*, 1–38. [CrossRef]
52. Silva, M.P.; Tulini, F.L.; Marinho, J.F.U.; Mazzocato, M.C.; De Martinis, E.C.P.; Luccas, V.; Favaro-Trindade, C.S. Semisweet chocolate as a vehicle for the probiotics *Lactobacillus acidophilus* LA3 and *Bifidobacterium animalis* subsp. *Lactis BLC1: Evaluation of chocolate stability and probiotic survival under in vitro simulated gastrointestinal conditions. LWT Food Sci. Technol.* **2017**, *75*, 640–647.
53. Pirouzian, H.R.; Konar, N.; Palabiyik, I.; Oba, S.; Toker, O.S. Pre-crystallization process in chocolate: Mechanism, importance and novel aspects. *Food Chem.* **2020**, *321*, 126718. [CrossRef] [PubMed]
54. Foong, Y.J.; Lee, S.T.; Ramli, N.; Tan, Y.N.; Ayob, M.K. Incorporation of potential probiotic *Lactobacillus plantarum* isolated from fermented cocoa beans into dark chocolate: Bacterial viability and physicochemical properties analysis. *J. Food Qual.* **2013**, *36*, 164–171. [CrossRef]
55. Toker, O.S.; Konar, N.; Palabiyik, I.; Pirouzian, H.R.; Oba, S.; Polat, D.G.; Poyrazoglu, E.S.; Sagdic, O. Formulation of dark chocolate as a carrier to deliver eicosapentaenoic and docosahexaenoic acids: Effects on product quality. *Food Chem.* **2018**, *254*, 224–231. [CrossRef] [PubMed]
56. Belščak-Cvitanović, A.; Komes, D.; Dujmović, M.; Karlović, S.; Biškić, M.; Brnčić, M.; Ježek, D. Physical, bioactive and sensory quality parameters of reduced sugar chocolates formulated with natural sweeteners as sucrose alternatives. *Food Chem.* **2015**, *167*, 61–70. [CrossRef] [PubMed]
57. Laličić-Petronijević, J.; Popov-Raljić, J.; Obradović, D.; Radulović, Z.; Paunović, D.; Petrušić, M.; Pezo, L. Viability of probiotic strains *Lactobacillus acidophilus* NCFM® and *Bifidobacterium lactis* HN019 and their impact on sensory and rheological properties of milk and dark chocolates during storage for 180 days. *J. Funct. Food* **2015**, *15*, 541–550. [CrossRef]
58. Żyżelewicz, D.; Nebesny, E.; Motyl, I.; Libudzisz, Z. Effect of milk chocolate supplementation with lyophilised lactobacillus cells on its attributes. *Czech J. Food Sci.* **2010**, *28*, 392–406. [CrossRef]
59. Glicerina, V.; Balestra, F.; Rosa, M.D.; Romani, S. Rheological, textural and calorimetric modifications of dark chocolate during process. *J. Food Eng.* **2013**, *119*, 173–179. [CrossRef]
60. Afoakwa, E.O. *Chocolate Science and Technology*, 1st ed.; Wiley-Blackwell Publishers: Oxford, UK, 2016; pp. 54–101.
61. Leung, K.S.; Galano, J.M.; Durand, T.; Lee, J.C.Y. Profiling of omega-polyunsaturated fatty acids and their oxidized products in salmon after different cooking methods. *Antioxidants* **2018**, *7*, 96. [CrossRef] [PubMed]
62. Mozuraityte, R.; Kristinova, V.; Standal, I.B.; Carvajal, A.K.; Aursand, M. Oxidative Stability and Shelf Life of Fish Oil. In *Oxidative Stability and Shelf Life of Foods Containing Oils and Fats*; AOCS Press: Rockville, MD, USA, 2016; pp. 209–231.
63. Goldfein, K.R.; Slavin, J.L. Why sugar is added to food: Food science 101. *Compr. Rev. Food Sci. Food Saf.* **2015**, *14*, 644–656. [CrossRef]
64. Faraji, H.; Lindsay, R.C. Antioxidant protection of bulk fish oils by dispersed sugars and polyhydric alcohols. *J. Agric. Food Chem.* **2005**, *53*, 736–744. [CrossRef] [PubMed]
65. Lagast, S.; De Steur, H.; Schouteten, J.J.; Gellynck, X. A comparison of two low-calorie sweeteners and sugar in dark chocolate on sensory attributes and emotional conceptualisations. *Int. J. Food Sci. Nutr.* **2017**, *69*, 344–357. [CrossRef] [PubMed]
66. Rad, A.H.; Pirouzian, H.R.; Konar, N.; Toker, O.S.; Polat, D.G. Effects of polyols on the quality characteristics of sucrose-free milk chocolate produced in a ball mill. *RSC Adv.* **2019**, *9*, 29676–29688. [CrossRef]

MDPI
St. Alban-Anlage 66
4052 Basel
Switzerland
Tel. +41 61 683 77 34
Fax +41 61 302 89 18
www.mdpi.com

Foods Editorial Office
E-mail: foods@mdpi.com
www.mdpi.com/journal/foods

www.ingramcontent.com/pod-product-compliance
Lightning Source LLC
Chambersburg PA
CBHW040744250926
51850CB00007BB/1401
9783036578545